"十三五"国家重点图书
当代化学学术精品译库

有机合成中的副反应（Ⅱ）
——芳香族分子取代反应

Side Reactions in Organic Synthesis Ⅱ：
Aromatic Substitutions

［瑞士］Florencio Zaragoza Dörwald 著

田伟生 史 勇 译

华东理工大学出版社
EAST CHINA UNIVERSITY OF SCIENCE AND TECHNOLOGY PRESS
·上海·

图书在版编目(CIP)数据

有机合成中的副反应.Ⅱ,芳香族分子取代反应／
(瑞士)佛罗伦西奥·萨拉戈萨·多沃德
(Florencio Zaragoza Dörwald)著；田伟生,史勇译.
—上海：华东理工大学出版社,2017.1
(当代化学学术精品译库)
书名原文：Side Reactions in Organic Synthesis
Ⅱ：Aromatic Substitutions
ISBN 978-7-5628-4881-3

Ⅰ.①有… Ⅱ.①佛… ②田… ③史… Ⅲ.①有机合成②芳香族化合物-取代反应 Ⅳ.①O621.3②O625

中国版本图书馆CIP数据核字(2016)第308627号

项目统筹／周　颖
责任编辑／周　颖
装帧设计／裘幼华
出版发行／华东理工大学出版社有限公司
　　　　　　地址：上海市梅陇路130号,200237
　　　　　　电话：021-64250306
　　　　　　网址：www.ecustpress.cn
　　　　　　邮箱：zongbianban@ecustpress.cn
印　　刷／山东鸿君杰文化发展有限公司
开　　本／710 mm×1000 mm　1/16
印　　张／17.5
字　　数／339千字
版　　次／2017年1月第1版
印　　次／2017年1月第1次
定　　价／88.00元

版权所有　侵权必究

译者前言

化学家的任务和兴趣是探索物质的化学组成以及它们的各种物理、化学和生物学性质，而有机化学家则把他们的兴趣主要集中在各类有机化合物分子的化学反应性能和转化方面的研究。

物质世界千变万化。每类化合物，甚至每个有机分子在不同条件下会产生不同的反应产物，即使是在相同反应条件下也有可能产生不完全相同的多种反应结果。全面了解每一类，甚至每个有机分子的反应性能是实现不同类型化合物和不同有机分子之间相互转化的基础，也被高效利用自然资源、洁净合成人类社会所必需的各种有机物质、促进人类社会可持续性发展所要求。但是，当今有机化学家在发表他们的研究结果时常常选择性地仅仅公布他们认为重要的反应结果，而很少公布全部的反应事实。社会的功利化发展使有机化学家在发表他们的研究结果时越来越"报喜不报忧"，或者仅仅是在为了显示自己所做的研究成果特别优秀，才会给出一些其他研究者未完全公开的有机化合物反应情报。如此，有机化学的同行、学生，特别是社会公众，无法从有机化学家发表的文章所涉及的有机反应中得到完整的结果。其现状是有机化学家利用社会资源不断地进行重复性研究工作，而社会公众仅能获得残缺不全的反应知识。丹麦化学家多沃德编著的《有机合成中的副反应》一书的价值是让社会公众真实地、系统地了解到有机化合物反应的复杂性、多样性。

有机合成化学发展已经有近 200 年的历史，在有机合成的发展历史中，有几件特别值得注意的事件：1828 年，德国化学家维勒首次实现尿素合成，开辟了人类有机合成新纪元；1965 年，美国化学家伍德沃德获得了诺贝尔化学奖，其标志着人类有机合成的能力在科学和艺术两方面已经达到了空前高的水平；2000 年，诺贝尔化学奖获得者美国化学家科里的"反合成分析"合成设计思想使有机合成由艺术真正变成为工具；高效、洁净、实用合成是当今有机合成化学家追求的新目标。有机合成化学不再是人类挑战自然界合成能力的艺术展示，而是为人类社会服务的工具。为了实现这一挑战性目标，特别需要人们全面了解每一类有机分子的反应性能，《有机合成中的副反应》和《有机合成中的副反应Ⅱ》正好能够适应广大读者的这一需求。

作为翻译者之一，我特别喜欢这本书著者的许多研究理念。探索自然规律不必过多在意世俗利害私念，服务人类社会必须全面客观了解自然规律。十一年前，我与同事彭逸华女士在我研究生期间的同学和好友荣国斌教授的支持下翻译出版了《有机合成中的副反应》。今年,我特别高兴与我现在的同事史勇副研究员(我以前的博士研究生)合作共同翻译《有机合成中的副反应Ⅱ》。愿这本翻译书对我国喜好有机化学的广大公众、从事有机化学及其相关学科的教学与研究人员及在校学生全面了解有机反应有所补益。鉴于我们的翻译水平有限,翻译本中难免会有瑕疵,敬请读者指正。

田伟生

2016年夏于中科院上海有机化学研究所

前　言

化学家的工作主要是解决问题。化学的最有趣之处不在于什么反应能发生，而在于什么反应不能发生以及为什么不能发生。困难的或"不可能"的反应、反应选择性差、收率低下、使用的催化剂昂贵或者产生过多废弃物等问题不会让化学家畏缩不前，反而正是它们为有关化学研究提供了重要的机遇。

十年前，我撰写了《有机合成中的副反应》一书，意在突出有机合成中一些最常见反应中的竞争反应及其局限性。尽管书名让一些读者感到困惑（有或无副反应一说），但同时我也收到很多正面的反馈意见。为此，我决定写一部续篇。

上一本书的重点是烷基化反应，即 sp^3 碳的取代反应。而本书的主题是芳香体系的取代反应，它在有机合成中具有同等的重要性。本书将展示这一常用合成转化中存在的主要问题与局限性，并希望有助于化学家们鉴定副产物以及设计更好的合成路线。正如之前的标题一样，本书主要目的在于鼓励大胆实验、激发灵感、接受挑战与激励的行动。

时间如此珍贵，我将尽量使书中文字简短（化学家从化学反应式中获取信息的速度比文字更快），并将所引文献以简短的代码列入反应式中，便于查阅。代码的格式是：年代-期刊名称-首页页码，例如 08joc4956 代表 J. Org. Chem., 2008:4956。所采用的期刊缩写可在"期刊名称缩写表"中找到，所有的专利可在 worldwide.espacenet.com 网站下载。

我衷心感谢保罗·汉塞尔曼（Paul Hanselmann）和马塞尔·苏哈托诺（Marcel Suhartono）对本书的校对和许多有益的讨论，同时感谢威立（WILEY-VCH）出版公司的编辑们，特别是安妮·布仁弗瑞（Anne Brennfuhrer）对本书的帮助和支持。

维斯普（Visp），瑞士
佛罗伦西奥·萨拉戈萨·多沃德
（Florencio Zaragoza Dörwald）
2014 年 5 月

术语与缩写
(Glossary and Abbreviations)

Ac	Acetyl, MeCO	乙酰基
acac	Acetylacetone, pentane-2,4-dione	乙酰丙酮基,2,4-戊二酮
Ada	Adamantyl	金刚烷基
AIBN	2,2'-Azobis(2-methylpropionitrile)	2,2'-偶氮二异丁腈
aq	Aqueous	水的
Ar	Undefined aryl group	未指明的芳香基团
BINAP	2,2'-Bis(diphenylphosphino)-1,1'-binaphthyl	2,2'-双(二苯基膦基)-1,1'-联萘
Boc	*tert*-Butyloxycarbonyl	叔丁氧基羰基
bpy	2,2'-Bipyridine	2,2'-联吡啶
CAN	Ceric ammonium nitrate, $(NH_4)_2Ce(NO_3)_6$	硝酸铈铵
cat	Catalyst or catalytic amount	催化剂,或催化量的
cod	1,5-Cyclooctadiene	1,5-环辛二烯
coe	*cis*-Cyclooctene	顺环辛烯
concd	Concentrated	浓的
cot	1,3,5-Cyclooctatriene	1,3,5-环辛三烯
Cp	Cyclopentadienyl	环戊二烯基
Cp*	Pentamethylcyclopentadienyl	五甲基环戊二烯基
Cy	Cyclohexyl	环己基
cym	Cymene, 4-sopropyltoluene	对异丙基甲苯
DABCO	1,4-Diazabicyclo[2.2.2]octane	1,4-二氮杂双环[2.2.2]辛烷
dba	1,5-Diphenyl-1,4-pentadien-3-one	1,5-二苯基-1,4-戊二烯-3-酮
DBU	1,8-Diazabicyclo[5.4.0]undec-5-ene	1,8-二氮杂双环[5.4.0]十一-5-烯

DCE	1,2 - Dichloroethane	1,2-二氯乙烷
DDQ	2,3 - Dichloro - 5,6 - dicyano - 1,4 - benzoquinone	2,3-二氯-5,6-二氰基-1,4-苯醌
DMA	N,N - Dimethylacetamide	N,N-二甲基乙酰胺
DME	1,2 - Dimethoxyethane	1,2-二甲氧基乙烷
DMF	N,N - Dimethylformamide	N,N-二甲基甲酰胺
DMI	1,3 - Dimethylimidazolidin - 2 - one	1,3-二甲基咪唑烷-2-酮
DMPU	1,3 - Dimethyltetrahydropyrimidin - 2 -one	1,3-二甲基四氢-2-嘧啶酮
DMSO	Dimethyl sulfoxide	二甲基亚砜
DPEphos	Bis [(2 - diphenylphosphino) phenyl]ether	双(2-二苯基膦基)苯基醚
dppb	1,4 - Bis(diphenylphosphino)butane	1,4-二(二苯基膦基)丁烷
dppf	1,1′ - Bis(diphenylphosphino) ferrocene	1,1′-二(二苯基膦基)二茂铁
dppp	1,3 - Bis(diphenylphosphino) propane	1,3-二(二苯基膦基)丙烷
dtbpy	2,6 - Di(*tert*-butyl)pyridine	2,6-二叔丁基吡啶
eq	Equivalent	当量
Fmoc	9 - Fluorenylmethyloxycarbonyl	9-芴基甲氧基羰基
GDP	Gross domestic product	国内生产总值
Hal	Undefined halogen	未指明的卤素
HFIP	1,1,1,3,3,3 - Hexafluoro - 2 - propanol	1,1,1,3,3,3-六氟-2-丙醇
HMPA	Hexamethylphosphoric triamide, $(Me_2N)_3PO$	六甲基磷酸三酰胺
L	Undefined ligand	未指明的配体
LTMP	Li - TMP	2,2,6,6-四甲基哌啶锂
Mes	Mesityl, 2,4,6 - trimethylphenyl	2,4,6-三甲基苯基
Ms	Methanesulfonyl	甲磺酰基
MS	Molecular sieves	分子筛
MW	Microwave	微波
NBS	N - Bromosuccinimide	N-溴代丁二酰亚胺
NCS	N - Chlorosuccinimide	N-氯代丁二酰亚胺
NIS	N - Iodosuccinimide	N-碘代丁二酰亚胺
NMP	N - Methylpyrrolidin - 2 - one	N-甲基吡咯烷基酮
Nu	Undefined nucleophile	未指明的亲核试剂

术语与缩写 (Glossary and Abbreviations)

PEGDM	Poly(ethylene glycol) dimethacrylate	聚(乙二醇)二甲基丙烯酸酯
phen	Phenanthroline	9,10-菲咯啉
pin	pinacolyl	频哪醇基,3,3-二甲基-2-丁基
Piv	Pivaloyl, 2,2-dimethylpropanoyl	新戊酰基
PPA	Polyphosphoric acid	多聚磷酸
pyr	Pyridine	吡啶
R	Undefined alkyl group	未指明的烷基
SET	Single electron transfer	单电子转移
S_NAr	Aromatic nucleophilic substitution	芳香亲核取代反应
S_N1	Monomolecular nucleophilic substitution	单分子亲核取代反应
S_N2	Bimolecular nucleophilic substitution	双分子亲核取代反应
S-phos	2-(2′,6′-Dimethoxybiphenyl) dicyclohexylphosphine	2-(2′,6′-二甲基联苯)二环己基膦
st. mat.	Starting material	起始原料
TBAB	Tetrabutylammonium bromide	四丁基溴化铵
TBAF	Tetrabutylammonium fluoride	四丁基氟化铵
TEMPO	(2,2,6,6-Tetramethyl-piperidin-1-yl)oxyl	2,2,6,6-四甲基-1-哌啶氧自由基
TFA	Trifluoroacetic acid	三氟乙酸
TFAA	Trifluoroacetic acid anhydride	三氟乙酸酐
TfOH	Triflic acid, F_3CSO_3H	三氟甲磺酸
THF	Tetrahydrofuran	四氢呋喃
TMEDA	N,N,N',N'-Tetramethyl-1,2-ethylenediamine	N,N,N',N'-四甲基-1,2-乙二胺
TMP	2,2,6,6-Tetramethylpiperidine	2,2,6,6-四甲基哌啶
Tol	Tolyl	对甲苯基
Ts	Tosyl, 4-toluenesulfonyl	甲磺酰基
wt	Weight	质量
xantphos	4,5-Bis(diphenylphosphino)-9,9-dimethylxanthene	4,5-双(二苯基膦)-9,9-二甲基氧杂蒽

期刊名称缩写表
(Journal Abbreviation List)

a	Arkivoc; Archive for Organic Chemistry	有机化学的档案文件
ac	Acta Crystallographica	晶体学报
acr	Accounts of Chemical Research	化学研究评论
ajc	Australian Journal of Chemistry	澳大利亚化学
ang	Angewandte Chemie, International Edition in English	德国应用化学(国际英文版)
asc	Advanced Synthesis & Catalysis	高等合成和催化
bcsj	Bulletin of the Chemical Society of Japan	日本化学会通报
bj	Biochemical Journal	生物化学
catc	Catalysis Communications	催化通信
catl	Catalysis Letters	催化快报
cb	Chemistry & Biology	化学与生物学
cc	Chemical Communications	化学通信
cej	Chemistry — A European Journal	欧洲化学
cjc	Canadian Journal of Chemistry	加拿大化学
cl	Chemistry Letters	化学快报
coc	Current Organic Chemistry	当代有机化学
cpb	Chemical & Pharmaceutical Bulletin	化学与药学通报
cr	Chemical Reviews	化学综述
ejoc	European Journal of Organic Chemistry	欧洲有机化学
hca	Helvetica Chimica Acta	瑞士化学学报
iec	Industrial & Engineering Chemistry	化学工艺
ja	Journal of the American Chemical Society	美国化学会志
jbcs	Journal of the Brazilian Chemical Society	巴西化学会志
jcat	Journal of Catalysis	催化学报
jcs(p1)	Journal of the Chemical Society, Perkin Transactions 1	英国化学会志
jmc	Journal of Medicinal Chemistry	药物化学

joc	Journal of Organic Chemistry	有机化学
jpc	Journal für Praktische Chemie	实用化学杂志
obmc	Organic & Biomolecular Chemistry	有机与生物分子化学
ol	Organic Letters	有机化学快报
oprd	Organic Process Research & Development	有机工艺研发
oscv(1)	Organic Syntheses, Collective Volume 1	有机合成,合订卷1
p	Pharmazie	药物学
pcs	Proceedings of the Chemical Society	化学学会论文集
rjoc	Russian Journal of Organic Chemistry	俄罗斯有机化学
sc	Synthetic Communications	合成通信
sl	Synlett	合成快报
syn	Synthesis	合成
tet	Tetrahedron	四面体
thl	Tetrahedron Letters	四面体快报
zok	Zhurnal Organicheskoi Khimii	俄罗斯有机化学（俄文版）

目　　录

1　芳烃的亲电烷基化反应 1
　　1.1　概述 1
　　　　1.1.1　过渡族金属络合物催化的反应 3
　　　　1.1.2　典型的副反应 3
　　1.2　芳烃的问题 6
　　　　1.2.1　缺电子芳烃 6
　　　　1.2.2　苯酚类 11
　　　　1.2.3　苯胺 14
　　　　1.2.4　唑类 17
　　1.3　亲电试剂的问题 18
　　　　1.3.1　甲基化 18
　　　　1.3.2　烯烃 18
　　　　1.3.3　烯丙基亲电试剂 19
　　　　1.3.4　环氧 21
　　　　1.3.5　α-卤代酮及相关的亲电体 22
　　　　1.3.6　硝基烷烃 25
　　　　1.3.7　酮类 26
　　　　1.3.8　醇 28
　　参考文献 30

2　芳烃的亲电烯基化反应 40
　　2.1　概述 40
　　2.2　与含离去基团取代的烯烃的烯基化反应 40
　　2.3　与未含离去基团取代的烯烃的烯基化反应 42
　　2.4　与炔烃的烯基化反应 47
　　参考文献 50

3　芳烃的亲电芳基化反应 54
　　3.1　概述 54
　　3.2　卤代芳烃的芳基化反应 55
　　　　3.2.1　通过阳离子中间体 55

  3.2.2 通过自由基 ... 56
  3.2.3 通过过渡金属螯合物 ... 59
  3.2.4 通过过渡金属催化 ... 59
 3.3 与重氮盐的芳基化反应 ... 63
 3.4 与其他官能团化芳烃的芳基化反应 ... 65
 3.5 未取代芳烃的芳基化反应 ... 67
 参考文献 ... 69

4 芳烃的亲电酰化反应 ... 75
 4.1 概述 ... 75
 4.2 芳烃的问题 ... 78
  4.2.1 芳烃的去烷基化/异构化反应 ... 78
  4.2.2 苯乙烯 ... 78
  4.2.3 苯胺、苯酚和苯硫酚 ... 80
  4.2.4 缺电子芳烃 ... 81
  4.2.5 唑类 ... 83
 4.3 亲电试剂的问题 ... 85
  4.3.1 酰卤的问题 ... 85
  4.3.2 羧酸酯类和内酯 ... 88
  4.3.3 碳酸衍生物 ... 90
  4.3.4 甲酸衍生物 ... 94
  4.3.5 混合羧酸酐和其他多重亲电试剂 ... 96
 参考文献 ... 98

5 芳烃的亲电卤代反应 ... 107
 5.1 概述 ... 107
 5.2 典型的副反应 ... 108
 5.3 区域选择性 ... 111
 5.4 催化 ... 113
 5.5 氟化反应 ... 114
 5.6 缺电子芳烃 ... 117
  5.6.1 吡啶 ... 118
  5.6.2 苯甲酸衍生物 ... 119
 5.7 富电子芳烃 ... 122
  5.7.1 酚类和芳基醚 ... 122
  5.7.2 苯胺 ... 123
  5.7.3 唑类 ... 126
 5.8 敏感的官能团 ... 130

5.8.1 烯烃	130
5.8.2 胺	131
5.8.3 醚	131
5.8.4 硫醇和硫醚	131
5.8.5 醛、酮和其他的 C—H 酸性化合物	133
5.8.6 酰胺	134
参考文献	134

6 通过亲电反应形成芳烃 C—N 键 … 143
6.1 芳烃的硝化反应 … 143
　　6.1.1　机理 … 143
　　6.1.2　区域选择性 … 146
　　6.1.3　催化 … 148
　　6.1.4　缺电子芳烃 … 148
　　6.1.5　富电子芳烃 … 150
6.2 芳烃的亲电胺化反应 … 155
　　6.2.1　典型的副反应 … 157
6.3 芳烃的亲电酰胺化反应 … 159
　　6.3.1　典型的副反应 … 160
参考文献 … 163

7 通过亲电反应形成芳烃 C—S 键 … 170
7.1 磺酰化反应 … 170
　　7.1.1　概述 … 170
　　7.1.2　典型的副反应 … 171
7.2 亚磺酰化反应 … 173
　　7.2.1　概述 … 173
　　7.2.2　典型的副反应 … 174
7.3 硫醚化(次磺酰化)反应 … 176
　　7.3.1　概述 … 176
　　7.3.2　典型的副反应 … 178
参考文献 … 178

8 芳烃的亲核取代反应 … 182
8.1 概述 … 182
　　8.1.1　机理 … 182
　　8.1.2　区域选择性 … 182
　　8.1.3　酸/碱催化 … 188

8.1.4　过渡族金属催化 ... 189
　8.2　亲电试剂的问题 ... 191
　　8.2.1　不兼容的官能团 ... 191
　　8.2.2　非活化芳烃 ... 192
　　8.2.3　硝基芳烃 ... 194
　　8.2.4　重氮盐 ... 200
　　8.2.5　酚类 ... 202
　　8.2.6　芳基醚和芳基硫醚 ... 203
　　8.2.7　其他苯酚衍生亲电 ... 204
　　8.2.8　芳炔 ... 205
　8.3　亲核试剂的问题 ... 206
　　8.3.1　烯醇盐 ... 206
　　8.3.2　有机镁及相关有机金属化合物 ... 208
　　8.3.3　氨 ... 212
　　8.3.4　伯胺和仲胺 ... 214
　　8.3.5　叔胺 ... 217
　　8.3.6　叠氮化物 ... 217
　　8.3.7　氢氧化物 ... 220
　　8.3.8　醇 ... 221
　　8.3.9　硫醇 ... 224
　　8.3.10　卤代物 ... 226
　参考文献 ... 228

后记：化学研究的质量 ... 244
　参考文献 ... 246

索引 ... 247

1 芳烃的亲电烷基化反应

1.1 概述

芳烃亲电烷基化反应是大规模工业化有机合成所必需的反应(图式1.1)。其引人注目的特征是:当以醇或烯烃作为亲电试剂进行反应时则无废弃物产生,而且原料来源极其丰富,通过一步简单反应即可获得结构复杂的化合物。其主要存在的问题是:反应区域选择性低,过度烷基化以及碳正离子中间体的异构化。通过芳烃亲电烷基化反应生产的重要产品包括异丙基苯(枯烯,起始原料为苯酚和丙酮)、乙基苯(起始原料为苯乙烯)、甲基苯酚、偕二芳基烷烃(制备聚合物的单体)、三苯甲基氯(原料为四氯化碳和苯[1])、二氯苯基三氯乙烷(DDT,原料为三氯乙醛和氯苯)以及三芳基甲烷类染料。

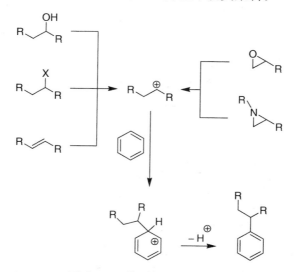

图式1.1 傅-克烷基化的机理

---

① 边栏数字为原版图书中文字对应的页码,与索引中的页码对应。

为获得可接受的收率,常常需要仔细优化大部分反应参数。由于在烷基化后芳烃的反应活性提高(每增加一个烷基提高约 2~3 倍),容易发生多烷基化的问题。多烷基化问题可以通过降低反应转化率或者调节反应温度、浓度、搅拌速度、所用溶剂(例如提供一个均相的反应体系)等参数来克服。如果起始原料可以回收,专业化的工厂通常使反应在低转化率下运行。而在实验室或使用复杂的高沸点原料的时候,芳烃的亲电烷基化反应就更加难以操作。

典型的芳烃亲电烷基化反应试剂包括脂族醇、烯烃、卤代物、羧酸和磺酸酯、醚、醛、酮以及亚胺。也有报道使用碳酸酯[2]、脲[3]、硝基烷烃[4]、叠氮化物[5]、重氮烷烃[6]、氨基醇[7]、环丙烷[8]和硫醚(图式 1.14)进行烷基化反应的例子。胺可通过转化成 N-烷基吡啶鎓盐中间体[9]或者瞬态脱氢为亚胺[10]的方式进行烷基化反应。图式 1.2 给出了一些傅-克(Friedel-Crafts)烷基化反应的实例。

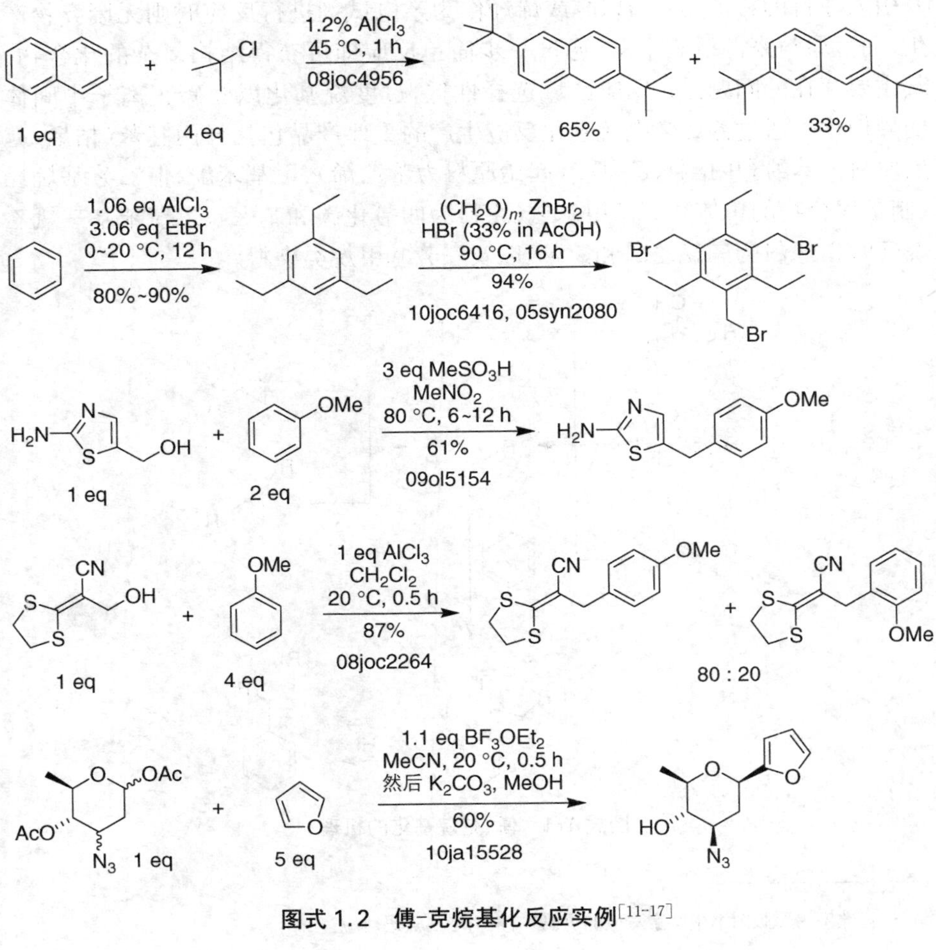

图式 1.2 傅-克烷基化反应实例[11-17]

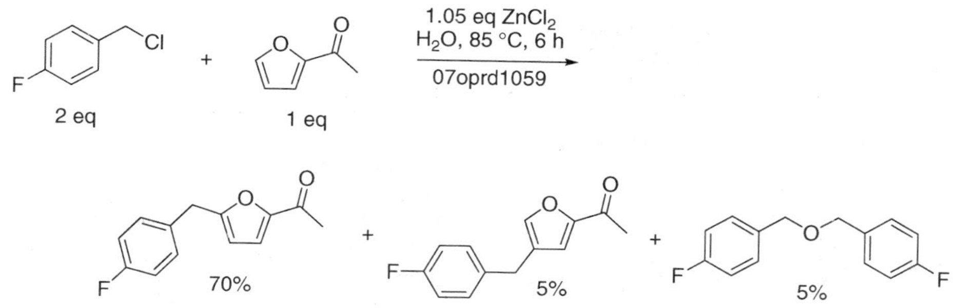

续图式1.2

大多数情况下,芳烃的亲电烷基化反应通过碳正离子中间体进行,手性仲卤代物或仲醇通常被完全消旋化。只有存在邻位基团并且可与之形成环状、构型稳定的碳正离子的时候,芳基化反应才有可能保持其构型[18]。

稳定的碳正离子(如叔碳正离子)容易生成,但它们与不稳定的碳正离子相比反应活性较低(选择性较高)。例如,三苯甲基或卓鎓($C_7H_7^+$)正离子可与苯甲醚反应,但不与苯反应。另一方面,碳正离子受临近的带正电荷基团的影响去稳定化,也可以提高其反应活性[7,19]。高度稳定化的阳离子甚至可以在中性反应条件下生成并进行芳基化反应[20]。

1.1.1 过渡族金属络合物催化的反应

钯、铑或钌等弱亲电过渡族金属络合物可以催化芳烃与烯烃或卤代烷的亲电烷基化反应。报道的多数实例是通过芳烃金属化形成螯合物进行的。当用高位阻要求的亲电试剂时,钌催化剂可以实现其间位烷基化反应(图式1.3第五个反应)。

对于碳正离子形成为最慢步骤(决速步骤)的反应,它可被任何能够稳定碳正离子中间体的化合物催化(促进碳正离子中间体的形成)。这种形式的催化作用在非极性溶剂中最为明显,因为非极性溶剂的溶剂化仅能轻微地稳定碳正离子。$IrCl_3$和$H_2[PtCl_6]$等一些过渡金属络合物可催化乙酸苄酯进行傅-克烷基化反应,反应可能是通过瞬间形成苄基金属络合物进行的(图式1.4)。由于在这些反应中观察到外消旋化,因此络合物中间体可能发生了快速转金属化。金属钌手性络合物催化剂已发展到能够从消旋醇制备对映体过量的烷基苯和烷基芳香杂环类化合物(图式1.18)。

1.1.2 典型的副反应

碳正离子中间体的重排是傅-克反应的常见副反应(图式1.5)。过渡金属络合形成的中间体的反应活性比未络合的碳正离子低,因此有时可借助过渡金属催化来避免碳正离子重排。

碳正离子还可作为氧化剂,从其他分子攫氢[31]。新形成的碳正离子也可进行芳烃烷基化,因此反应体系较为复杂(图式1.6)。

图式1.3 过渡族金属催化的芳烃烷基化反应[21-26]

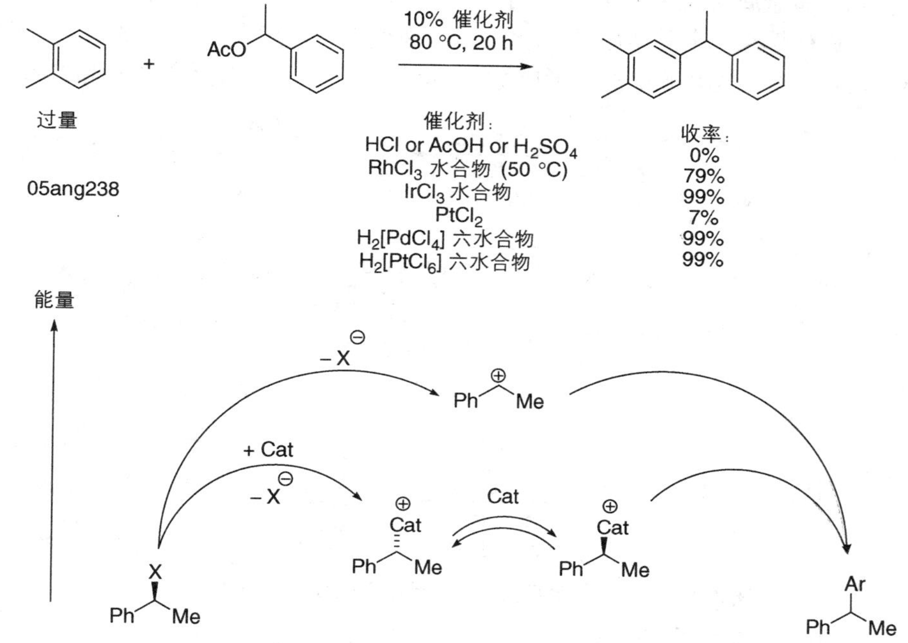
10% 催化剂
80 °C, 20 h
过量
05ang238
催化剂:
HCl or AcOH or H_2SO_4
$RhCl_3$ 水合物 (50 °C)
$IrCl_3$ 水合物
$PtCl_2$
$H_2[PdCl_4]$ 六水合物
$H_2[PtCl_6]$ 六水合物
收率:
0%
79%
99%
7%
99%
99%
能量
+ Cat
Cat
Ar
Ph
Me

图式 1.4 傅-克烷基化催化过程[28]

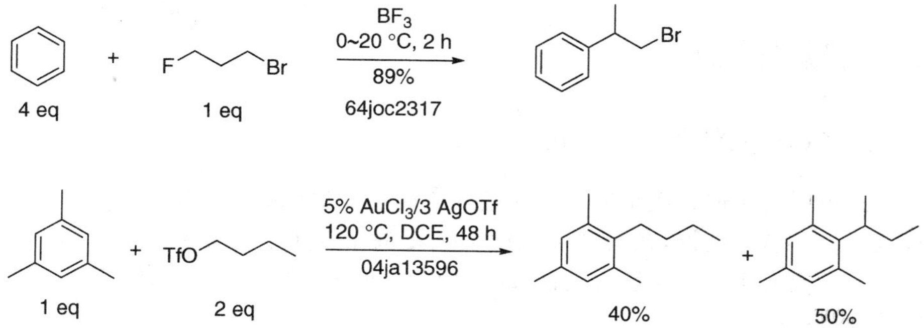
BF_3
0~20 °C, 2 h
89%
64joc2317
4 eq
1 eq
5% $AuCl_3$/3 AgOTf
120 °C, DCE, 48 h
04ja13596
1 eq
2 eq
40%
50%

图式 1.5 傅-克烷基化中的碳正离子重排反应[29,30]

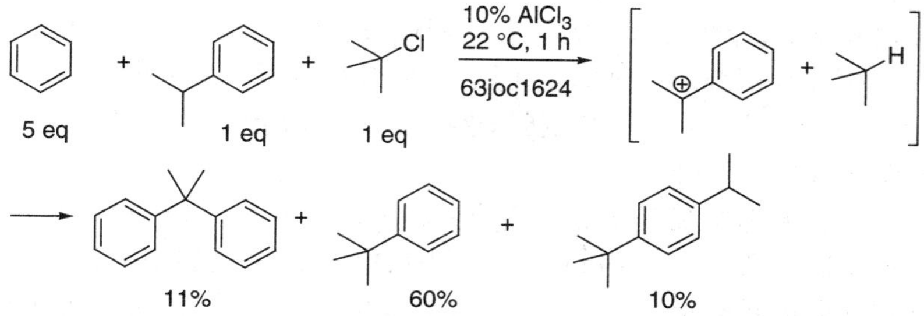
10% $AlCl_3$
22 °C, 1 h
63joc1624
5 eq
1 eq
1 eq
11%
60%
10%

图式 1.6 傅-克反应中碳正离子攫氢引发的副反应[32]

当以贵金属卤代物作为催化剂,或者以 α-卤代酮、α-卤代酯(见第 1.3.5 节)、全卤代烷作为亲电试剂时,芳烃会进行卤代而非烷基化反应(图式 1.7)。卤素与良好的离去基团相链接(可形成稳定的碳阴离子)的卤代烷是亲电卤代试剂。

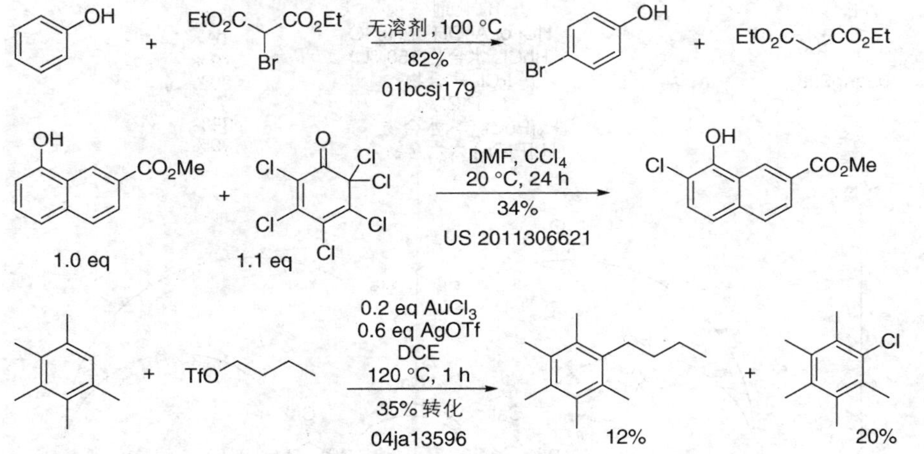

图式 1.7 三氯化金与卤代烷引起的芳烃卤化反应[30,33,34]

如果烷基化试剂的浓度太低,芳烃会发生酸催化的氧化二聚(Scholl 反应)[35]。该反应在酚类和苯胺等富电子芳烃中特别容易发生。

1.2 芳烃的问题

1.2.1 缺电子芳烃

缺电子芳烃通过碳正离子方式进行的烷基化反应收率通常低下。其主要原因是反应速度太慢,在与芳烃反应前碳正离子发生了重排和聚合。如果碳正离子不会发生其他副反应的话,缺电型芳烃的傅-克烷基化反应就能够实现高收率(图式 1.8)。

缺电子芳烃可以跟烯烃或卤代烷通过芳烃金属化中间体进行烷基化反应。螯合物的形成对过渡金属催化反应的区域选择性至关重要(图式 1.9)。在钌和铑催化下,烯烃与苯乙酮和苯乙酮亚胺的邻位烷基化甚至可在室温下进行[39]。对于位阻要求苛刻的卤代烃,钌络合物可以引发缺电子芳烃的间位烷基化[24]。当反应中存在氧化剂时,这些反应可以得到苯乙烯类化合物,而不是烷基苯化合物[40-42](见第 2.3 节)。

用作苯乙酮邻位烷基化催化剂的金属不但可以插入 C—H 键中,还能以相近的速率插入到 C—O 键和 C—N 键中(图式 1.10),其选择性有时可以通过精确选择催化剂来改善[47]。上述烷基化反应的一个潜在副反应是芳烃羟化反应,当反应体系中存在氧化剂时,此副反应容易发生[48,49]。

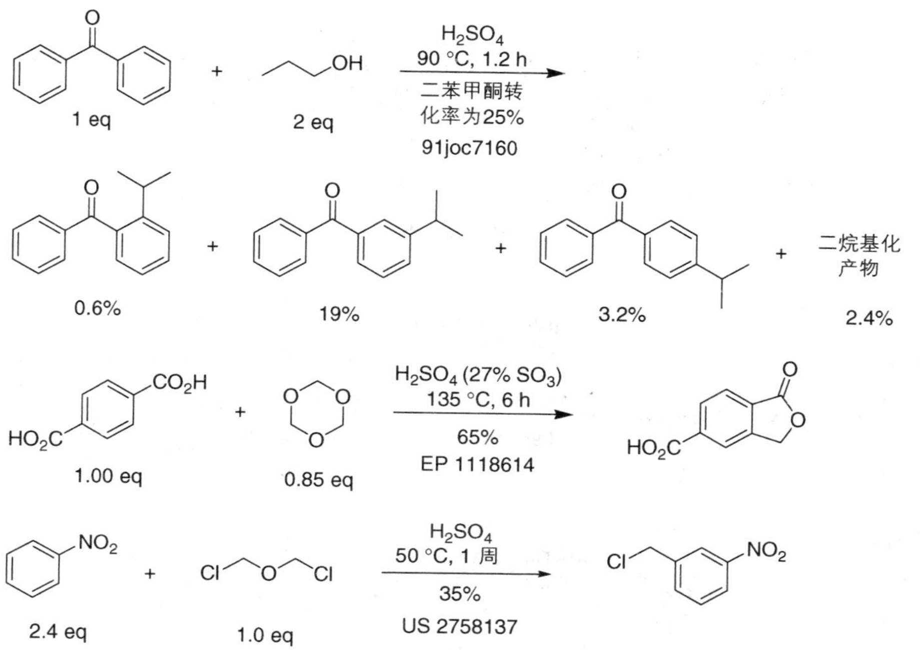

图式 1.8 缺电子芳烃的傅-克烷基化反应[36-38]

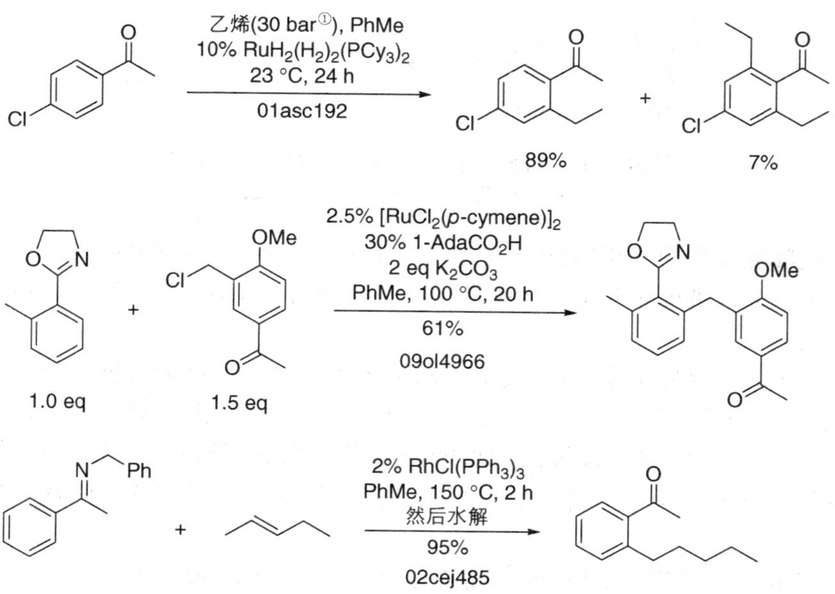

图式 1.9 铑、钌和钯催化经螯合介导的缺电子芳烃烷基化反应[43-46]

---

① 1 bar = 10^5 Pa。

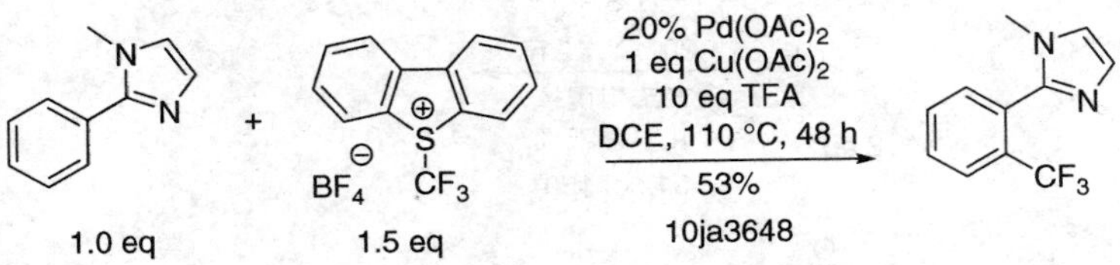

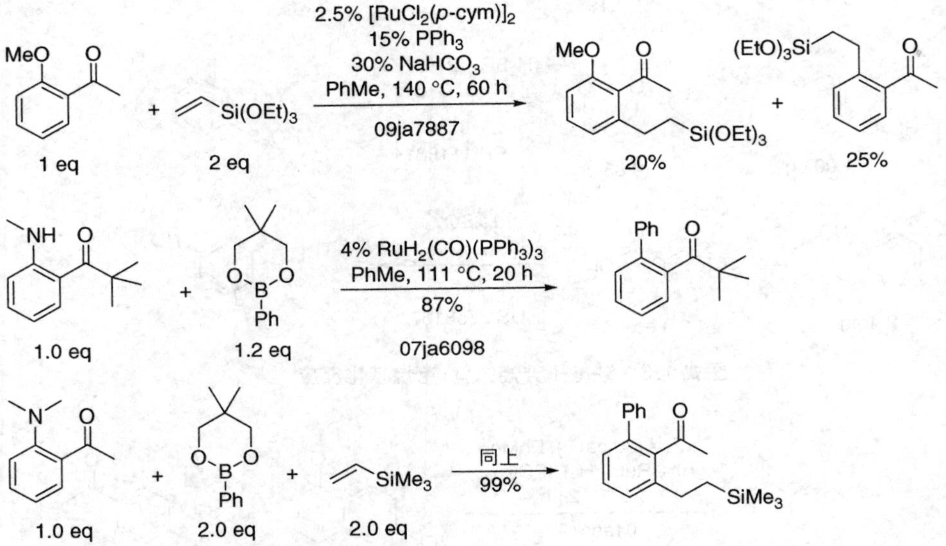

图式 1.10 钌催化下苯乙酮的邻位烷基化和芳基化反应[50,51]（更多实例：[52,53]）

吡啶氮氧化物、噻唑或咪唑等杂芳烃的 C—H 具有较强酸性，即使不形成螯合物也可以被催化金属化。在图式 1.11 的例子中，中间体实际是金属卡宾络合物。

在强烈的反应条件下，氟代苯或硝基苯也可以不通过形成螯合物而被金属化，并且被醛、酮等亲电试剂原位捕获（图式 1.12）。由于竞争性卡尼扎罗（Cannizzaro）反应和强亲核试剂可能导致发生酮的断裂反应[如哈勒-鲍尔（Haller - Bauer）反应]，这些反应需要使用过量的亲电试剂并仔细优化反应条件。

缺电子芳烃和杂芳烃，如吡啶盐等，可与富电子的碳自由基反应。这些自由基可以从烷烃、烷基卤化物、羧酸和一些二酰基过氧化物[58]（图式 1.13）获得，或通过硼烷的氧化[59]生成。然而，这类烷基化的区域选择性往往较差。

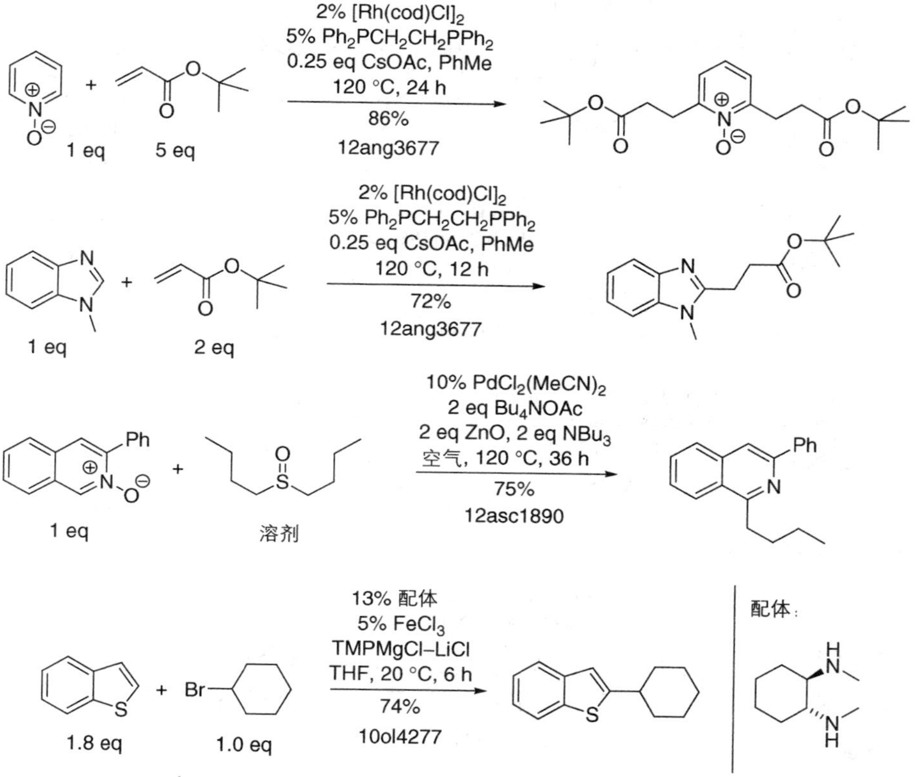

图式 1.11 具有 C—H 酸性的杂芳烃的金属化和烷基化反应[54-56]

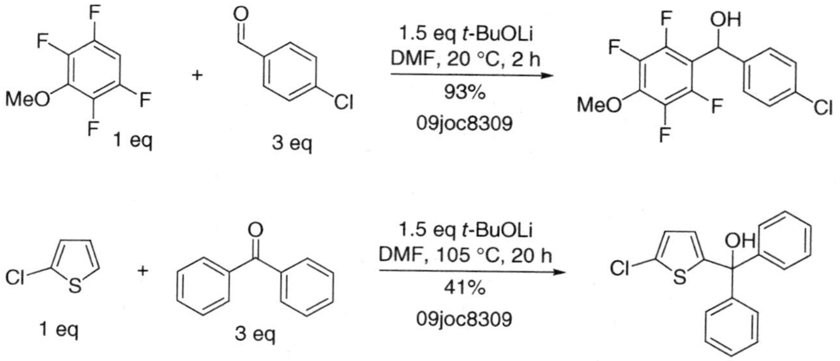

图式 1.12 具有 C—H 酸性的芳烃的金属化和烷基化反应[57]

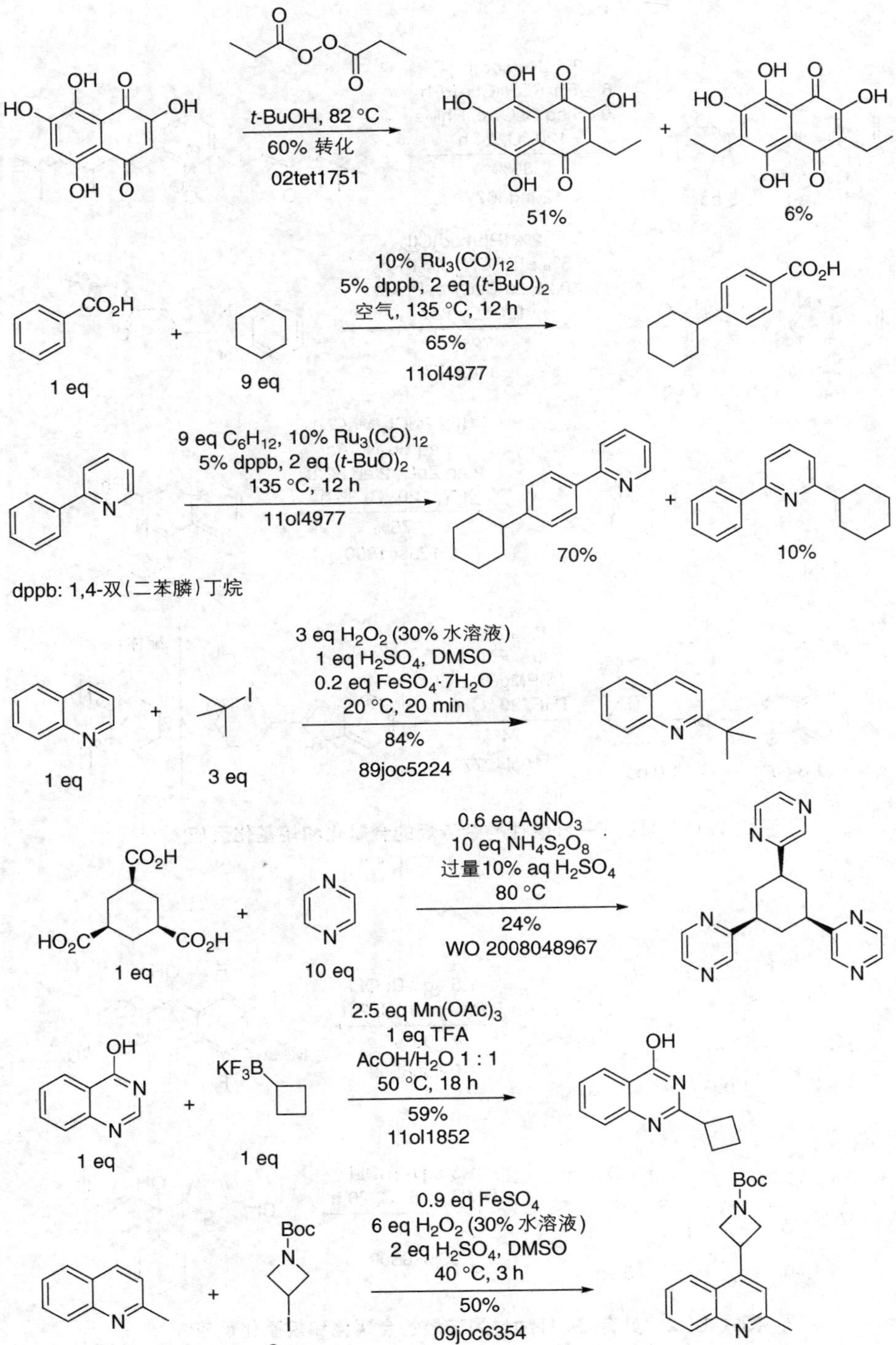

dppb: 1,4-双(二苯膦)丁烷

图式1.13 芳烃与自由基的烷基化反应[59-64] (更多实例: [65])

1.2.2 苯酚类

由于游离的酚羟基可使路易斯酸失活,并且苯酚存在烯酮的互变异构,其本身也可作为亲电试剂(见下文),因此在傅-克型反应中苯酚类是一类自身存在问题的亲核试剂。此外,在氧化剂存在下酚类容易二聚形成联芳烃。

在适当的反应条件下,酚类可进行碳烷基化而不发生大量的氧烷基化。稳定化的碳正离子是软亲电试剂,可与芳烃或烯烃等软亲核试剂优先反应。在酸性条件下,酚类化合物的氧烷基化只有在与硬的烷基化试剂(如重氮甲烷、碳酸二甲酯、甲醇、甲酯、烷氧基鏻盐或者缩醛等)反应时,才能被观察到。酚的氧烷基化产物在酸存在下有时可以重排成为碳烷基化产物[66](图式 1.14)。

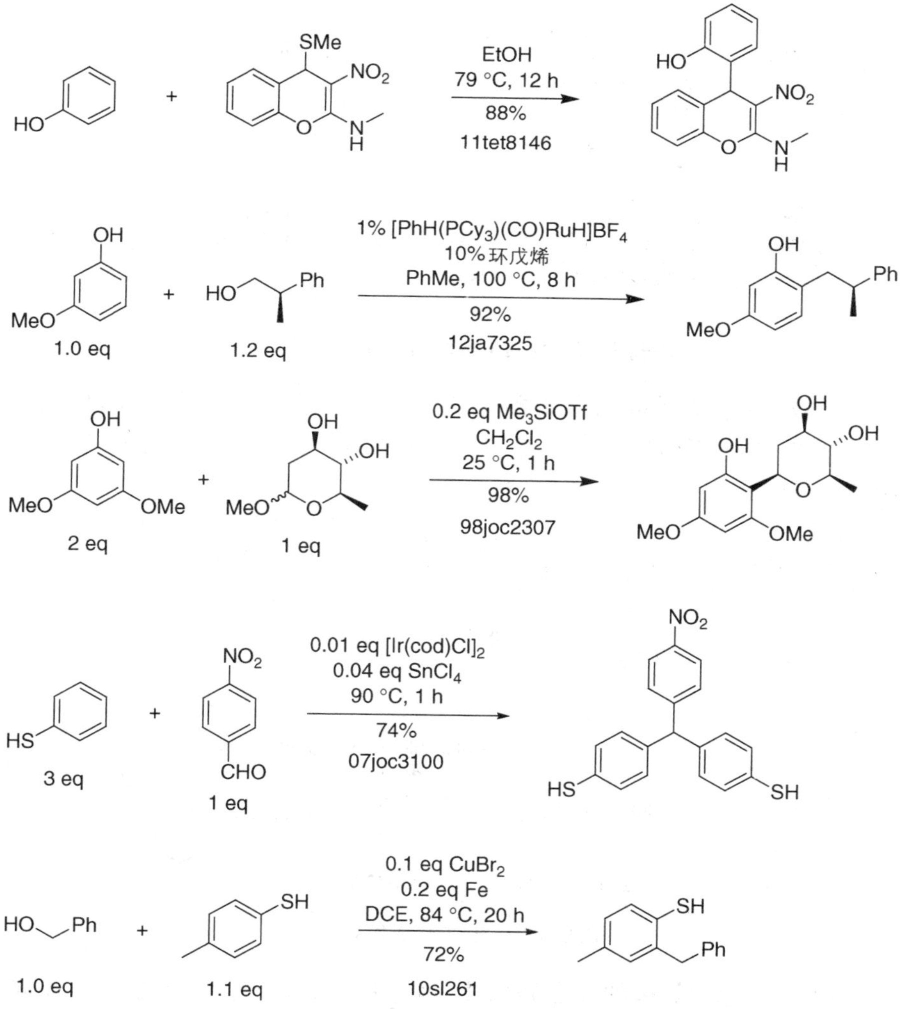

图式 1.14 酸性条件下酚和苯硫酚的碳烷基化反应[67-71]

在高温下，苯酚和酚铝可被烯烃碳烷基化（图式1.15）。与酰苯胺铝的反应相比，该反应比较困难且反应应用范围较窄（见1.2.3节）。虽然烷基化首先发生在邻位，但在大量烯烃存在下可得到2,4,6-三烷基化和更高烷基化的酚类产物[72,73]。在高压条件下，甚至会与烯烃发生狄尔斯-阿尔德（Diels-Alder）反应[74]。如今，许多重要的烷基酚都是在非均相催化剂存在下，通过烯烃与苯酚在高温下进行烷基化来制备的[73,75]。

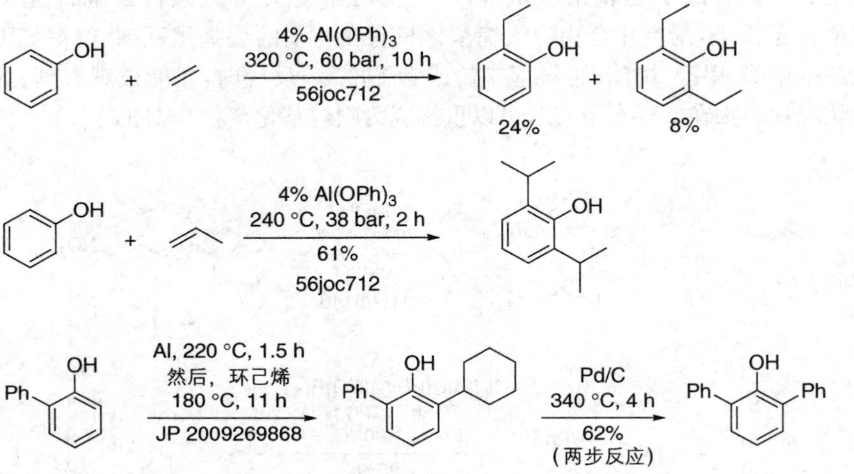

图式1.15 酚铝与烯烃的烷基化反应[76,77]

某些双亲电试剂可以在苯酚的碳和氧原子上同时进行烷基化反应。例如，在酸[78]或铑催化剂[79]存在下，1,3-二烯与酚反应生成色满化合物（图式1.16）。

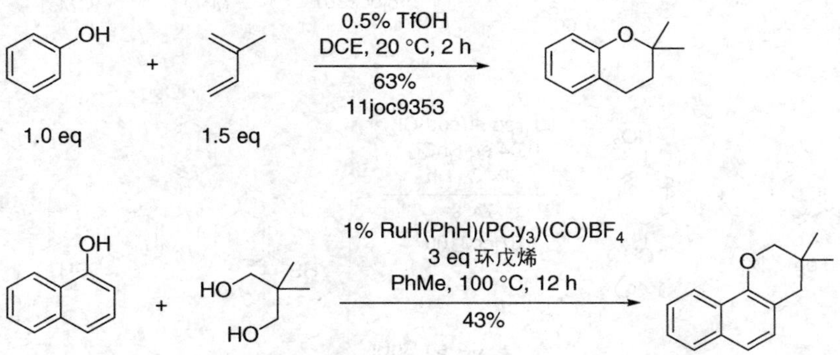

图式1.16 从酚形成色满[68,80]

苯酚是环己二烯酮的互变异构体，因此其也能以环己二烯酮的方式进行反应。特别是1-（或2-）萘酚、1,3-苯二酚和1,3,5-苯三酚显示出强的环己烯酮特征。在卤化铝或者氟化氢/五氟化锑的存在下，酚和芳基醚与芳烃反应得到 3-(或 4-)芳基环己烯酮[81-83]。根据酸的用量和酚的碱性，它们可以发生烯酮共轭加成，也可以进行双阳离子的芳基化反应，其结果很难被精确地预测(图式 1.17)。此外，形成的 4,4-二取代的环己烯酮可以经酸引发重排反应而生成 3,4-二取代的环己酮。含离去基团(卤化物、羟基)取代的酚可在芳基化反应后发生消除反应生成 3-(或 4-)芳基苯酚。

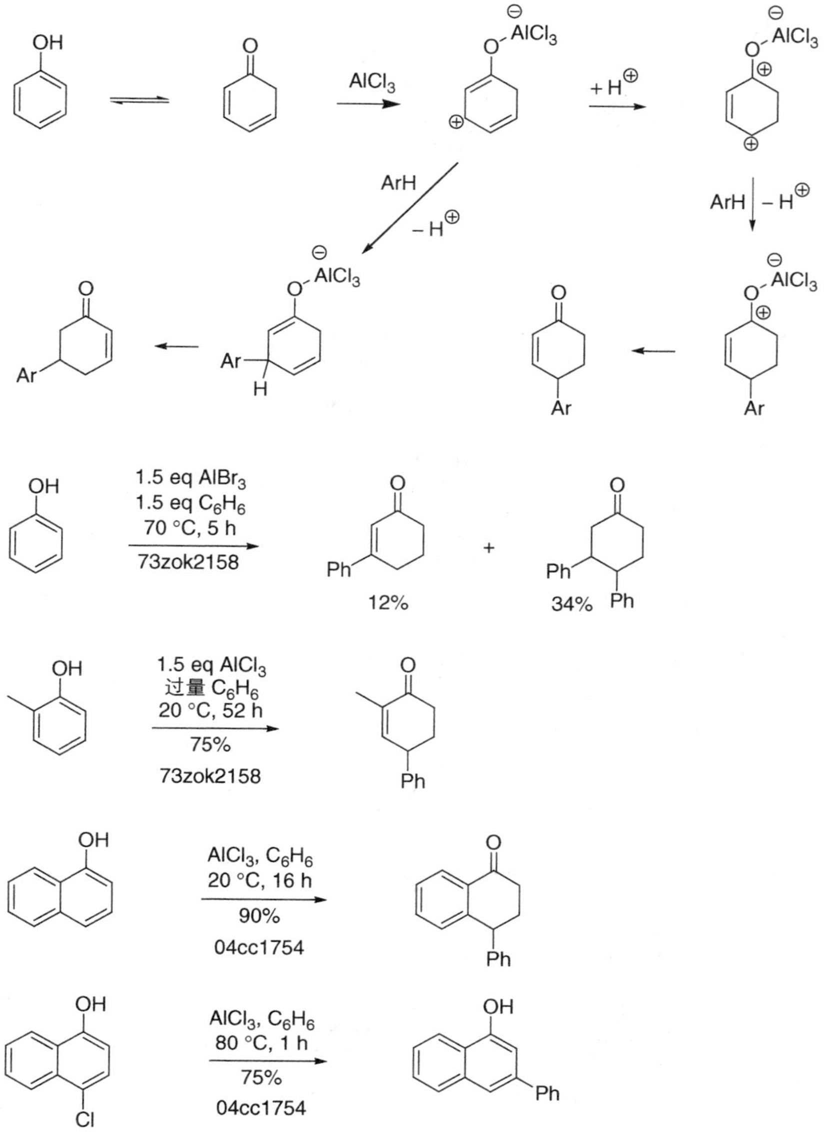

图式 1.17 酸引发的酚芳基化反应[84,85]

1.2.3 苯胺

氮原子尽管能被酸质子化,但是在酸性条件下苯胺的碳原子和氮原子上都能够进行烷基化反应。合适的烷基化试剂包括醇、醚、烯烃、醛、酮和卤代烷。

尽管铵盐是吸电子基团,但是苯胺的傅-克烷基化反应通常还是遵循邻位和对位选择性,并且其比相应苯的反应更容易进行。如盐酸苯胺在乙酸存在的条件下可以进行对位三苯甲基化反应,但是苯与三苯甲基正离子却不反应(图式 1.18 中第一个例子)。

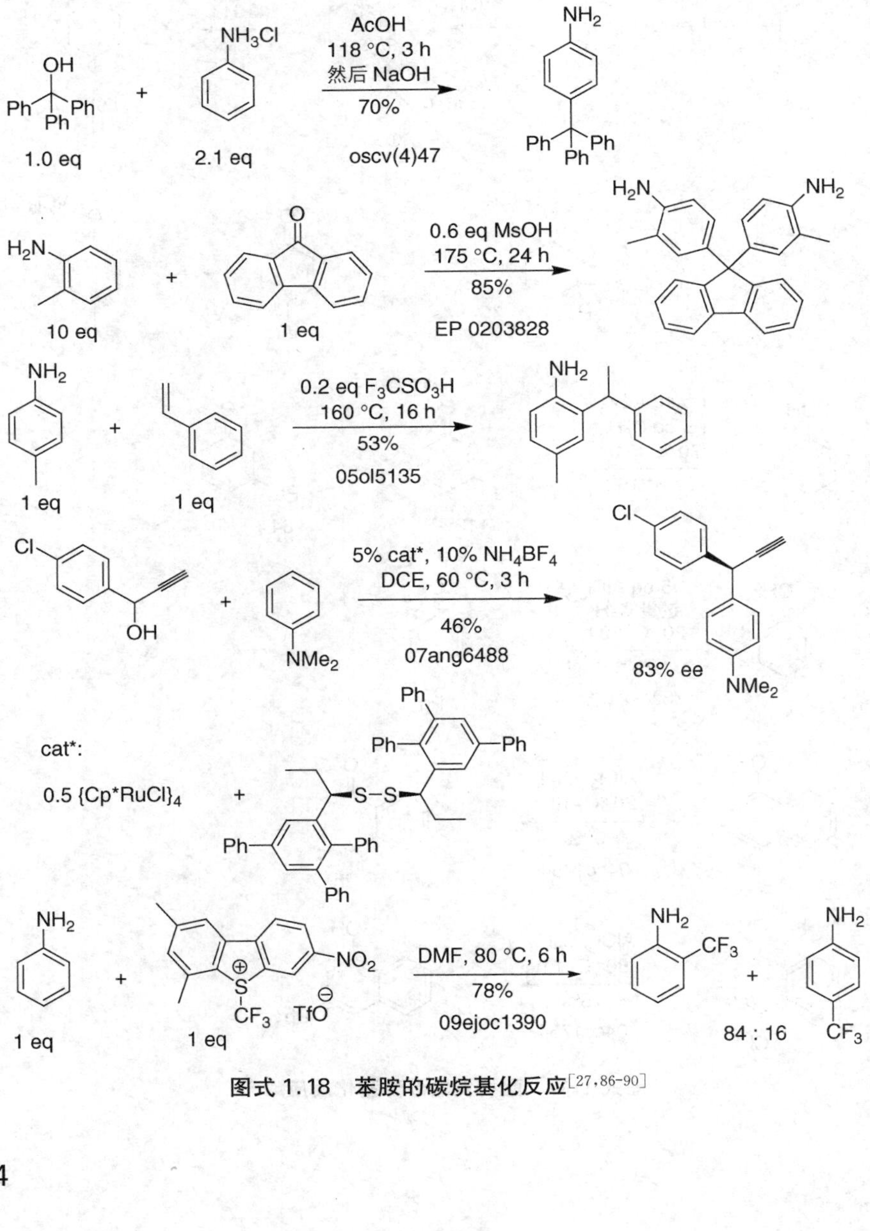

图式 1.18 苯胺的碳烷基化反应[27,86-90]

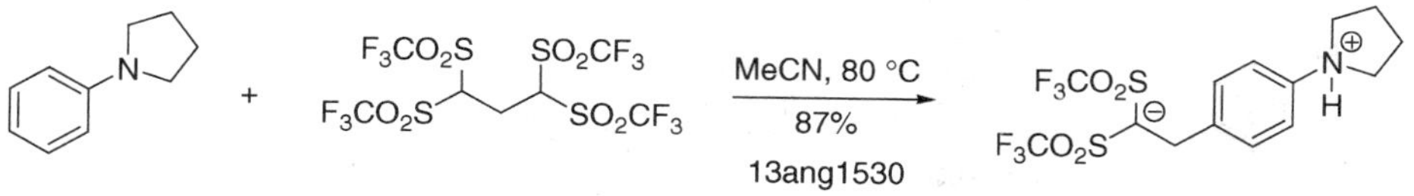

续图式 1.18

苯胺与烷基化试剂反应的结果难以精确预测。化学计量的强酸通常有利于碳烷基化。在高温下或在酸存在下，N-烷基苯胺能够发生脱烷基化反应，并且自身可以作为烷基化试剂[91-93]。反应偶尔也会得到氮烷基化和碳烷基化产物的混合物（图式 1.19）。

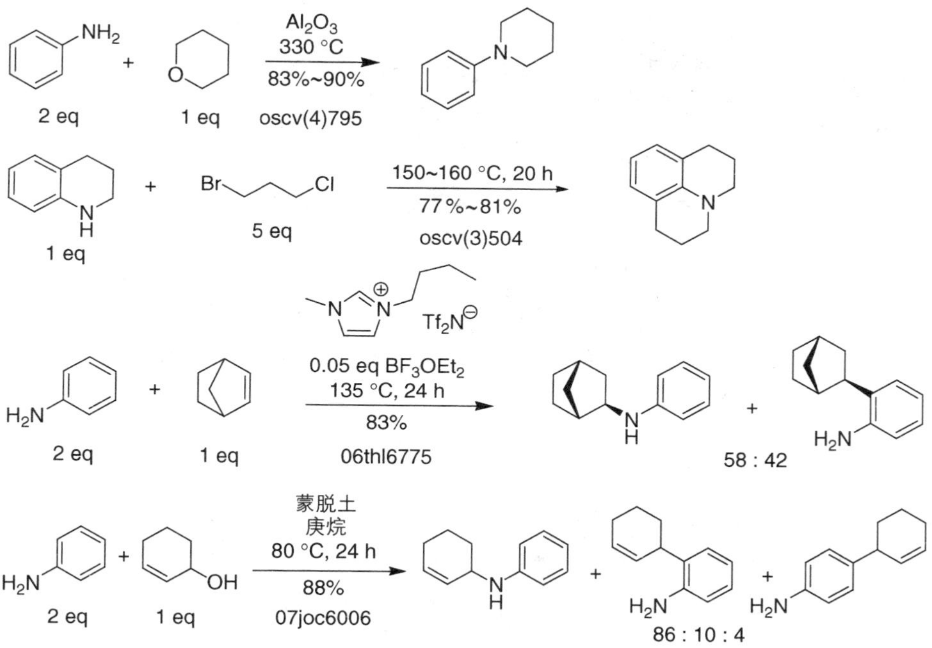

图式 1.19　苯胺的碳和氮烷基化实例[94-97]（更多实例：[98,99]）

在酸存在下，醛或酮与苯胺在室温下能可逆地形成缩醛胺、亚胺或者烯胺。加热则导致在碳上发生不可逆的烷基化。因此，甲醛与苯胺在低温下反应仅仅形成缩醛胺、苄胺或者特洛格尔（Tröger）碱；而在高温下，二芳基甲烷是主要产物（图式 1.20）。氢原子从醛或苯胺转移到亚胺盐中间体，导致了氮烷基化苯胺副产物的形成。

从苯胺制备二芳基甲烷的过程中常见的一个副反应是形成三芳基甲烷类染料。空气是合适的氧化剂，并且氧化作用可被钒酸盐催化（图式 1.21）。即使严格无氧的情况下，通过亚胺盐中间体的氧化也能产生少量的染料副产物。

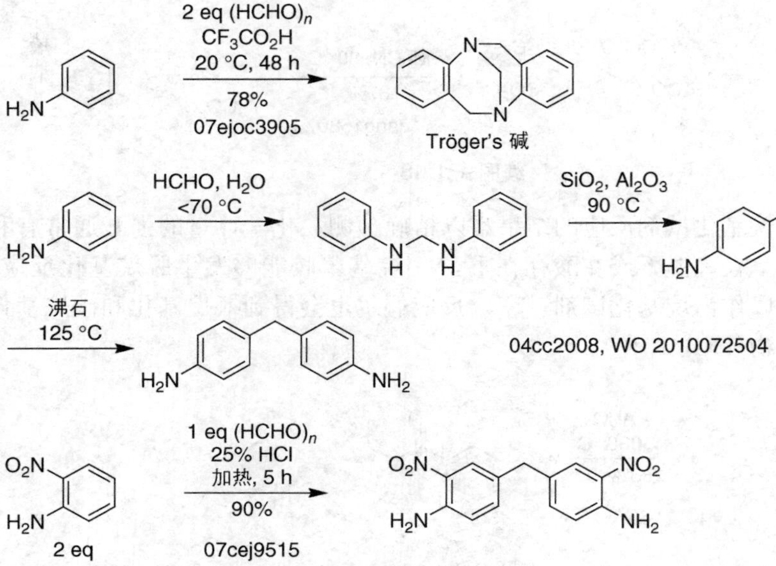

图式 1.20 由苯胺和甲醛形成二芳基甲烷[100-103]

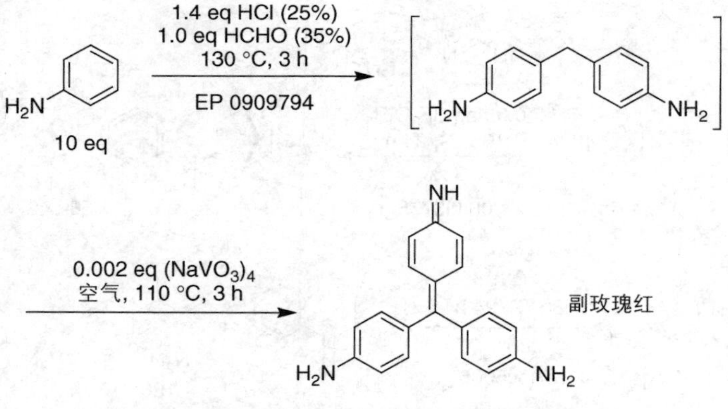

图式 1.21 从二芳基甲烷形成三芳基甲烷染料[104]

在碱性条件下,苯胺与烯烃能够发生选择性邻位烷基化反应,反应需要首先用铝/氯化铝处理苯胺,使其转化成为苯胺铝(图式 1.22)。然而这个有趣的反应适用范围较小,不适合酚的烷基化[76]。

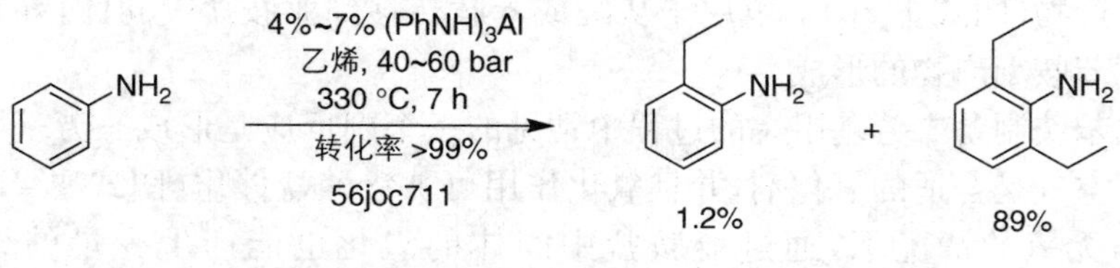

图式 1.22 苯胺与烯烃的烷基化反应[105-107] (更多实例:[108])

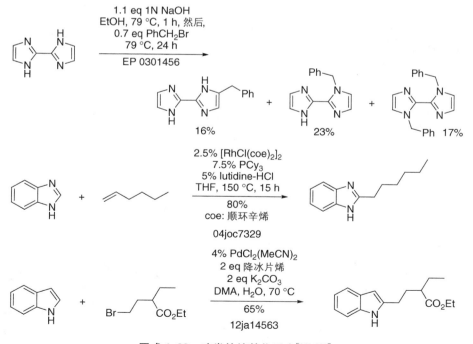

1.2.4 唑类

含有游离 N—H 基团的唑类在其氮或碳原子上能被烷基化。这种反应,特别对于吲哚和苯并咪唑等含有芳烃的底物,其反应结果几乎无法预测。唑类可以在化学计量的金属化后进行烷基化,这将使反应的区域选择性得到进一步提高。唑类化合物与硬亲电试剂(如甲基化试剂等)易于发生氮烷基化,而与软亲电试剂(如烯烃等)有时可发生高选择性的碳烷基化。图式 1.23 给出了一些非金属化的唑类化合物进行烷基化的例子。

图式 1.23 唑类的烷基化反应[109-111]

1.3 亲电试剂的问题

1.3.1 甲基化

由于傅-克烷基化需要形成自由的碳正离子或类碳正离子中间体,所以甲基化不容易进行。酚类化合物需要在高温下才能与甲醇发生 C-甲基化(图式1.24)。酸催化的甲基化反应可能无法形成自由的甲基正离子,所以反应中间体更可能是催化剂与甲基化试剂形成的络合物[112]。

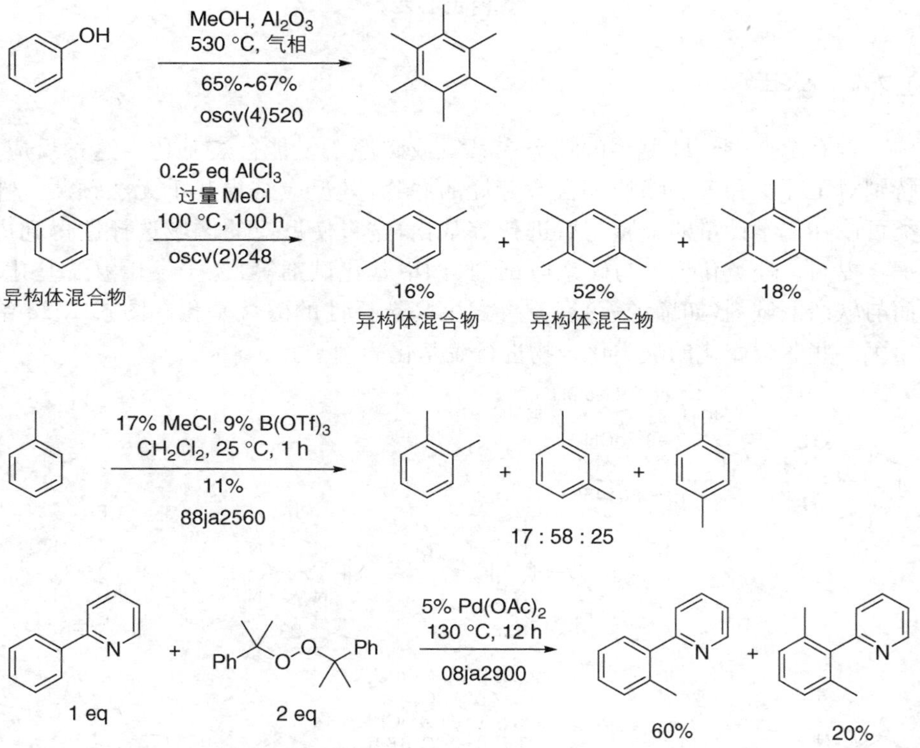

图式 1.24 芳烃与甲醇、氯甲烷和甲基自由基的甲基化反应[112-115]

1.3.2 烯烃

酸性条件下,非对称的烯烃与芳烃反应可得到两种不同产物——来自稳定碳正离子的马氏产物,或者来自较不稳定但活性较高的碳正离子的反马氏产物。与其他酸引发的烯烃加成反应一样,芳烃通常被更稳定的碳正离子烷基化。过渡族金属催化的芳烃烷基化也是如此[116]。无论如何,随着催化剂的发展,人们已经能够从末端烯烃制备线性烷基化芳烃[117,118](图式1.3)。

被吸电子基团取代的烯烃[迈克尔（Michael）反应受体]可以通过强亲电性的 β-碳原子对芳烃进行烷基化（例如[119]）。硝基烯烃也是如此，但硝基烯烃在强酸性水溶液中能够被水解成酮羰基（图式 1.25）。

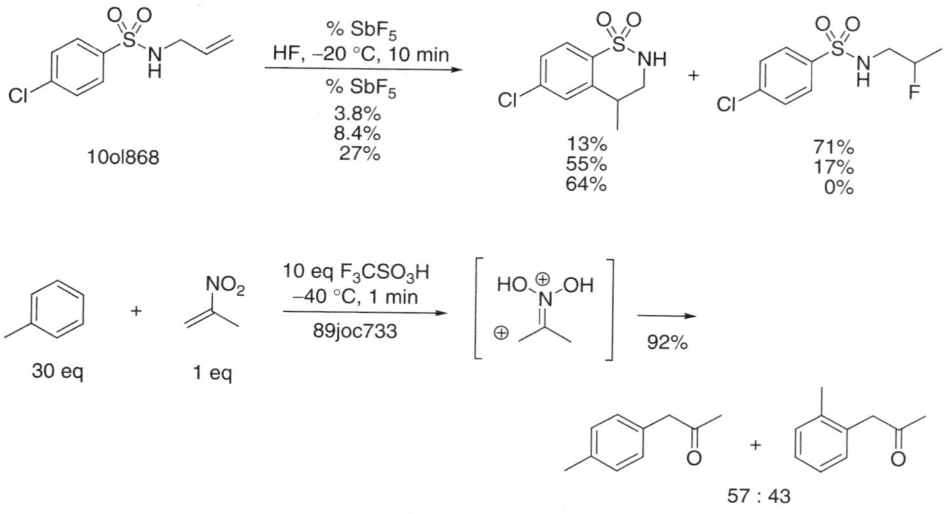

图式 1.25 烯烃的芳香烷基化反应[120,121]

酸引发的烯烃烷基化反应的一个典型副反应是烯烃的聚合反应。苯乙烯和丙烯酸酯的聚合反应特别容易进行。由于聚合反应通常需要一个最低浓度，有时可以通过保持烯烃低浓度来抑制聚合反应。在氧化剂或过渡金属的存在下，芳烃与烯烃反应得到苯乙烯类分子而非烷基芳烃（见第 2.3 节）。

1.3.3 烯丙基亲电试剂

芳烃与烯丙基亲电试剂反应经常生成几个异构体的混合物。优势的共振式（更稳定但活性低）并非总是能够控制反应区域选择性，立体效应也会影响反应过程（图式 1.26）。反应结果可以这样进行合理化解释，但其预测价值却受到限制。

在酸存在下，烯丙基亲电试剂是 1,3-丙烯双正离子的合成等价物。正因如此，产物经环化反应生成二氢化茚是一个潜在的副反应。环化反应有时能够通过过量使用芳烃来阻止。如果使用了钯催化剂，则预计赫克（Heck）反应（代替烯丙基取代反应）（图式 1.27）成为进一步的副反应。

丙烯酸酯是另一种类型的 1,3-双亲电体，在酸性条件下它与芳烃反应形成双环化合物（图式 1.28）。

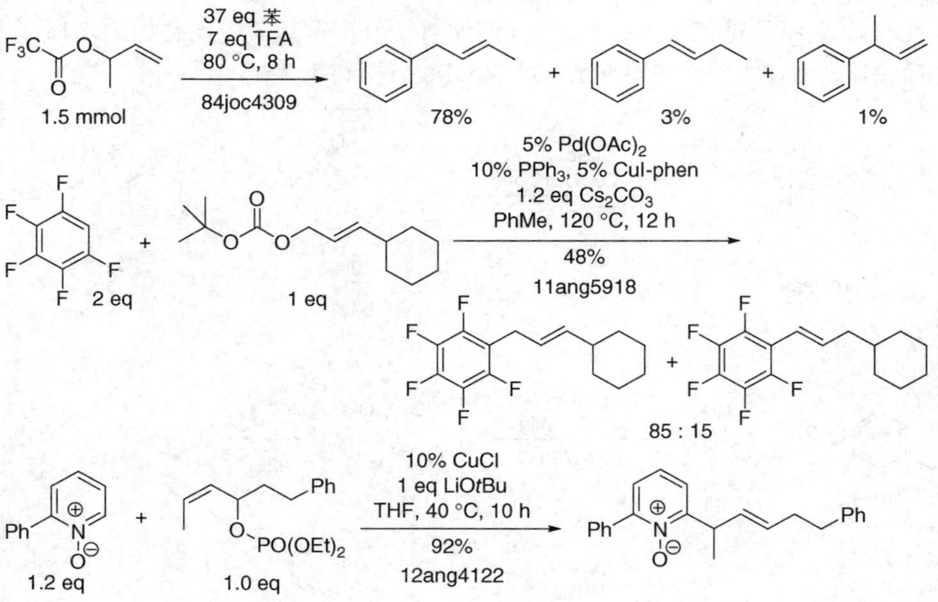

图式 1.26 芳烃与烯丙基亲电试剂发生烷基化反应的例子[122-124]（更多实例：[125,126]）

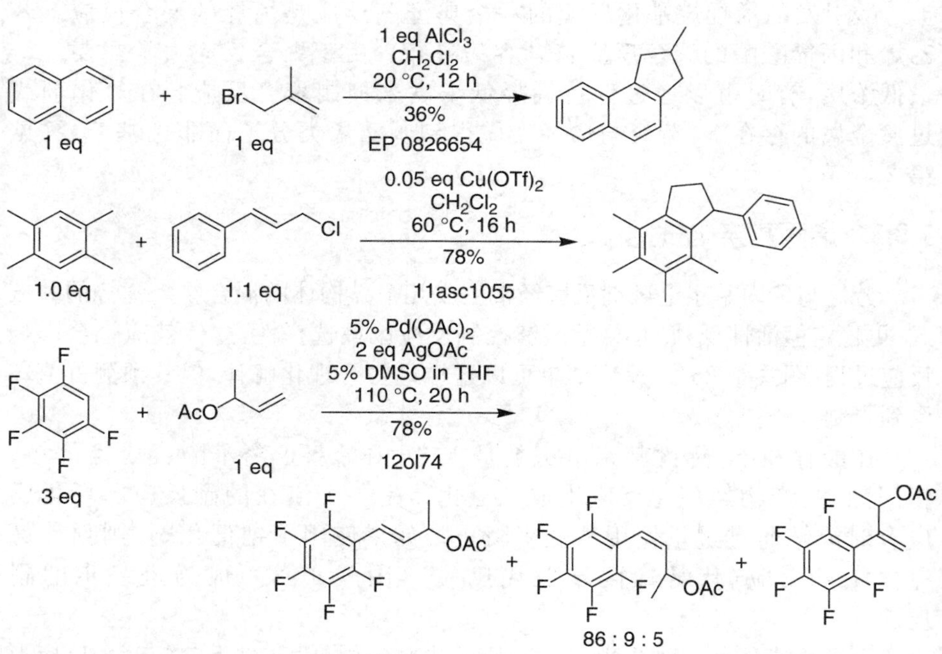

图式 1.27 烯丙基亲电试剂环化和 Heck 反应[126-128]

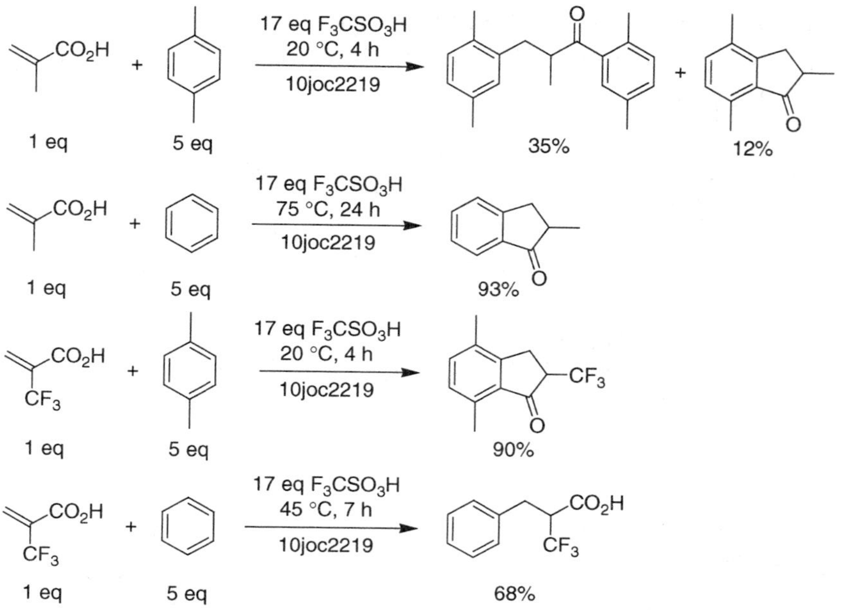

图式 1.28 酸引发的丙烯酸与芳烃的反应[129]

1.3.4 环氧

芳烃通常通过在能够形成稳定性较高的碳正离子的碳原子上环氧化而进行烷基化。因此，芳烃与烷基、芳基或者烯基环氧的反应主要生成伯醇，而与吸电子基团取代的环氧反应则主要得到仲醇。酸引发的芳烃与环氧氯丙烷和缩水甘油醚反应也易于生成仲醇（图式 1.29）。

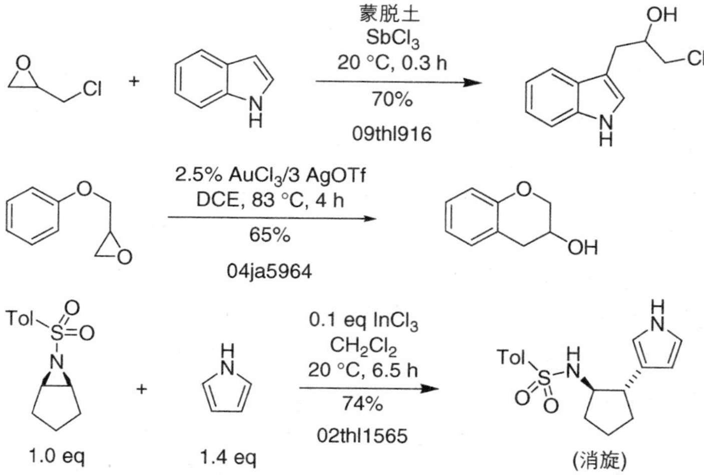

图式 1.29 芳烃与环氧化物和氮丙啶进行烷基化的例子[130-132]（更多实例：[133]）

环氧为活泼中间体，反应条件选择不当常常导致多种副反应的发生。典型的副反应包括环氧重排成为醛或酮、环氧的二聚或低聚以及芳烃被新生成的醇进一步烷基化等（图式1.30）。

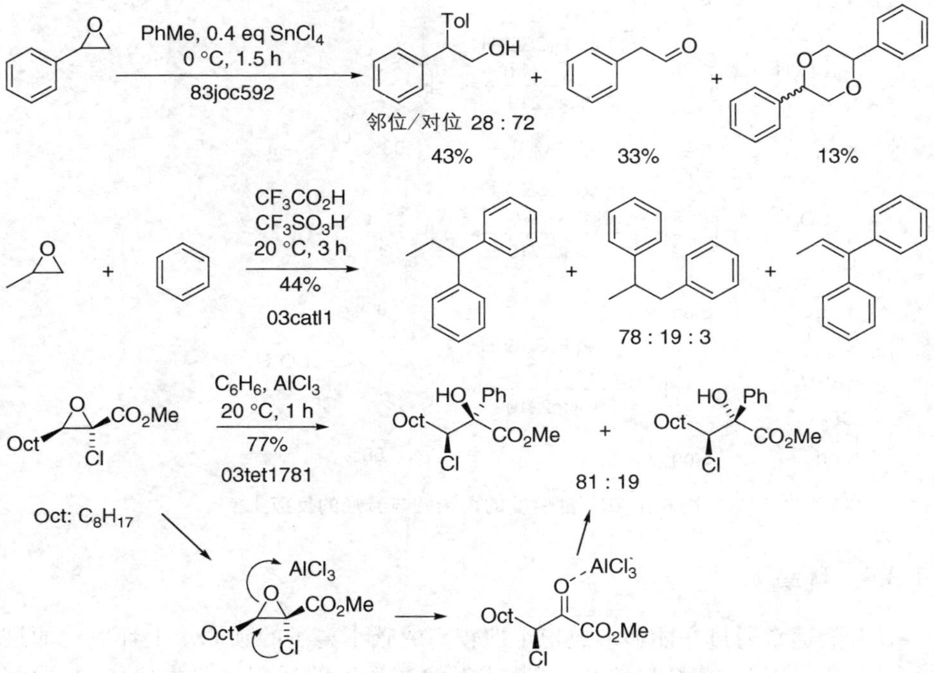

图式1.30 芳烃与环氧化物烷基化过程中的副反应[134-136]

1.3.5 α-卤代酮及相关的亲电体

卤素原子与 C—H 酸性碳原子相连的烷基卤化物(α-卤代酮、α-卤代酯、α-卤代腈等)具有独特的反应性质。它们很难通过失去卤素原子生成碳正离子(去稳定化)，仅有几个酸催化的芳烃与此类亲电体烷基化反应的报道[137,138]（图式1.31）。然而，这些卤代烷的亲核取代反应反而更容易进行。亲核试剂首先对羰基加成可能是这类亲电试剂反应活性被增强的一个原因[139]。

芳烃也能与酮、酯和腈反应，因此当 α-卤代酮及相关亲电试剂与芳烃进行烷基化反应时，此类副反应就在预料之中（图式1.32）。此外，α-卤代酮也可作为卤化试剂或氧化剂[144]，并且在碱存在下它们还可以发生二聚或三聚反应。

酮和酯也可被转化成自由基，然后加成到芳烃或杂芳烃上。产生这种自由基最常见的策略是 α-卤代酮或 α-卤代酯的光解以及酮的氧化（图式1.33）。因为脂肪族 α-卤代酯仅仅吸收短波长的紫外光，而芳烃会阻断短波长的紫外光与脂肪族 α-卤代酯的作用，所以芳烃通常不能用作反应溶剂（图式1.33中第二个实例）。

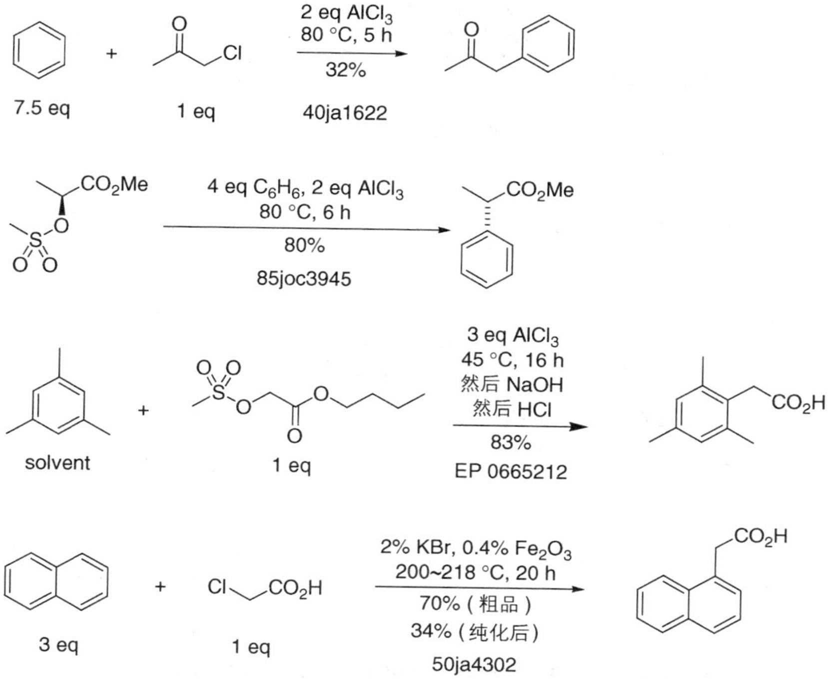

图式 1.31　芳烃与 α-卤代酮及相关亲电体的亲电烷基化反应[140-143]

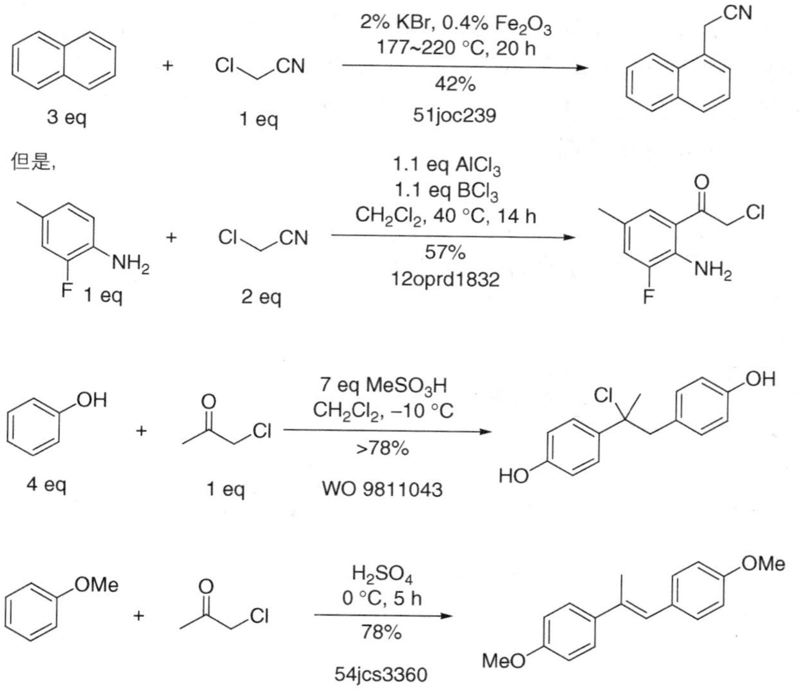

图式 1.32　α-氯代酮和氯代腈与芳烃的烷基化或酰化[145-148]

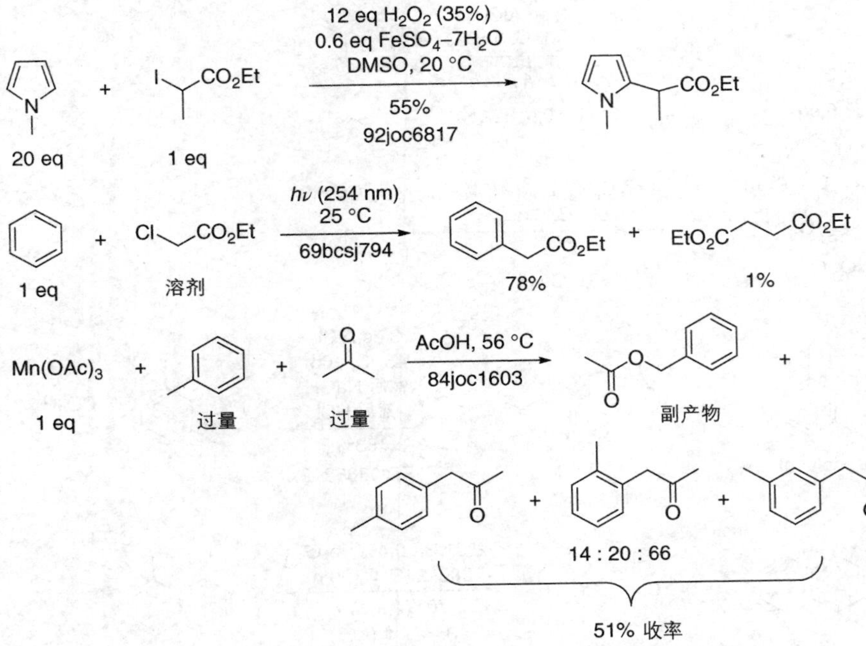

图式 1.33　α-卤代酯和卤代酮通过自由基进行芳基化[149-151]

α-重氮酮或α-重氮酯是金属卡宾络合物的前体，金属卡宾络合物可以直接插入到芳烃的 C—H 键中（图式 1.34）。然而，卡宾络合物中间体具有高反应活性和亲电性，可以跟很多官能团发生烷基化，可以攫氢，还可以跟烯烃、炔烃乃至芳烃发生环丙烷化反应。因此，重氮羰基化合物（或重氮烷[6]）很少用作芳烃的亲电烷基化试剂。

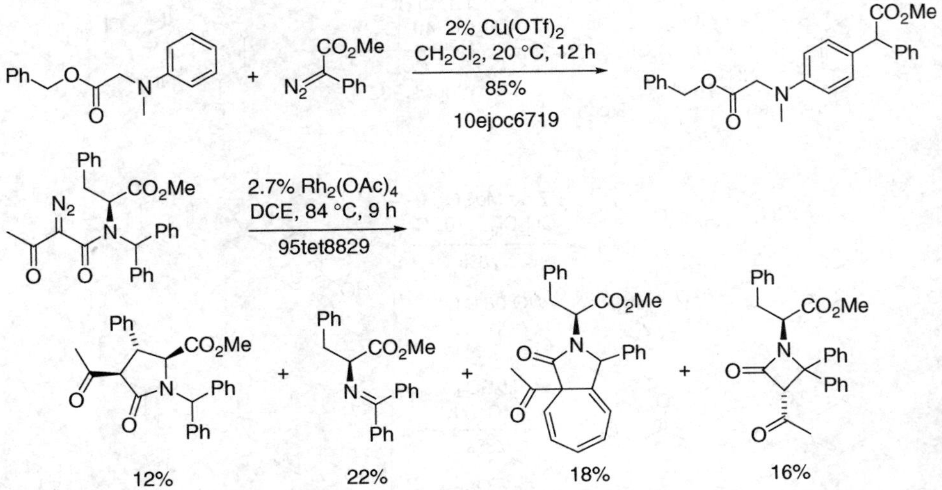

图式 1.34　α-重氮酯与芳烃的反应[152,153]

1 芳烃的亲电烷基化反应

α-卤代酮及相关亲电试剂通过间接亲核取代方式的芳基化在第 8.2.3 节讨论。

1.3.6 硝基烷烃

硝基烷烃的硝基作为离去基团对芳烃进行烷基化仅有少数几个例子被报道[4]（图式 1.35）。这个反应因有许多潜在的副反应而变得复杂。硝基烷烃可以在不失去硝基的情况下作为碳亲电体；此外，在强酸的存在下，硝基可以与芳烃在氧原子上发生反应。例如，用三氟甲磺酸处理 2-芳基-1-硝基乙烷可以得到氧-芳基化的肟（图式 1.35）。在此类型的反应中，硝基的氧原子具有亲电性。硝基进行亲电芳基胺化也有报道（图式 1.35 中最后一个例子）。

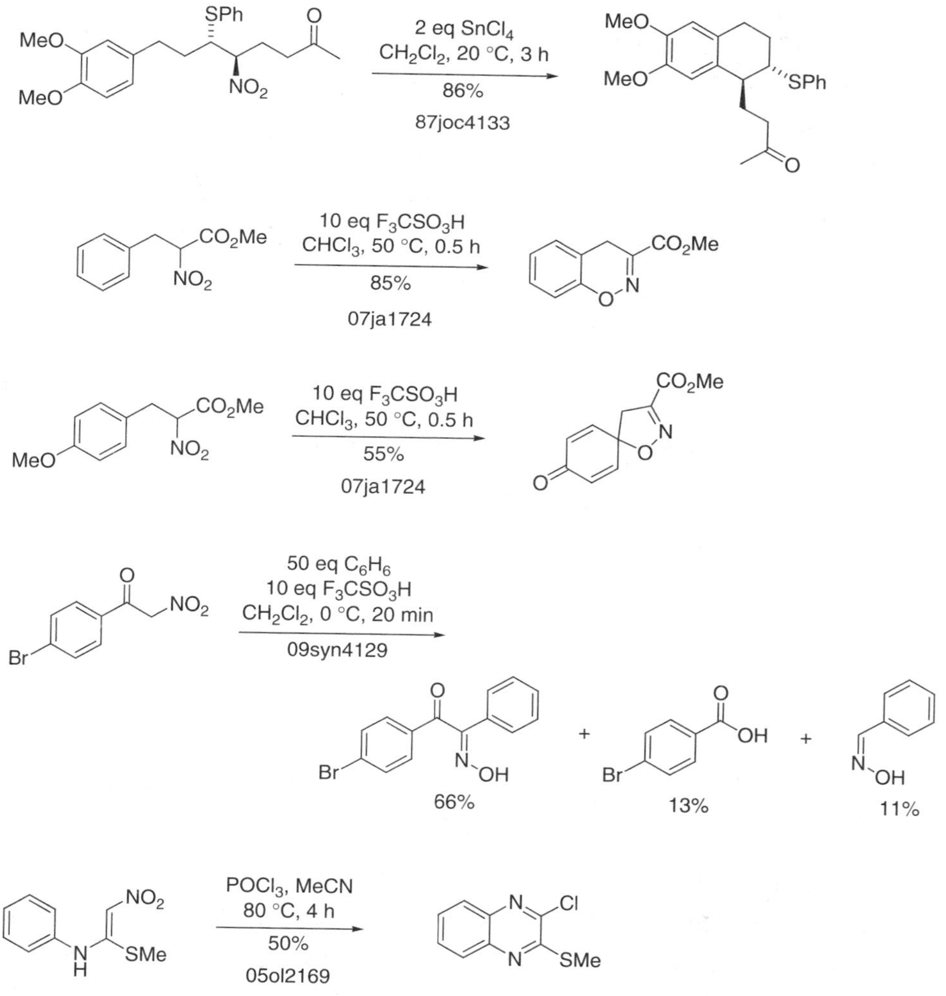

图式 1.35　硝基烷烃与芳烃的反应[154-157]（更多实例：[158]）

在脱水试剂存在下,伯硝基烷烃(RCH_2NO_2)能被转化成为腈氧化物。后者非常活泼,容易二聚、多聚、重排为异氰酸酯,也能与亲核试剂反应或进行1,3-偶极环加成反应。

1.3.7 酮类

在酸催化下,简单的二烷基酮可以与酚、苯胺或吡咯等富电子芳烃反应,但不与苯或甲苯反应。反应产生的叔苄醇常常会与第二分子芳烃发生烷基化,生成偕二芳基烷烃。偶尔也可观察到中间体醇的脱水反应及其脱水产物烯烃进一步进行低聚。如醇是目标产物,则需要选择温和的酸性催化剂并且仔细优化反应条件。

例如,异丙烯基苯不能直接通过丙酮和苯反应制备(最近的研究见文献[159]),原因是该反应很易形成的枯基阳离子会与苯反应[160]。从丙酮直接制备异丙烯基苯是很有价值的转化,因为在由异丙苯过氧醇生产苯酚的过程中有1当量丙酮生成,而它目前尚不能直接用于制备异丙苯。现有工艺是将丙酮氢化为异丙醇,然后将其脱水为丙烯,再用于苯的烷基化(图式1.36)。尽管苯与异丙醇的直接烷基化是可能的[161,162],但大多数傅-克烷基化的催化剂遇水失活,因此使用丙烯来进行异丙烯化要比用异丙醇更加方便。

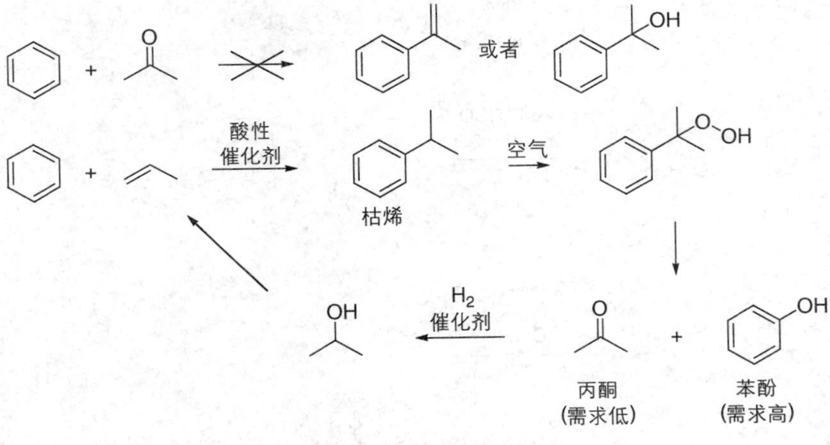

图式 1.36 苯与丙酮合成苯酚

只有三氟甲基酮、1,2-二酮或者α-酮酸酯等被吸电子基团取代的酮能够与未活化的芳烃反应。芴酮也很活泼,原因是O—质子化后的芴酮具有反芳香性。最初形成的醇如果不快速形成碳正离子的话,常常能够被分离出来(图式1.37)。

酮类进行傅-克烷基化可能的副反应是形成二芳基甲烷、产物的低度聚合以及起始原料酮的羟醛缩合反应。此外,在氧化剂的存在下,酮还可以通过形成自由基中间体进行α-芳基化[151]。当芳烃与酮的傅-克烷基化是在氢供体存在下进行时,能够发生芳烃的还原型烷基化反应(图式1.38)。

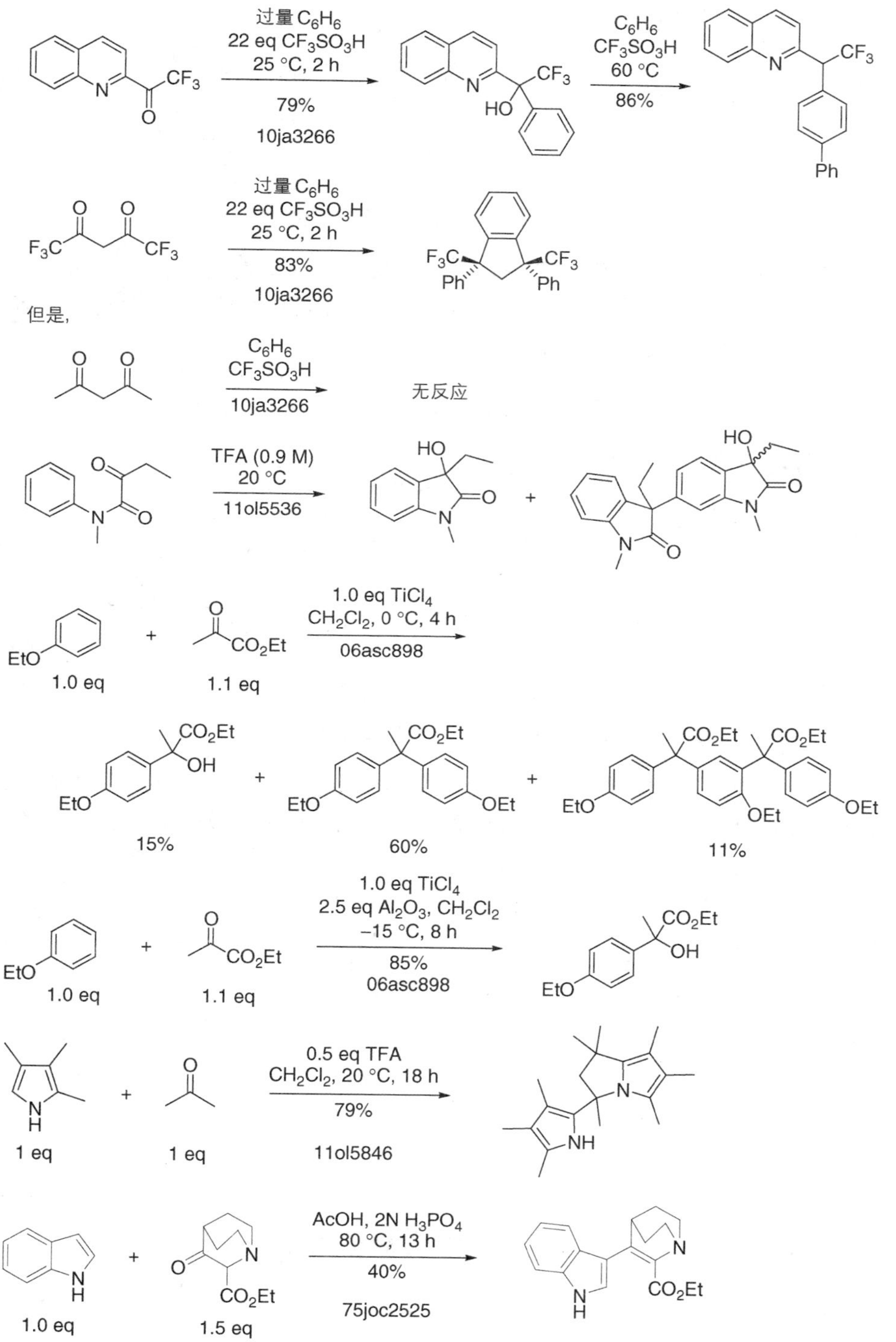

1 芳烃的亲电烷基化反应

图式 1.37　酮与芳烃和杂芳烃进行烷基化[163-167]

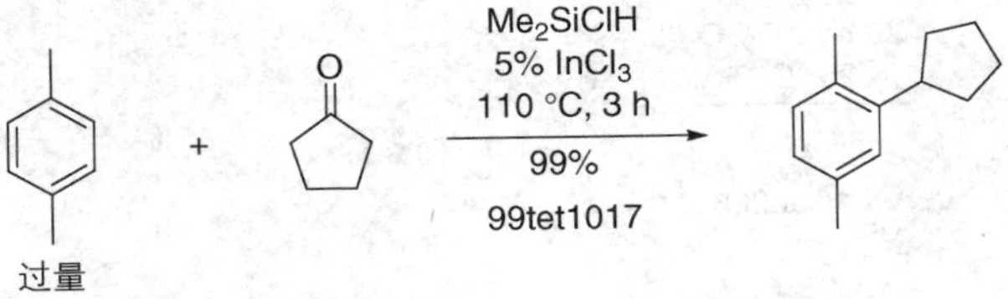

图式 1.38　酮与芳烃的还原型烷基化[168]

含强酸性 C—H 的酮,如 β-酮酸酯,容易在其碳上钯化。所得中间体可经历 β-氢消除得到 α,β-不饱和酮。后者是迈克尔反应受体,能够对富电子芳烃进行烷基化反应(图式 1.39)。

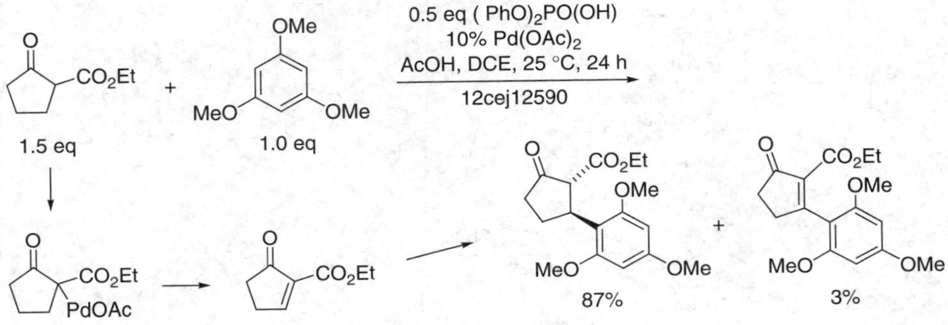

图式 1.39　脱氢是钯催化酮的芳基化反应中的副反应[169]

有些情况下,亲核试剂不是进攻苄基亲电试剂的苄位,而是进攻芳环位置(图式 1.37 中第一个反应)。无取代芳烃与四氢萘酮及相关芳酮亲电芳基化反应的例子也有报道(图式 1.40)。

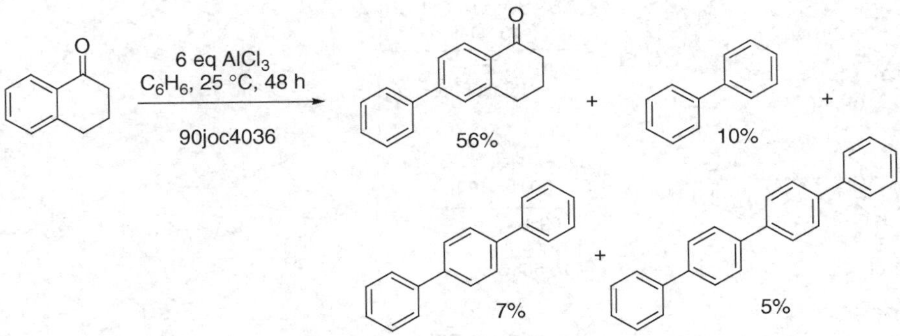

图式 1.40　苯和四氢萘酮的芳基化[83]

1.3.8　醇

醇被广泛用作傅-克烷基化的亲电试剂,通常它比卤代烷更活泼,但在芳烃烷基化时需要更多的酸。伯醇和非苄型醇很少被作为烷基化试剂,原因是它们能快速重排成为更稳定的二级或三级阳离子。

与其他类型的亲电试剂一样,不能快速形成碳正离子的醇不适合用于芳烃烷基化。例如,目前未发现过用2,2,2-三卤代乙醇或氰醇进行阳离子型芳烃烷基化反应的实例,仅有几个α-羟基羧酸或α-羟基酮的芳烃烷基化反应被报道,大部分实例中还是使用具有能够稳定碳正离子的α-取代醇(即苄醇)。

在强碱性条件下,吲哚的C-3位能被乙醇酸烷基化,但反应是通过醇被氧化成醛后进行的(图式1.41)。芴与醇在苄位亚甲基上发生类似的烷基化也有报道[170,171]。

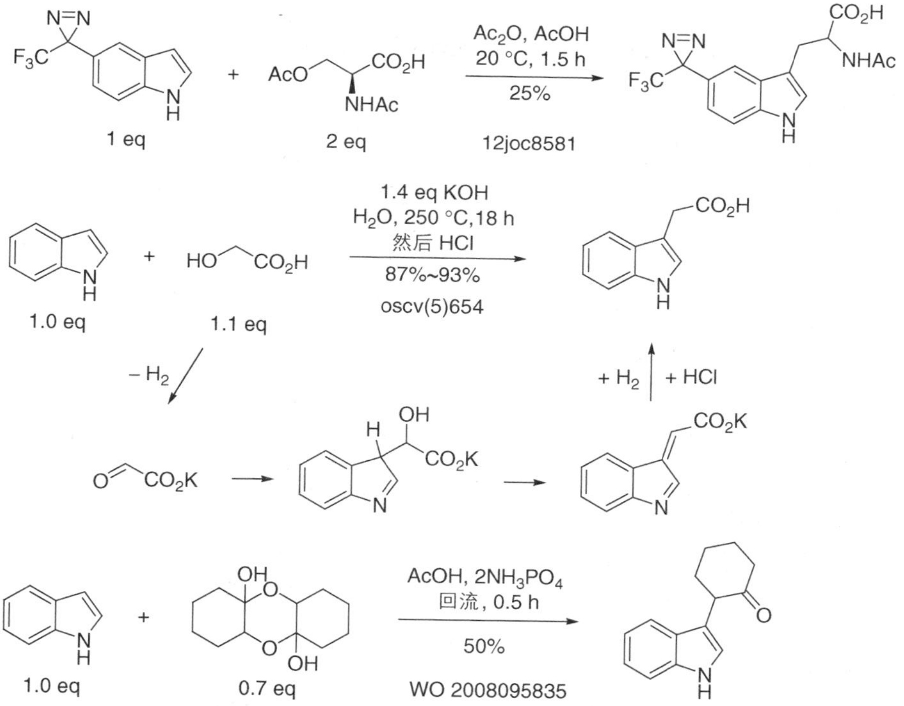

图式1.41 吲哚与醇和酯的烷基化[172-174]

醇或者酯可通过脱水形成迈克尔(Michael)受体,它们在反应中作为软亲电试剂并且非常适合富电子芳烃的烷基化(图式1.41中第一个反应)。

2-氨基和2-烷氧基醇是在酸性条件下不容易对芳烃进行烷基化的醇类型。氧和氮原子的电负性比碳原子大,诱导效应导致相应的碳正离子去稳定化。此外,酸会使胺和醚质子化,进一步降低了所需双正离子形成的可能性。换言之,仅有被活化的醇(即苄醇或烯丙醇)或分子内烷基化能以可接受的收率进行芳烃烷基化反应(图式1.42)。

图式 1.42　芳烃与 2-氨基醇和乙二醇的烷基化[175-177]

参考文献[①]

1. Bachmann, W.E. (1955) Triphenyl-chloromethane. *Org. Synth.*, Coll. Vol. **3**, 841–845.
2. Xu, X., Xu, X., Li, H., Xie, X., and Li, Y. (2010) Iron-catalyzed, microwave-promoted, one-pot synthesis of 9-substituted xanthenes by a cascade benzylation-cyclization process. *Org. Lett.*, **12**, 100–103.
3. Chung, K.H., Kim, J.N., and Ryu, E.K. (1994) Friedel–Crafts alkylation reactions of benzene with amide bond containing compounds. *Tetrahedron Lett.*, **35**, 2913–2914.
4. Bonvino, V., Casini, G., Ferappi, M., Cingolani, G.M., and Pietroni, B.R. (1981) Nitro compounds as alkylating reagents in Friedel–Crafts conditions: reaction of 2-nitropropane with benzene. *Tetrahedron*, **37**, 615–620.
5. Margosian, D., Speier, J., and Kovacic, P. (1981) Formation of (1-adamantylcarbinyl)arenes from 3-azidohomoadamantane–aluminum chloride–aromatic substrates. *J. Org. Chem.*, **46**, 1346–1350.
6. Zhao, X., Wu, G., Zhang, Y., and Wang, J. (2011) Copper-catalyzed direct benzylation or allylation of 1,3-azoles with *N*-tosylhydrazones. *J. Am. Chem. Soc.*, **133**, 3296–3299.
7. Klumpp, D.A., Aguirre, S.L., Sanchez, G.V. Jr., and de Leon, S.J. (2001) Reactions of amino alcohols in superacid: the direct observation of dicationic intermediates and their application in synthesis. *Org. Lett.*, **3**, 2781–2784.
8. Ohwada, T., Kasuga, M., and Shudo, K. (1990) Direct observation of an intermediate in the oxygen atom rearrangement of 2-cyclopropylnitrobenzene in a strong acid. *J. Org. Chem.*, **55**, 2717–2719.
9. Katritzky, A.R., Lopez Rodriguez, M.L., Keay, J.G., and King, R.W. (1985) Nucleophilic displacement with heterocycles as leaving groups. Part 16. Reactions of secondary alkyl primary amines with 5,6,8,9-tetrahydro-7-phenyldibenzo[*c,h*]xanthylium trifluoromethanesulfonate to give intermediates solvolysing without rearrangement. *J. Chem. Soc., Perkin Trans. 2*, 165–169.
10. Imm, S., Bähn, S., Tillack, A., Mevius, K., Neubert, L., and Beller, M. (2010) Selective ruthenium-catalyzed alkylation of indoles by using amines. *Chem. Eur. J.*, **16**, 2705–2709.
11. Quast, H., Nüdling, W., Klemm, G., Kirschfeld, A., Neuhaus, P., Sander, W., Hrovat, D.A., and Borden, W.T. (2008) A perimidine-derived non-Kekulé triplet diradical. *J. Org. Chem.*, **73**, 4956–4961.
12. Mazik, M. and Sonnenberg, C. (2010) Isopropylamino and isobutylamino groups as recognition sites for carbohydrates: acyclic receptors with enhanced binding affinity toward β-galactosides. *J. Org. Chem.*, **75**, 6416–6423.
13. Wallace, K.J., Hanes, R., Anslyn, E., Morey, J., Kilway, K.V., and Siegel, J. (2005) Preparation of

---

[①]　为方便读者查阅,本书原样复制原著参考文献。

13. 1,3,5-tris(aminomethyl)-2,4,6-triethylbenzene from two versatile 1,3,5-tri(halosubstituted) 2,4,6-triethylbenzene derivatives. *Synthesis*, 2080–2083.
14. Saulnier, G., Dodier, M., Frennesson, D.B., Langley, D.R., and Vyas, D.M. (2009) Nucleophilic capture of the imino-quinone methide type intermediates generated from 2-aminothiazol-5-yl carbinols. *Org. Lett.*, **11**, 5154–5157.
15. Piao, C., Zhao, Y., Han, X., and Liu, Q. (2008) $AlCl_3$-mediated direct carbon–carbon bond-forming reaction of α-hydroxyketene-S,S-acetals with arenes and synthesis of 3,4-disubstituted dihydrocoumarin derivatives. *J. Org. Chem.*, **73**, 2264–2269.
16. O'Keefe, B.M., Mans, D.M., Kaelin, D.E., and Martin, S.F. (2010) Total synthesis of isokidamycin. *J. Am. Chem. Soc.*, **132**, 15528–15530.
17. Izumi, K., Kabaki, M., Uenaka, M., and Shimizu, S. (2007) One-step synthesis of 5-(4-fluorobenzyl)-2-furyl methyl ketone: a key intermediate of HIV-integrase inhibitor S-1360. *Org. Process Res. Dev.*, **11**, 1059–1061.
18. Piccolo, O., Azzena, U., Melloni, G., Delogu, G., and Valoti, E. (1991) Stereospecific Friedel–Crafts alkylation of aromatic compounds: synthesis of optically active 2- and 3-arylalkanoic esters. *J. Org. Chem.*, **56**, 183–187.
19. Klumpp, D.A., Garza, M., Jones, A., and Mendoza, S. (1999) Synthesis of aryl-substituted piperidines by superacid activation of piperidones. *J. Org. Chem.*, **64**, 6702–6705.
20. Hofmann, M., Hampel, N., Kanzian, T., and Mayr, H. (2004) Electrophilic alkylations in neutral aqueous or alcoholic solutions. *Angew. Chem. Int. Ed.*, **43**, 5402–5405.
21. Liu, C. and Widenhoefer, R.A. (2006) Scope and mechanism of the Pd^{II}-catalyzed arylation/carboalkoxylation of unactivated olefins with indoles. *Chem. Eur. J.*, **12**, 2371–2382.
22. Zhao, Y. and Chen, G. (2011) Palladium-catalyzed alkylation of ortho-C(sp^2)–H bonds of benzylamide substrates with alkyl halides. *Org. Lett.*, **13**, 4850–4853.
23. Ackermann, L., Novák, P., Vicente, R., and Hofmann, N. (2009) Ruthenium-catalyzed regioselective direct alkylation of arenes with unactivated alkyl halides through C–H bond cleavage. *Angew. Chem. Int. Ed.*, **48**, 6045–6048.
24. Hofmann, N. and Ackermann, L. (2013) Meta-selective C–H bond alkylation with secondary alkyl halides. *J. Am. Chem. Soc.*, **135**, 5877–5884.
25. Tsai, A.S., Brasse, M., Bergman, R.G., and Ellman, J.A. (2011) Rh(III)-catalyzed oxidative coupling of unactivated alkenes via C–H activation. *Org. Lett.*, **13**, 540–542.
26. Yu, D., Lee, S., Sum, Y.N., and Zhang, Y. (2012) Selective formation of formamidines or 7-aminomethylbenzoxazoles from unprecedented couplings between benzoxazoles and amines. *Adv. Synth. Catal.*, **354**, 1672–1678.
27. Matsuzawa, H., Miyake, Y., and Nishibayashi, Y. (2007) Ruthenium-catalyzed enantioselective propargylation of aromatic compounds with propargylic alcohols via allenylidene intermediates. *Angew. Chem. Int. Ed.*, **46**, 6488–6491.
28. Mertins, K., Iovel, I., Kischel, J., Zapf, A., and Beller, M. (2005) Transition-metal-catalyzed benzylation of arenes and heteroarenes. *Angew. Chem. Int. Ed.*, **44**, 238–242.
29. Olah, G.A. and Kuhn, S.J. (1964) Selective Friedel–Crafts reactions. I. Boron halide catalyzed haloalkylation of benzene and alkylbenzenes with fluorohaloalkanes. *J. Org. Chem.*, **29**, 2317–2320.
30. Shi, Z. and Chuan, H. (2004) Direct functionalization of arenes by primary alcohol sulfonate esters catalyzed by gold(III). *J. Am. Chem. Soc.*, **126**, 13596–13597.
31. Mayr, H., Lang, G., and Ofial, A.R. (2002) Reactions of carbocations with unsaturated hydrocarbons: electrophilic alkylation or hydride abstraction? *J. Am. Chem. Soc.*, **124**, 4076–4083.
32. Serres, C. and Fields, E.K. (1963) Synthesis of 2,2-diarylpropanes by hydride transfer. *J. Org. Chem.*, **28**, 1624–1627.

33. Coumbarides, G.S., Dingjan, M., Eames, J., and Weerasooriya, N. (2001) Investigations into the bromination of substituted phenols using diethyl bromomalonate and diethyl dibromomalonate. *Bull. Chem. Soc. Jpn.*, **74**, 179–180.

34. Kinoyama, I., Miyazaki, T., Koganemaru, Y., Shiraishi, N., Kawamoto, Y., and Washio, T. (2011) Acylguanidine derivatives. US Patent 2011306621.

35. Pradhan, A., Dechambenoit, P., Bock, H., and Durola, F. (2013) Twisted polycyclic arenes by intramolecular Scholl reactions of C3-symmetric precursors. *J. Org. Chem.*, **78**, 2266–2274.

36. Shen, Y., Liu, H., Wu, M., Du, W., Chen, Y., and Li, N. (1991) Friedel–Crafts alkylation of benzenes substituted with *meta*-directing groups. *J. Org. Chem.*, **56**, 7160–7162.

37. Dall'Asta, L., Casazza, U., and Cotticelli, G. (2001) Process for the preparation of 5-carboxyphthalide. EP Patent 1118614.

38. Buc, S.R. (1956) Preparation of *meta*-nitrobenzyl chlorides. US Patent 2758137.

39. Grellier, M., Vendier, L., Chaudret, B., Albinati, A., Rizzato, S., Mason, S., and Sabo-Etienne, S. (2005) Synthesis, neutron structure, and reactivity of the bis(dihydrogen) complex $RuH_2(\eta^2-H_2)_2(PCyp_3)_2$ stabilized by two tricyclopentylphosphines. *J. Am. Chem. Soc.*, **127**, 17592–17593.

40. Mochida, S., Hirano, K., Satoh, T., and Miura, M. (2011) Rhodium-catalyzed regioselective olefination directed by a carboxylic group. *J. Org. Chem.*, **76**, 3024–3033.

41. Park, S.H., Kim, J.Y., and Chang, S. (2011) Rhodium-catalyzed selective olefination of arene esters via C–H bond activation. *Org. Lett.*, **13**, 2372–2375.

42. Rakshit, S., Grohmann, C., Besset, T., and Glorius, F. (2011) Rh(III)-catalyzed directed C–H olefination using an oxidizing directing group: mild, efficient, and versatile. *J. Am. Chem. Soc.*, **133**, 2350–2353.

43. Busch, S. and Leitner, W. (2001) Ruthenium-catalysed Murai-type couplings at room temperature. *Adv. Synth. Catal.*, **343**, 192–195.

44. Ackermann, L. and Novák, P. (2009) Regioselective ruthenium-catalyzed direct benzylations of arenes through C–H bond cleavages. *Org. Lett.*, **11**, 4966–4969.

45. Jun, C., Moon, C.W., Hong, J., Lim, S., Chung, K., and Kim, Y. (2002) Chelation-assisted Rh^I-catalyzed *ortho*-alkylation of aromatic ketimines, or ketones with olefins. *Chem. Eur. J.*, **8**, 485–492.

46. Wang, X., Truesdale, L., and Yu, J. (2010) Pd(II)-catalyzed *ortho*-trifluoromethylation of arenes using TFA as a promoter. *J. Am. Chem. Soc.*, **132**, 3648–3649.

47. Ueno, S., Kochi, T., Chatani, N., and Kakiuchi, F. (2009) Unique effect of coordination of an alkene moiety in products on ruthenium-catalyzed chemoselective C–H alkenylation. *Org. Lett.*, **11**, 855–858.

48. Yang, Y., Lin, Y., and Rao, Y. (2012) Ruthenium(II)-catalyzed synthesis of hydroxylated arenes with ester as an effective directing group. *Org. Lett.*, **14**, 2874–2877.

49. Yadav, M.R., Rit, R.K., and Sahoo, A.K. (2012) Sulfoximines: a reusable directing group for chemo- and regioselective *ortho* C–H oxidation of arenes. *Chem. Eur. J.*, **18**, 5541–5545.

50. Martinez, R., Simon, M., Chevalier, R., Pautigny, C., Genet, J.-P., and Darses, S. (2009) C–C bond formation via C–H bond activation using an in situ-generated ruthenium catalyst. *J. Am. Chem. Soc.*, **131**, 7887–7895.

51. Ueno, S., Chatani, N., and Kakiuchi, F. (2007) Ruthenium-catalyzed C–C bond formation via the cleavage of an unreactive aryl C–N bond in aniline derivatives with organoboronates. *J. Am. Chem. Soc.*, **129**, 6098–6099.

52. Sonoda, M., Kakiuchi, F., Chatani, N., and Murai, S. (1997) Ruthenium-catalyzed addition of C–H bonds in aromatic ketones to olefins. The effect

of various substituents at the aromatic ring. *Bull. Chem. Soc. Jpn.*, **70**, 3117–3128.

53. Ueno, S., Mizushima, E., Chatani, N., and Kakiuchi, F. (2006) Direct observation of the oxidative addition of the aryl C–O bond to a ruthenium complex and elucidation of the relative reactivity between aryl C–O and aryl C–H bonds. *J. Am. Chem. Soc.*, **128**, 16516–16517.

54. Ryu, J., Cho, S.H., and Chang, S. (2012) A versatile rhodium(I) catalyst system for the addition of heteroarenes to both alkenes and alkynes by a C–H bond activation. *Angew. Chem. Int. Ed.*, **51**, 3677–3681.

55. Yao, B., Song, R., Liu, Y., Xie, Y., Li, J., Wang, M., Tang, R., Zhang, X., and Deng, C. (2012) Palladium-catalyzed C–H oxidation of isoquinoline N-oxides: selective alkylation with dialkyl sulfoxides and halogenation with dihalo sulfoxides. *Adv. Synth. Catal.*, **354**, 1890–1896.

56. Tran, L.D. and Daugulis, O. (2010) Iron-catalyzed heterocycle and arene deprotonative alkylation. *Org. Lett.*, **12**, 4277–4279.

57. Popov, I., Do, H., and Daugulis, O. (2009) In situ generation and trapping of aryllithium and arylpotassium species by halogen, sulfur, and carbon electrophiles. *J. Org. Chem.*, **74**, 8309–8313.

58. Linhardt, R.J., Montgomery, B.L.E., Osby, J., and Sherbine, J. (1982) Mechanism for diacyl peroxide decomposition. *J. Org. Chem.*, **47**, 2242–2251.

59. Molander, G.A., Colombel, V., and Braz, V.A. (2011) Direct alkylation of heteroaryls using potassium alkyl- and alkoxymethyltrifluoroborates. *Org. Lett.*, **13**, 1852–1855.

60. Glazunov, V.P., Tchizhova, A.Y., Pokhilo, N.D., Anufriev, V.P., and Elyakov, G.B. (2002) First direct observation of tautomerism of monohydroxynaphthazarins. *Tetrahedron*, **58**, 1751–1757.

61. Guo, X. and Li, C.-J. (2011) Ruthenium-catalyzed *para*-selective oxidative cross-coupling of arenes and cycloalkanes. *Org. Lett.*, **13**, 4977–4979.

62. Minisci, F., Vismara, E., and Fontana, F. (1989) Homolytic alkylation of protonated heteroaromatic bases by alkyl iodides, hydrogen peroxide, and dimethyl sulfoxide. *J. Org. Chem.*, **54**, 5224–5227.

63. Miller, B.L., Palde, P.B., and Gareiss, P.C. (2008) Tripodal cyclohexane derivatives and their use as carbohydrate receptors., WO Patent 2008048967.

64. Duncton, M.A.J., Estiarte, M.A., Johnson, R.J., Cox, M., O'Mahony, D.J.R., Edwards, W.T., and Kelly, M.G. (2009) Preparation of heteroaryloxetanes and heteroarylazetidines by use of a Minisci reaction. *J. Org. Chem.*, **74**, 6354–6357.

65. Yoshida, M., Amemiya, H., Kobayashi, M., Sawada, H., Hagii, H., and Aoshima, K. (1985) Perfluoropropylation of aromatic compounds with bis(heptafluorobutyryl) peroxide. *J. Chem. Soc., Chem. Commun*, 234–236.

66. Firth, B.E. and Rosen, T.J. (1985) Preparation of *ortho*-alkylated phenols. US Patent 4538008.

67. Rao, H.S.P., Geetha, K., and Kamalraj, M. (2011) Synthesis of 4-(2-hydroxyaryl)-3-nitro-4*H*-chromenes. *Tetrahedron*, **67**, 8146–8154.

68. Lee, D.-H., Kwon, K.-H., and Yi, C.S. (2012) Dehydrative C–H alkylation and alkenylation of phenols with alcohols: expedient synthesis for substituted phenols and benzofurans. *J. Am. Chem. Soc.*, **134**, 7325–7328.

69. Toshima, K., Matsuo, G., Ishizuka, T., Ushiki, Y., Nakata, M., and Matsumura, S. (1998) Aryl and allyl *C*-glycosidation methods using unprotected sugars. *J. Org. Chem.*, **63**, 2307–2313.

70. Podder, S., Choudhury, J., Roy, U.K., and Roy, S. (2007) Dual-reagent catalysis within Ir–Sn domain: highly selective alkylation of arenes and heteroarenes with aromatic aldehydes. *J. Org. Chem.*, **72**, 3100–3103.

71. Zhang, C., Gao, X., Zhang, J., and Peng, X. (2010) Fe/CuBr$_2$-catalyzed benzylation of arenes and thiophenes with benzyl alcohols. *Synlett*, 261–265.

72. Firth, B.E. (1981) Preparation of 2,4,6-triisopropylphenol. US Patent 4275248.

73. Klemm, L.H. and Taylor, D.R. (1980) Alumina-catalyzed reactions of hydroxyarenes and hydroaromatic ketones. 9. Reaction of phenol with l-propanol. *J. Org. Chem.*, **45**, 4320–4326.

74. Kealy, T.J. and Coffman, D.D. (1961) Thermal addition reactions of monocyclic phenols with ethylene. *J. Org. Chem.*, **26**, 987–992.

75. Zhao, C., Camaioni, D.M., and Lercher, J.A. (2012) Selective catalytic hydroalkylation and deoxygenation of substituted phenols to bicycloalkanes. *J. Catal.*, **288**, 92–103.

76. Kolka, J., Napolitano, J.P., and Ecke, G.G. (1956) The *ortho*-alkylation of phenols. *J. Org. Chem.*, **21**, 712–713.

77. Nakagawa, Y. and Sato, T. (2009) Preparation of 2,6-diphenylphenols. JP Patent 2009269868.

78. Youn, S.W. and Eom, J.I. (2006) Ag(I)-catalyzed sequential C–C and C–O bond formations between phenols and dienes with atom economy. *J. Org. Chem.*, **71**, 6705–6707.

79. Ancel, J.E., Bienayme, H., and Meilland, P. (1996) Procédé des préparations de phénols substitués. EP Patent 0742194.

80. Dang, T.T., Boeck, F., and Hintermann, L. (2011) Hidden Brønsted acid catalysis: pathways of accidental or deliberate generation of triflic acid from metal triflates. *J. Org. Chem.*, **76**, 9353–9361.

81. Repinskaya, I.B., Barkhutova, D.D., Makarova, Z.S., Alekseeva, A.V., and Koptyug, V.A. (1985) Condensation of phenols and their derivatives with aromatic compounds in the presence of acidic agents. VII. Comparison of the condensation of phenols with aromatic compounds in HF–SbF$_5$ and in the presence of aluminum halides. *Zh. Org. Khim.*, **21**, 836–845 (translation: 759–767).

82. Repinskaya, I.B. and Koptyug, V.A. (1980) Condensation of substituted resorcinols with benzene and chlorobenzene. *Zh. Org. Khim.*, **16**, 1508–1514 (translation: 1298–1303).

83. El-Zohry, M.F. and El-Khawaga, A.M. (1990) Nonconventional Friedel–Crafts chemistry. 1. Reaction of α-tetralone and anthrone with arenes under Friedel–Crafts conditions. *J. Org. Chem.*, **55**, 4036–4039.

84. Koptyug, V.A. and Golounin, A.V. (1973) Reaction of phenols with Lewis acids VIII. Reaction of phenol and its methylated derivatives with benzene and aluminum halides. *Zh. Org. Khim.*, **9**, 2158–2163 (translation: 2172–2176).

85. Koltunov, K.Y., Walspurger, S., and Sommer, J. (2004) Superelectrophilic activation of polyfunctional organic compounds using zeolites and other solid acids. *Chem. Commun.*, 1754–1755.

86. Witten, B. and Reid, E.E. (1963) p-Aminotetraphenylmethane. *Org. Synth.*, Coll. Vol. **4**, 47–48.

87. Schultz, W.J., Portelli, G.B., and Tane, J.P. (1986) Epoxy resin curing agent, curing process and composition containing it. EP Patent 0203828.

88. Cherian, A.E., Domski, G.J., Rose, J.M., Lobkovsky, E.B., and Coates, G.W. (2005) Acid-catalyzed *ortho*-alkylation of anilines with styrenes: an improved route to chiral anilines with bulky substituents. *Org. Lett.*, **7**, 5135–5137.

89. Macé, Y., Raymondeau, B., Pradet, C., Blazejewski, J.-C., and Magnier, E. (2009) Benchmark and solvent-free preparation of sulfonium salt based electrophilic trifluoromethylating reagents. *Eur. J. Org. Chem.*, **2009**, 1390–1397.

90. Yanai, H., Yoshino, T., Fujita, M., Fukaya, H., Kotani, A., Kusu, F., and Taguchi, T. (2013) Synthesis, characterization, and applications of zwitterions containing a carbanion moiety. *Angew. Chem. Int. Ed.*, **52**, 1560–1563.

91. Hickinbottom, W.J. (1933) The elimination of tertiary alkyl groups from alkylanilines by hydrolysis. *J. Chem. Soc.*, 1070–1073.

92. Pillai, R.B.C. (1999) Isomerisation of *N*-isopropylaniline and *N*-n-propylaniline over zeolites. *J. Indian Chem. Soc.*, **76**, 157–158.

93. Beyer, H., Gerencsérné, A., Palkovics, I., Gémes, I., Rátosi, E., Sebestyén,

B., Horváth, J., Czágler, I., Perger, J., Hegedüs, I., Arányi, P., Torkos, L., Hódossy, L., Borbéli, G., and Halász, I. (1992) Process for the Preparation of Ring-Alkylated Anilines by Isomerization of *N*-alkylanilines Over Acidic Zeolites as Catalysts. Ger. Offen. 4023652.
94. Bourns, A.N., Embleton, H.W., and Hansuld, M.K. (1963) 1-Phenylpiperidine. *Org. Synth.*, Coll. Vol. **4**, 795–798.
95. Glass, D.B. and Weissberger, A. (1955) Julolidine. *Org. Synth.*, Coll. Vol. **3**, 504–505.
96. Lapis, A.A.M., DaSilveira Neto, B.A., Scholten, J.D., Nachtigall, F.M., Eberlin, M.N., and Dupont, J. (2006) Intermolecular hydroamination and hydroarylation reactions of alkenes in ionic liquids. *Tetrahedron Lett.*, **47**, 6775–6779.
97. Motokura, K., Nakagiri, N., Mizugaki, T., Ebitani, K., and Kaneda, K. (2007) Nucleophilic substitution reactions of alcohols with use of montmorillonite catalysts as solid Brønsted acids. *J. Org. Chem.*, **72**, 6006–6015.
98. Tsuji, Y., Huh, K.-T., Ohsugi, Y., and Watanabe, Y. (1985) Ruthenium complex catalyzed *N*-heterocyclization. Syntheses of *N*-substituted piperidines, morpholines, and piperazines from amines and 1,5-diols. *J. Org. Chem.*, **50**, 1365–1370.
99. Wei, H., Qian, G., Xia, Y., Li, K., Li, Y., and Li, W. (2007) BiCl$_3$-catalyzed hydroamination of norbornene with aromatic amines. *Eur. J. Org. Chem.*, **2007**, 4471–4474.
100. Didier, D. and Sergeyev, S. (2007) Bromination and iodination of 6*H*,12*H*-5,11-methanodibenzo[*b,f*][1,5]diazocine: a convenient entry to unsymmetrical analogs of Tröger's base. *Eur. J. Org. Chem.*, **2007**, 3905–3910.
101. Corma, A., Botella, P., and Mitchell, C. (2004) Replacing HCl by solid acids in industrial processes: synthesis of diamino diphenyl methane (DADPM) for producing polyurethanes. *Chem. Commun.*, 2008–2010.
102. Mitchell, C.J., Corma Canos, A., Carr, R.H., and Botella Asunción, P. (2010) Process for production of methylene-bridged polyphenyl polyamines. WO Patent 2010072504.
103. Chauvin, A.-S., Comby, S., Song, B., Vandevyver, C.D.B., Thomas, F., and Bünzli, J.-C.G. (2007) A polyoxyethylene-substituted bimetallic europium helicate for luminescent staining of living cells. *Chem. Eur. J.*, **13**, 9515–9526.
104. Ellis, G.D., Dimarcello, B.J., and Bradshaw, D.J. (1999) Preparation of a dye for coloring protein-based fibers and cellulose-based materials from the oxidation byproducts of the manufacture of a triphenylmethane dye. EP Patent 0909794.
105. Ecke, G.G., Napolitano, J.P., and Kolka, A.J. (1956) The *ortho*-alkylation of aromatic amines. *J. Org. Chem.*, **21**, 711–712.
106. Chockalingam, K. (2009) Preparation of 2-(1,3-dimethylbutyl)aniline and other branched alkyl-substituted anilines. WO Patent 2009029383.
107. Diamond, S.E., Szalkiewicz, A., and Mares, F. (1979) Reactions of aniline with olefins catalyzed by group 8 metal complexes: *N*-alkylation and heterocycle formation. *J. Am. Chem. Soc.*, **101**, 490–491.
108. Ecke, G.G., Napolitano, J.P., Filbey, A.H., and Kolka, A.J. (1957) Ortho-alkylation of aromatic amines. *J. Org. Chem.*, **22**, 639–642.
109. Matthews, D.P., McCarthy, J.R., and Whitten, J.P. (1989) Novel (aryl or heteroaromatic methyl)-2,2′-bi-1*H*-imidazoles, EP Patent 0301456.
110. Tan, K.L., Park, S., Ellman, J.A., and Bergman, R.G. (2004) Intermolecular coupling of alkenes to heterocycles via C–H bond activation. *J. Org. Chem.*, **69**, 7329–7335.
111. Jiao, L., Herdtweck, E., and Bach, T. (2012) Pd(II)-catalyzed regioselective 2-alkylation of indoles via a norbornene-mediated C–H activation: mechanism and applications. *J. Am. Chem. Soc.*, **134**, 14563–14572.
112. Olah, G.A., Farooq, O., Farnia, S.M.F., and Olah, J.A. (1988)

Boron, aluminum, and gallium tris(trifluoromethanesulfonate) (triflate): effective new Friedel–Crafts catalysts. *J. Am. Chem. Soc.*, **110**, 2560–2565.

113. Cullinane, N.M., Chard, S.J., and Dawkins, C.W.C. (1963) Hexamethylbenzene. *Org. Synth.*, Coll. Vol. **4**, 520–521.

114. Smith, L.I. (1943) Durene. *Org. Synth.*, Coll. Vol. **2**, 248–253.

115. Zhang, Y., Feng, J., and Li, C.-J. (2008) Palladium-catalyzed methylation of aryl C–H bonds by using peroxides. *J. Am. Chem. Soc.*, **130**, 2900–2901.

116. Nakao, Y., Kashihara, N., Kanyiva, K.S., and Hiyama, T. (2010) Nickel-catalyzed hydroheteroarylation of vinylarenes. *Angew. Chem. Int. Ed.*, **49**, 4451–4454.

117. Watson, A.J.A., Maxwell, A.C., and Williams, J.M.J. (2010) Ruthenium-catalyzed aromatic C–H activation of benzylic alcohols via remote electronic activation. *Org. Lett.*, **12**, 3856–3859.

118. Matsumoto, T., Taube, D.J., Periana, R.A., Taube, H., and Yoshida, H. (2000) Anti-Markovnikov olefin arylation catalyzed by an iridium complex. *J. Am. Chem. Soc.*, **122**, 7414–7415.

119. Sun, Z.-M., Zhang, J., Manan, R.S., and Zhao, P. (2010) Rh(I)-catalyzed olefin hydroarylation with electron-deficient perfluoroarenes. *J. Am. Chem. Soc.*, **132**, 6935–6937.

120. Liu, F., Martin-Mingot, A., Jouannetaud, M., Zunino, F., and Thibaudeau, S. (2010) Superelectrophilic activation in superacid HF/SbF$_5$ and synthesis of benzofused sultams. *Org. Lett.*, **12**, 868–871.

121. Okabe, K., Ohwada, T., Ohta, T., and Shudo, K. (1989) Novel electrophilic species equivalent to α-keto cations. Reactions of O,O-diprotonated nitro olefins with benzenes yield arylmethyl ketones. *J. Org. Chem.*, **54**, 733–734.

122. Fujiwara, Y., Kuromaru, H., and Taniguchi, H. (1984) Trifluoroacetic acid catalyzed allylic phenylation of α-methylallyl acetate, α-methylallyl trifluoroacetate, and α-methylallyl alcohol with benzene. *J. Org. Chem.*, **49**, 4309–4310.

123. Fan, S., Chen, F., and Zhang, X. (2011) Direct palladium-catalyzed intermolecular allylation of highly electron-deficient polyfluoroarenes. *Angew. Chem. Int. Ed.*, **50**, 5918–5923.

124. Makida, Y., Ohmiya, H., and Sawamura, M. (2012) Regio- and stereocontrolled introduction of secondary alkyl groups to electron-deficient arenes through copper-catalyzed allylic alkylation. *Angew. Chem. Int. Ed.*, **51**, 4122–4127.

125. Yao, T., Hirano, K., Satoh, T., and Miura, M. (2011) Stereospecific copper-catalyzed C–H allylation of electron-deficient arenes with allyl phosphates. *Angew. Chem. Int. Ed.*, **50**, 2990–2994.

126. Zhang, Y., Chen, L., and Lu, T. (2011) A copper(II) triflate-catalyzed tandem Friedel–Crafts alkylation/cyclization process towards dihydroindenes. *Adv. Synth. Catal.*, **353**, 1055–1060.

127. Müller, H., Trübenbach, P., Rieger, B., Wagner, J.M., and Dietrich, U. (1998) Verfahren zur Herstellung von Indenderivaten. EP Patent 0826654.

128. Li, Z., Zhang, Y., and Liu, Z.-Q. (2012) Pd-catalyzed olefination of perfluoroarenes with allyl esters. *Org. Lett.*, **14**, 74–77.

129. Prakash, G.K.S., Paknia, F., Vaghoo, H., Rasul, G., Mathew, T., and Olah, G.A. (2010) Preparation of trifluoromethylated dihydrocoumarins, indanones, and arylpropanoic acids by tandem superacidic activation of 2-(trifluoromethyl)acrylic acid with arenes. *J. Org. Chem.*, **75**, 2219–2226.

130. Liu, Y.-H., Liu, Q.-S., and Zhang, Z.-H. (2009) An efficient Friedel–Crafts alkylation of nitrogen heterocycles catalyzed by antimony trichloride/montmorillonite K-10. *Tetrahedron Lett.*, **50**, 916–921.

131. Shi, Z. and He, C. (2004) An Au-catalyzed cyclialkylation of electron-rich arenes with epoxides to prepare 3-chromanols. *J. Am. Chem. Soc.*, **126**, 5964–5965.

132. Yadav, J.S., Reddy, B.V.S., Abraham, S., and Sabitha, G. (2002) InCl$_3$-catalyzed regioselective opening of aziridines

with heteroaromatics. *Tetrahedron Lett.*, **43**, 1565–1567.

133. Thirupathi, B., Srinivas, R., Prasad, A.N., Kumar, J.K.P., and Reddy, B.M. (2010) Green progression for synthesis of regioselective β-amino alcohols and chemoselective alkylated indoles. *Org. Process Res. Dev.*, **14**, 1457–1463.

134. Taylor, S.K., Clark, D.L., Heinz, K.L., Schramm, S.B., Westermann, C.D., and Barnell, K.K. (1983) Friedel–Crafts reactions of some conjugated epoxides. *J. Org. Chem.*, **48**, 592–596.

135. Molnar, A., Ledneczki, I., Bucsi, I., and Bartok, M. (2003) Alkylation of benzene with cyclic ethers in superacidic media. *Catal. Lett.*, **89**, 1–9.

136. Lin, J.-R., Gubaidullin, A.T., Mamedovb, V.A., and Tsuboi, S. (2003) Nucleophilic addition reaction of aromatic compounds with α-chloroglycidates in the presence of Lewis acid. *Tetrahedron*, **59**, 1781–1790.

137. Schultz, E.M. and Mickey, S. (1955) α,α-Diphenylacetone. *Org. Synth.*, Coll. Vol. **3**, 343–346.

138. Robb, C.M. and Schultz, E.M. (1955) Diphenylacetonitrile. *Org. Synth.*, Coll. Vol. **3**, 347–349.

139. Itoh, S., Yoshimura, N., Sato, M., and Yamataka, H. (2011) Computational study on the reaction pathway of α-bromoacetophenones with hydroxide ion: possible path bifurcation in the addition/substitution mechanism. *J. Org. Chem.*, **76**, 8294–8299.

140. Mason, J.P. and Terry, L.I. (1940) Preparation of phenylacetone. *J. Am. Chem. Soc.*, **62**, 1622.

141. Piccolo, O., Spreafico, F., Visentin, G., and Valoti, E. (1985) Alkylation of aromatic compounds with optically active lactic acid derivatives: synthesis of optically pure 2-arylpropionic acid and esters. *J. Org. Chem.*, **50**, 3945–3946.

142. Stelzer, U. (1995) Verfahren zur Herstellung von substituierten Phenylessigsäurederivaten und Zwischenprodukten. EP Patent 0665212.

143. Ogata, Y. and Ishiguro, J. (1950) Preparation of α-naphthaleneacetic acid by the condensation of naphthalene with chloroacetic acid. *J. Am. Chem. Soc.*, **72**, 4302.

144. Kawai, D., Kawasumi, K., Miyahara, T., Hirashita, T., and Araki, S. (2009) Cyclopropanation of electron-deficient alkenes with activated dibromomethylene compounds mediated by lithium iodide or tetrabutylammonium salts. *Tetrahedron*, **65**, 10390–10394.

145. Ogata, Y., Ishiguro, J., and Kitamura, Y. (1951) Condensation of naphthalenes with α-halo fatty acids and related reactions. *J. Org. Chem.*, **21**, 239–242.

146. Chung, J.Y.L., Steinhübel, D., Krska, S.W., Hartner, F.W., Cai, C., Rosen, J., Mancheno, D.E., Pei, T., DiMichele, L., Ball, R.G., Chen, C., Tan, L., Alorati, A.D., Brewer, S.E., and Scott, J.P. (2012) Asymmetric synthesis of a glucagon receptor antagonist via Friedel–Crafts alkylation of indole with chiral α-phenyl benzyl cation. *Org. Process Res. Dev.*, **16**, 1832–1845.

147. Clement, K., Tasset, E.L., and Walker, L.L. (1998) Process for the preparation of stilbene diols. WO Patent 9811043.

148. Zaheer, S.H., Singh, B., Bhushan, B., Bhargava, P.M., Kacker, I.K., Ramachandran, K., Sastri, V.D.N., and Rao, N.S. (1954) Reactions of α-halogeno-ketones with aromatic compounds. Part I. Reactions of chloroacetone and 3-chlorobutanone with phenol and its ethers. *J. Chem. Soc.*, 3360–3362.

149. Baciocchi, E., Muraglia, E., and Sleiter, G. (1992) Homolytic substitution reactions of electron-rich pentatomic heteroaromatics by electrophilic carbon-centered radicals. Synthesis of α-heteroarylacetic acids. *J. Org. Chem.*, **57**, 6817–6820.

150. Ogata, Y., Itoh, T., and Izawa, Y. (1969) Photochemical ethoxycarbonylmethylation of benzene with ethyl haloacetates. *Bull. Chem. Soc. Jpn.*, **42**, 794–797.

151. Kurz, M.E., Baru, V., and Nguyen, P.-N. (1984) Aromatic acetonylation promoted by manganese(III) and cerium(IV) salts. *J. Org. Chem.*, **49**, 1603–1607.

152. Tayama, E., Yanaki, T., Iwamoto, H., and Hasegawa, E. (2010) Copper(II) triflate catalyzed intermolecular aromatic substitution of *N,N*-disubstituted

anilines with diazo esters. *Eur. J. Org. Chem.*, **2010**, 6719–6721.
153. Zaragoza, F. (1995) Remarkable substituent effects on the chemoselectivity of rhodium(II) carbenoids derived from *N*-(2-diazo-3-oxobutyryl)-L-phenylalanine esters. *Tetrahedron*, **51**, 8829–8834.
154. Ono, N., Kamimura, A., Sasatani, H., and Kaji, A. (1987) Lewis acid induced nucleophilic substitution reactions of β-nitro sulfides via episulfonium ions. *J. Org. Chem.*, **52**, 4133–4135.
155. Nakamura, S., Sugimoto, H., and Ohwada, T. (2007) Formation of 4*H*-1,2-benzoxazines by intramolecular cyclization of nitroalkanes. Scope of aromatic oxygen-functionalization reaction involving a nitro oxygen atom and mechanistic insights. *J. Am. Chem. Soc.*, **129**, 1724–1732.
156. Takamoto, M., Kurouchi, H., Otani, Y., and Ohwada, T. (2009) Phenylation reaction of α-acylnitromethanes to give 1,2-diketone monooximes: involvement of carbon electrophile at the position α to the nitro group. *Synthesis*, 4129–4136.
157. Venkatesh, C., Singh, B., Mahata, P.K., Ila, H., and Junjappa, H. (2005) Heteroannulation of nitroketene *N*,*S*-arylaminoacetals with POCl$_3$: a novel highly regioselective synthesis of unsymmetrical 2,3-substituted quinoxalines. *Org. Lett.*, **7**, 2169–2172.
158. Yao, C.-F., Kao, K.-H., Liu, J.-T., Chu, C.-M., Wang, Y., Chen, W.-C., Lin, Y.-M., Lin, W.-W., Yan, M.-C., Liu, J.-Y., Chuang, M.-C., and Shiue, J.-L. (1998) Generation of nitroalkanes, hydroximoyl halides and nitrile oxides from the reactions of β-nitrostyrenes with Grignard or organolithium reagents. *Tetrahedron*, **54**, 791–822.
159. Shutkina, O.V., Ponomareva, O.A., and Ivanova, I.I. (2013) Catalytic synthesis of cumene from benzene and acetone. *Pet. Chem.*, **53**, 20–26.
160. Zimmerman, H.E. and Zweig, A. (1961) Carbanion rearrangements. II. *J. Am. Chem. Soc.*, **83**, 1196–1213.
161. Girotti, G., Rivetti, F., Ramello, S., and Carnelli, L. (2003) Alkylation of benzene with isopropanol on β-zeolite: influence of physical state and water concentration on catalyst performances. *J. Mol. Catal. A*, **204–205**, 571–579.
162. Ohkubo, T., Aoki, S., Ishibashi, M., Imai, M., Fujita, T., and Fujiwara, K. (2012) Process for producing alkylated aromatic compounds and process for producing phenols. US Patent 2012004471.
163. O'Connor, M.J., Boblak, K.N., Topinka, M.J., Kindelin, P.J., Briski, J.M., Zheng, C., and Klumpp, D.A. (2010) Superelectrophiles and the effects of trifluoromethyl substituents. *J. Am. Chem. Soc.*, **132**, 3266–3267.
164. Gorokhovik, I., Neuville, L., and Zhu, J. (2011) Trifluoroacetic acid-promoted synthesis of 3-hydroxy, 3-amino and spirooxindoles from α-keto-*N*-anilides. *Org. Lett.*, **13**, 5536–5539.
165. Si, Y.-G., Chen, J., Li, F., Li, J.-H., Qin, Y.-J., and Jiang, B. (2006) Highly regioselective Friedel–Crafts reactions of electron-rich aromatic compounds with pyruvate catalyzed by Lewis acid-base: efficient synthesis of pesticide cycloprothrin. *Adv. Synth. Catal.*, **348**, 898–904.
166. Smithen, D.A., Cameron, T.S., and Thompson, A. (2011) One-pot synthesis of asymmetric annulated bis(pyrrole)s. *Org. Lett.*, **13**, 5846–5849.
167. Freter, K. (1975) 3-Cycloalkenylindoles. *J. Org. Chem.*, **40**, 2525–2529.
168. Miyai, T., Onishi, Y., and Baba, A. (1999) Novel reductive Friedel–Crafts alkylation of aromatics catalyzed by indium compounds: chemoselective utilization of carbonyl moieties as alkylating reagents. *Tetrahedron*, **55**, 1017–1026.
169. Yip, K.-T., Nimje, R.Y., Leskinen, M.V., and Pihko, P.M. (2012) Palladium-catalyzed dehydrogenative β′-arylation of β-keto esters under aerobic conditions: interplay of metal and Brønsted acids. *Chem. Eur. J.*, **18**, 12590–12594.
170. Schoen, K.L. and Becker, E.I. (1963) 9-Methylfluorene. *Org. Synth.*, Coll. Vol. **4**, 623–626.
171. Fleckenstein, C.A., Kadyrov, R., and Plenio, H. (2008) Efficient large-scale synthesis of 9-alkylfluorenyl phosphines for Pd-catalyzed cross-coupling

172. Murai, Y., Masuda, K., Sakihama, Y., Hashidoko, Y., Hatanaka, Y., and Hashimoto, M. (2012) Comprehensive synthesis of photoreactive (3-trifluoromethyl)diazirinyl indole derivatives from 5- and 6-trifluoroacetylindoles for photoaffinity labeling. *J. Org. Chem.*, **77**, 8581–8587.
173. Johnson, H.E. and Crosby, D.G. (1973) Indole-3-acetic acid. *Org. Synth.*, Coll. Vol. **5**, 654–656.
174. Weinbrenner, S., Dunkern, T., Marx, D., Schmidt, B., Stengel, T., Flockerzi, D., Kautz, U., Hauser, D., Diefenbach, J., Christiaans, J.A.M., and Menge, W.M.P.B. (2008) 6-Benzyl-2,3,4,7-tetrahydroindolo[2,3-*c*]quinoline compounds useful as PDE5 inhibitors. WO Patent 2008095835.
175. Davoust, M., Kitching, J.A., Fleming, M.J., and Lautens, M. (2010) Diastereoselective benzylic arylation of tetralins. *Chem. Eur. J.*, **16**, 50–54.
176. Hoefgen, B., Decker, M., Mohr, P., Schramm, A.M., Rostom, S.A.F., El-Subbagh, H., Schweikert, P.M., Rudolf, D.R., Kassack, M.U., and Lehmann, J. (2006) Dopamine/serotonin receptor ligands. 10: SAR studies on azecine-type dopamine receptor ligands by functional screening at human cloned D_1, D_{2L}, and D_5 receptors with a microplate reader based calcium assay lead to a novel potent D_1/D_5 selective antagonist. *J. Med. Chem.*, **49**, 760–769.
177. Johnson, H.E. (1965) Process for producing 3-substituted indoles. US Patent 3197479.

reactions. *Org. Process Res. Dev.*, **12**, 475–479.

2 芳烃的亲电烯基化反应

2.1 概述

本章将着重介绍能在未官能团化芳环的 C—H 上引入烯烃的反应。这类反应中最常见的例子是富电子芳烃的亲电烯基化和通过原位金属化的烯基化（图式 2.1）。原则上，缺电子芳烃也能够被富电子烯烃或炔烃烯基化，但这样的例子鲜有报道。

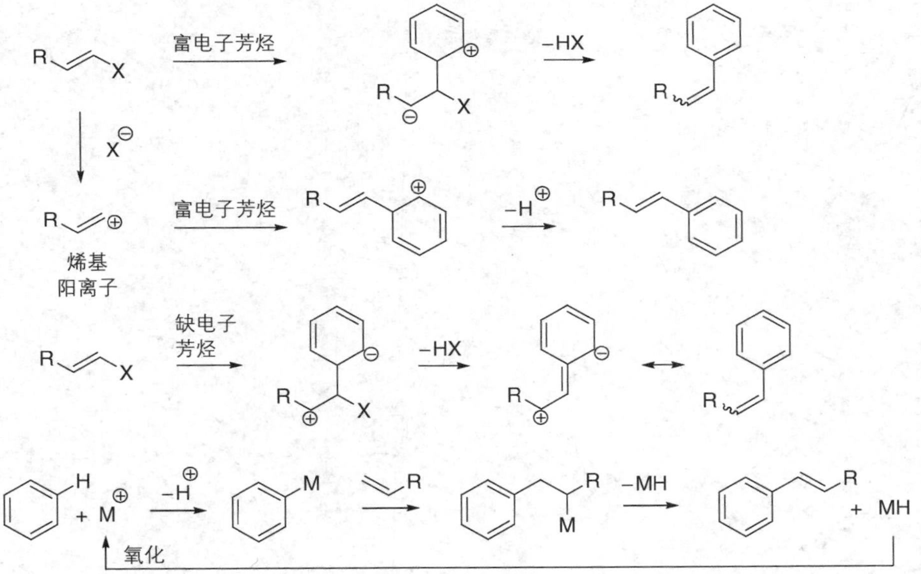

图式 2.1　未官能化的芳烃的烯基化反应机理（X：亲核取代反应的离去基团；M：金属，钯是典型）

2.2 与含离去基团取代的烯烃的烯基化反应

富电子芳烃与烯基卤或相关亲电试剂的亲电烯基化反应范围较窄。烯基碳正离子是高活性的中间体，容易发生重排、还原以及无选择性的亲电反应。为了获得可以接受的反应收率，通常需要使用显著过量的亲核试剂(图式 2.2)。

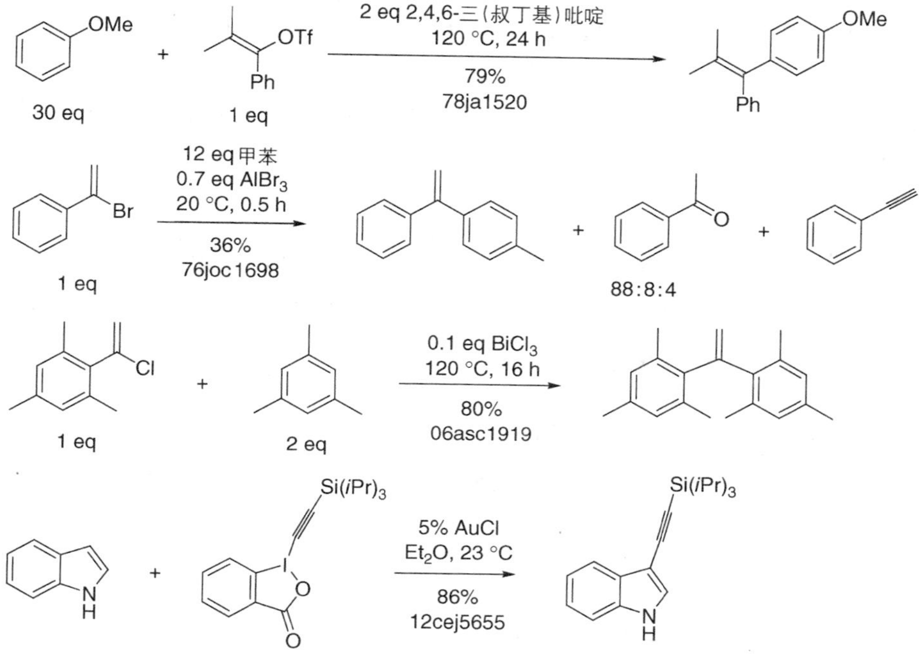

图式 2.2 芳烃与含离去基团取代的烯烃和炔烃的烯基化和炔基化反应[1-4]

在一价铜盐的存在下，带酸性 C—H(pK_a<27)的芳烃或杂芳烃能被烯卤烯基化。反应首先形成一个芳基铜中间体，它被烯卤进一步烯基化(图式 2.3)。芳基铜中间体的烯基化反应类似于乌尔曼(Ullmann)反应中由卤代芳烃形成的芳基铜中间体的芳基化反应[5-8]，很可能是通过芳基铜中间体对烯烃进行亲核加成，随后消除卤化铜实现的。

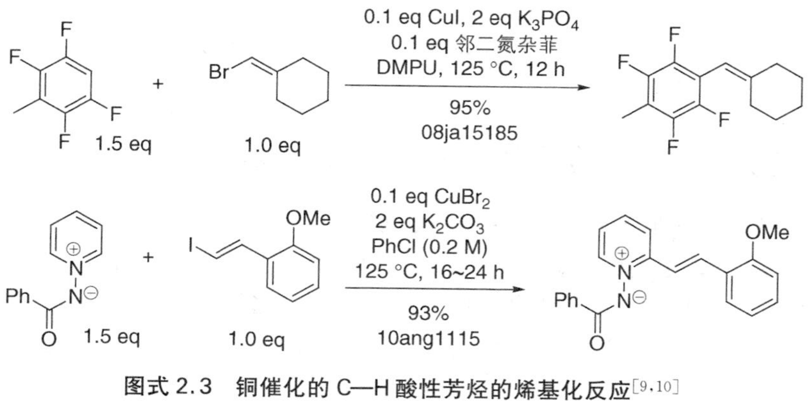

图式 2.3 铜催化的 C—H 酸性芳烃的烯基化反应[9,10]

芳烃也可能通过形成金属螯合物被金属化,由此产生的反应中间体能够与含有离去基团取代的烯烃反应(图式2.4)。报道最多的例子是钌、铑和钯催化的烯基化反应。合适的烯化试剂包括烯基乙酸酯、烯基硼酸酯、烯基酸酐以及β-卤代丙烯酸酯[11,12]。这些反应的机理尚未完全阐明,可能是通过芳基(乙烯基)金属络合物进行的,该络合物由原料经两步反应形成。如图式2.4中倒数第二个例子所示,除氢之外,其他基团在金属螯合物形成过程中也能被置换。

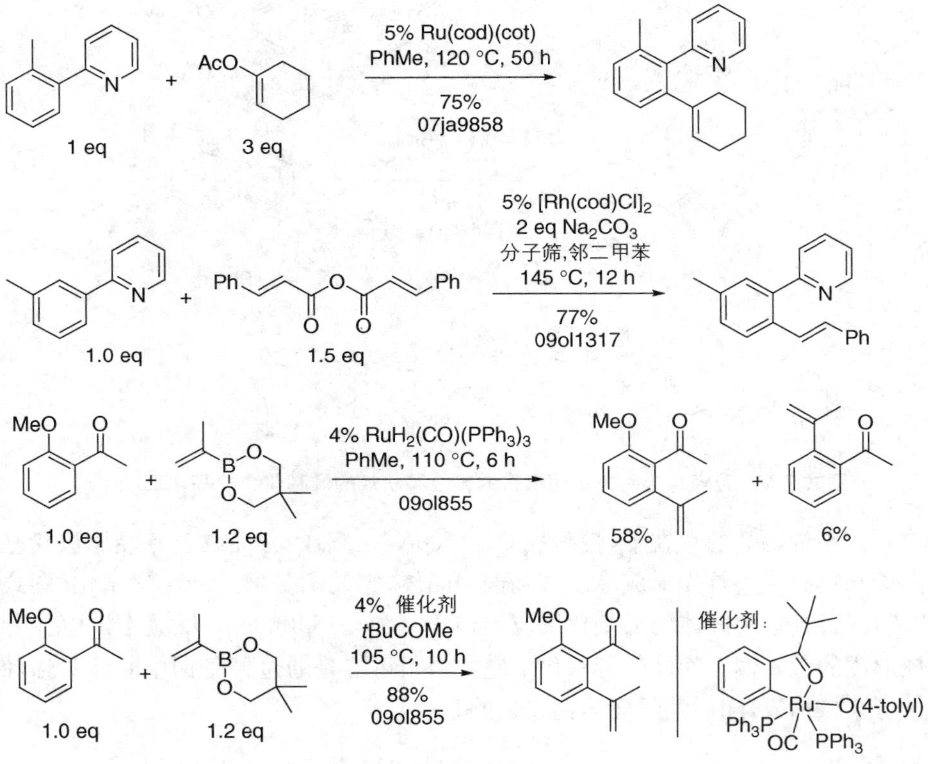

图式2.4 通过形成螯合物金属化的芳烃烯基化反应[13-15]

2.3 与未含离去基团取代的烯烃的烯基化反应

卤代芳烃能够通过赫克反应与无离去基团取代的烯烃发生烯基化反应[16]。未含离去基团的芳烃或烯烃的烯基化形式上是一个氧化反应,为了再生催化剂(例如,零价钯氧化到二价钯),需要使用化学计量的氧化剂。常见氧化剂有氧气、二价铜盐、苯醌、过氧化物或者价格十分昂贵的一价银盐[17](图式2.5和图式2.6)。

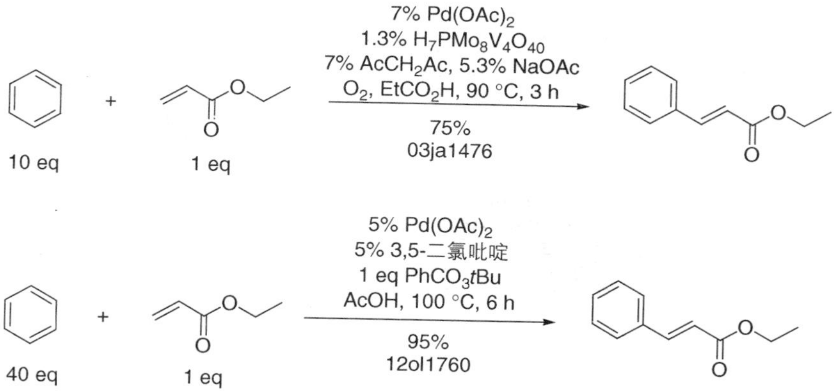

图式 2.5　苯与丙烯酸酯的氧化乙烯化反应[18,19]（更多实例：[20-23]）

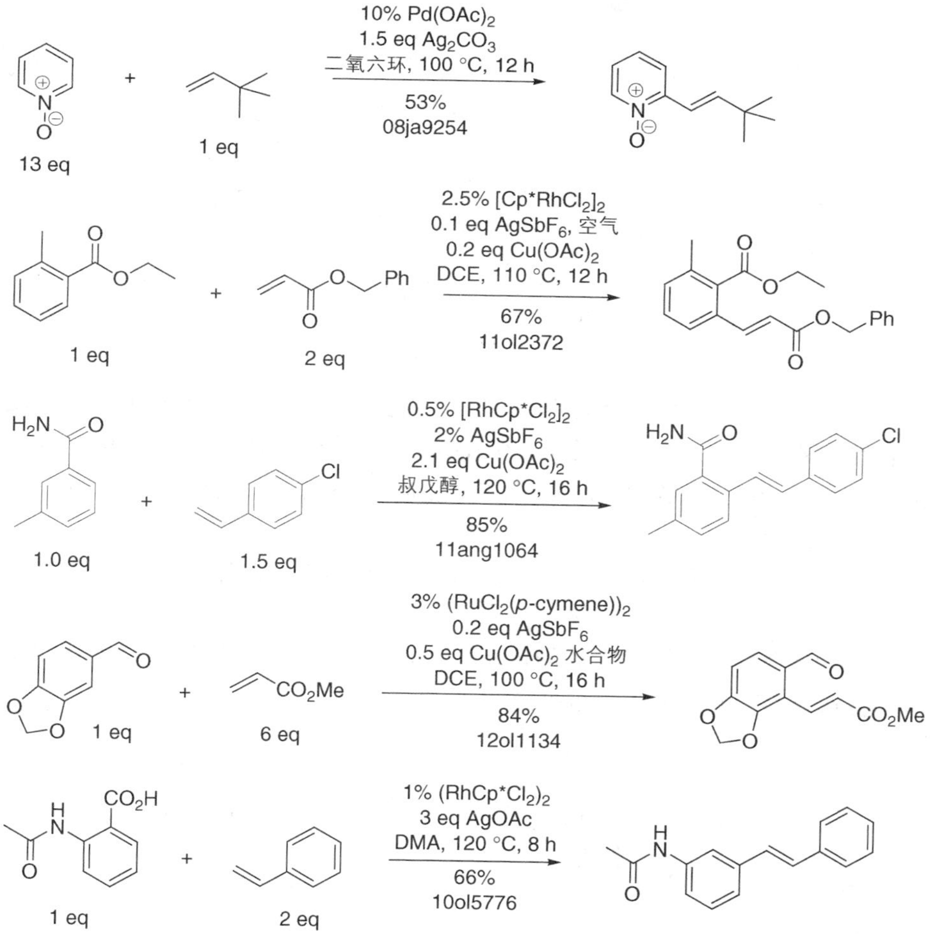

图式 2.6　取代芳烃的邻位烯基化反应[24-31]（更多实例：[32-36]）

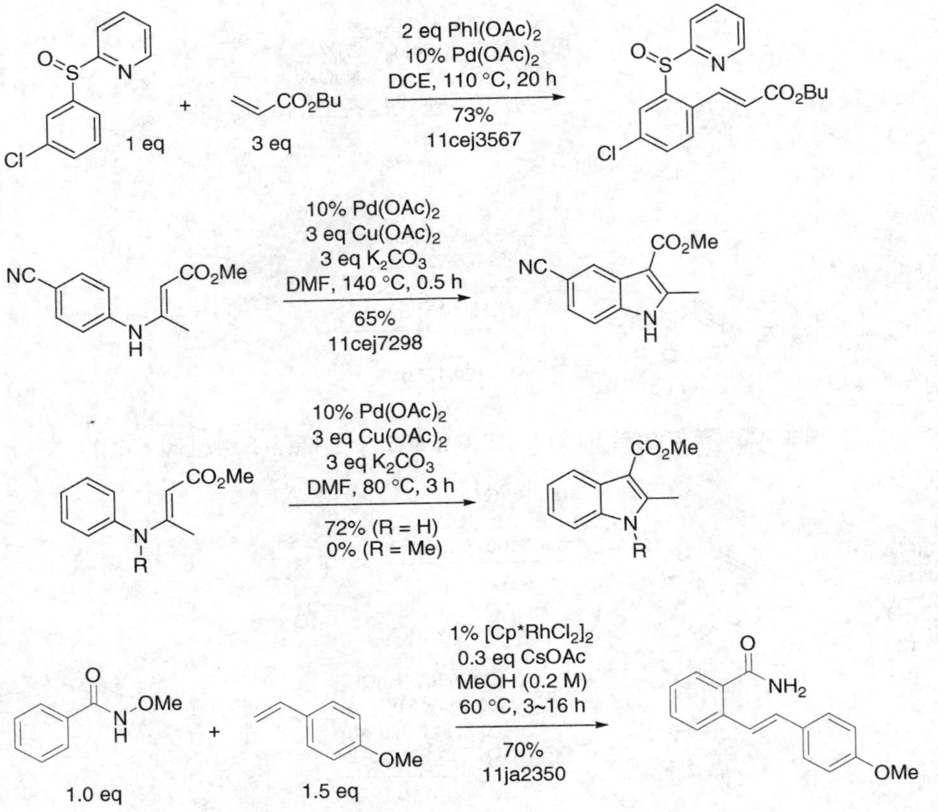

续图式 2.6

芳烃金属化的区域选择性受其 C—H 基团的酸性、螯合物形成能力以及空间相互排斥作用控制。许多官能团能够通过螯合作用导向金属化到其邻位，这些导向基团包括羰基、吡咯、吡啶、嘧啶、吡嗪、叔胺、磺胺类、酰胺、醇、硅醇以及三氮烯[37]。

任何一个含能够被还原基团的起始原料也可作为氧化剂，来替代外加的氧化剂（如图式 2.6 中最后一个例子）。

没有螯合基团的取代芳烃的烯基化反应，其区域选择性通常较差。很少有间位选择性烯基化的例子被报道（图式 2.7）。

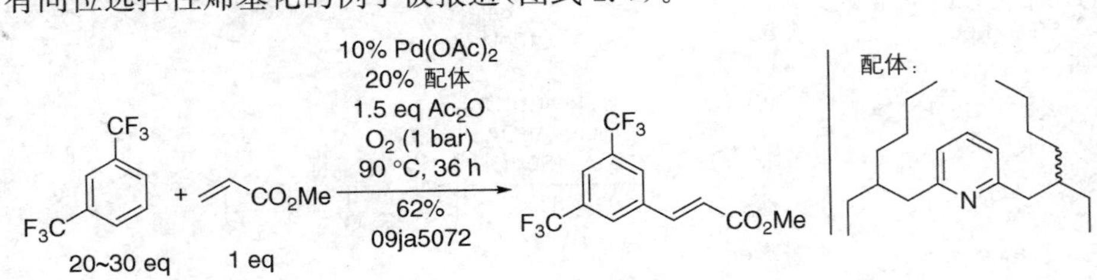

图式 2.7 芳烃的间位烯基化[38,39]

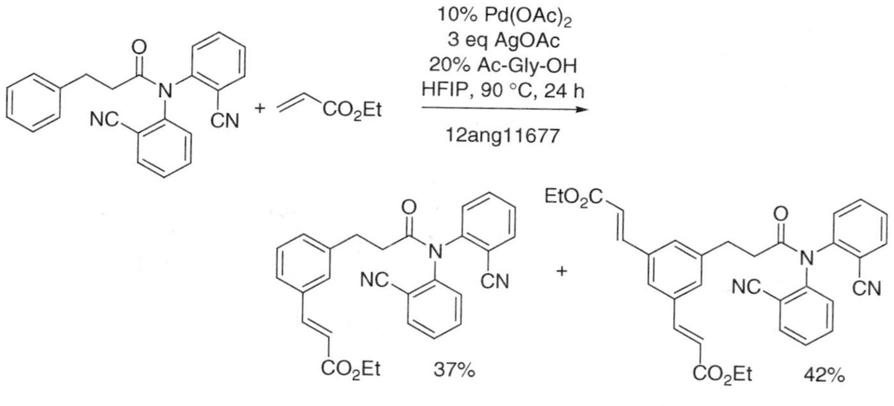

续图式 2.7

芳烃的氧化烯基化所需反应条件有时会导致烷基的脱氢。对于某些起始原料,可以在生成烯烃的同时对其进行芳基化(图式 2.8)。当烯烃中间体不稳定或者处理困难(由于它的毒性)时,这一策略就具有了价值。

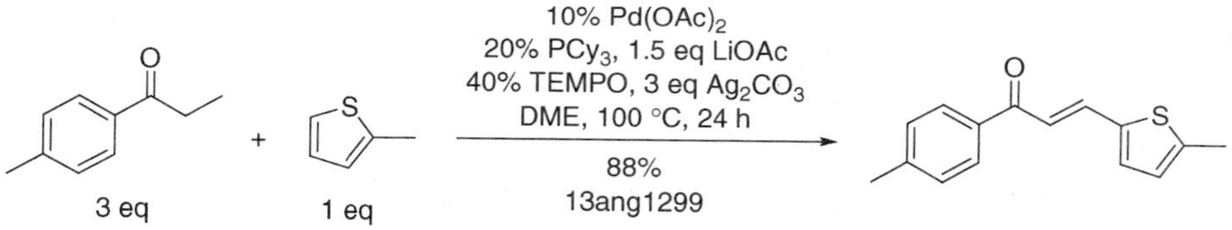

图式 2.8 脱氢原位生成烯烃[40]

芳烃与烯烃直接烯基化的典型副反应包括烯烃聚合以及形成螯合的官能团对烯烃发生分子内加成反应。五元杂环的形成特别容易而且很难抑制。此外,不带吸电子或供电子基团的非对称烯烃对芳烃的烯基化反应通常得到的产物是异构体混合物(图式 2.9)。

由于烯基取代后不会显著改变芳烃的电性,芳烃烯基化的一个常见副反应是多次烯基化。卤代芳烃生成氧化烯基化和赫克反应的混合物。在氧化剂不存在时,一些催化剂会导致形成烷基芳烃,而不是烯基芳烃[44]。另外,具有络合或螯合能力的官能团或杂质(氰、咪唑类、吡啶或含硫化合物)会使催化剂失活。另一个潜在副反应是芳烃的氧化自偶联,酚类和苯胺类的富电子芳烃很容易发生此类反应。在氧化剂和酸或碱的存在下,高温会导致富电子芳烃的脱酰化和其他官能团转化(图式 2.10)。

化学计量氧化剂能增强这些反应固有的危险性。可使用仅仅具抗氧化性能的溶剂,如卤代烷、乙腈、丙酮或乙酸等。为避免反应物的大量积聚,在大规模制备时,氧化剂应随着反应的进行逐渐加入。

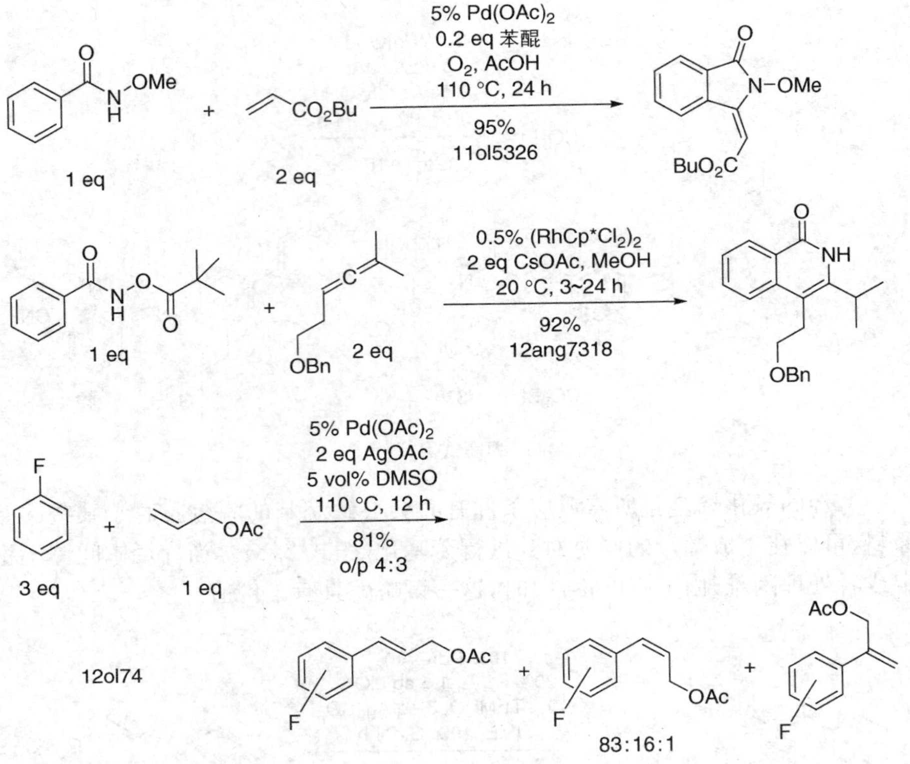

图式 2.9　芳烃烯基化中形成双环产物和区域(位置)异构体[41-43]

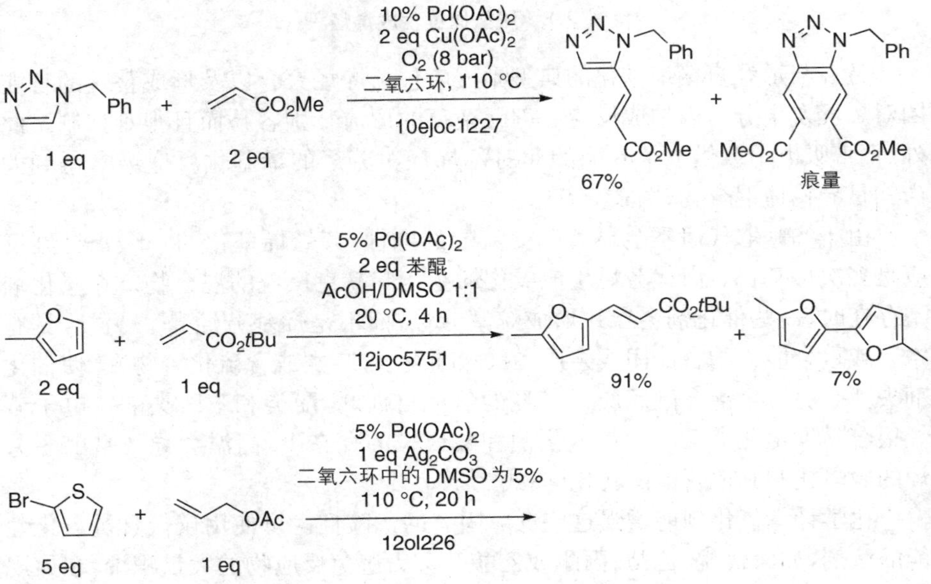

图式 2.10　芳烃氧化烯基化反应中的副反应[24,45-48]

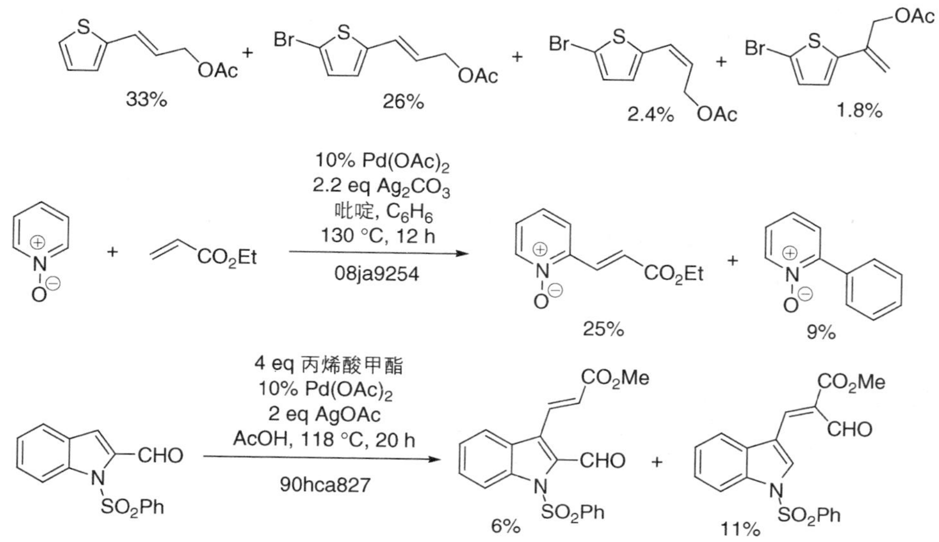

续图式 2.10

富电子芳烃或杂芳烃与缺电子烯烃的烯基化反应的另一个潜在的副反应是环加成反应。萘、蒽和呋喃特别容易与缺电子烯烃发生狄尔斯-阿尔德反应（图式 2.11）。

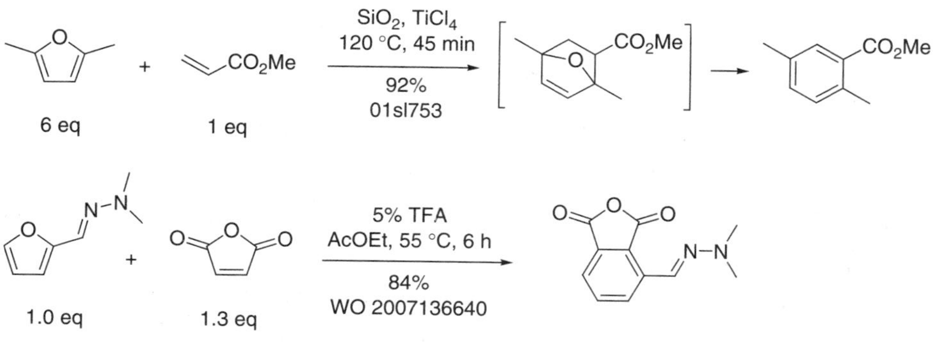

图式 2.11 呋喃的狄尔斯-阿尔德反应[50,51]

2.4 与炔烃的烯基化反应

炔烃是烯基卤的合成等价体，如在钯或铑催化剂存在下能直接对芳烃进行烯基化（图式 2.12）。反应通过氢芳基化（芳基加成到炔烃上）方式进行，不发生氧化反应，因此无须氧化剂。

末端炔烃和非末端炔烃常常表现出不同的反应性，有些反应条件或仅适用于末端或非末端炔烃,但非二者皆可。因此,在图式 2.12 中第一个例子的条件仅适用于非末端炔烃,而第二个例子的条件仅适用于末端炔烃。遗憾的是,最重要的炔烃,即乙炔,很少在这些新反应中进行试验。

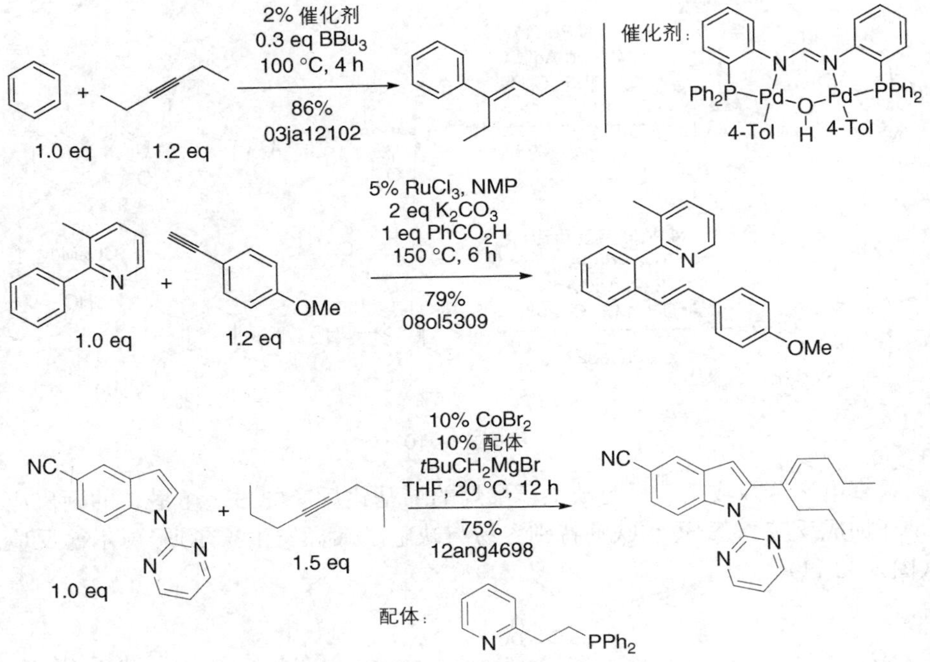

图式 2.12　与炔烃的芳香烯基化反应[52-54]

与烯烃对芳烃的烯基化相似,炔烃烯基化的典型副反应也是多烯化、环化反应及异构体混合物的形成(图式 2.13)。此外,所得的苯乙烯可对芳烃进行烷基化,生成偕二芳基烷烃[55]。一些路易斯酸也能催化水、醇、胺或者硫醇对炔烃的加成反应。

缺电子的杂芳烃可以用(易爆炸的)乙炔基铜或乙炔基银来进行炔化反应。这种反应的第一个中间体是炔基二氢芳烃。在氧化剂不存在时,炔基可充当氢受体,并引起杂芳烃的去芳构化。现已发展出允许缺电子芳烃与末端炔烃通过"一锅法"烯基化的催化剂和反应条件(图式 2.14)。

已发展出芳烃氧化炔化反应的方法(图式 2.15),反应以富电子芳烃和缺电子炔烃的反应效果最佳。后者是优秀的迈克尔受体,因此,大多数亲核试剂容易对该反应的起始原料及产物进行加成。

当准备放大上述反应规模时,炔烃容易发生聚合或在高浓度下容易爆炸,这些应当牢记在心,过渡金属存在时尤其如此。此外,在空气和铜盐存在下末端炔烃容易二聚生成 1,3-丁二炔[艾格林顿(Eglinton)反应]。

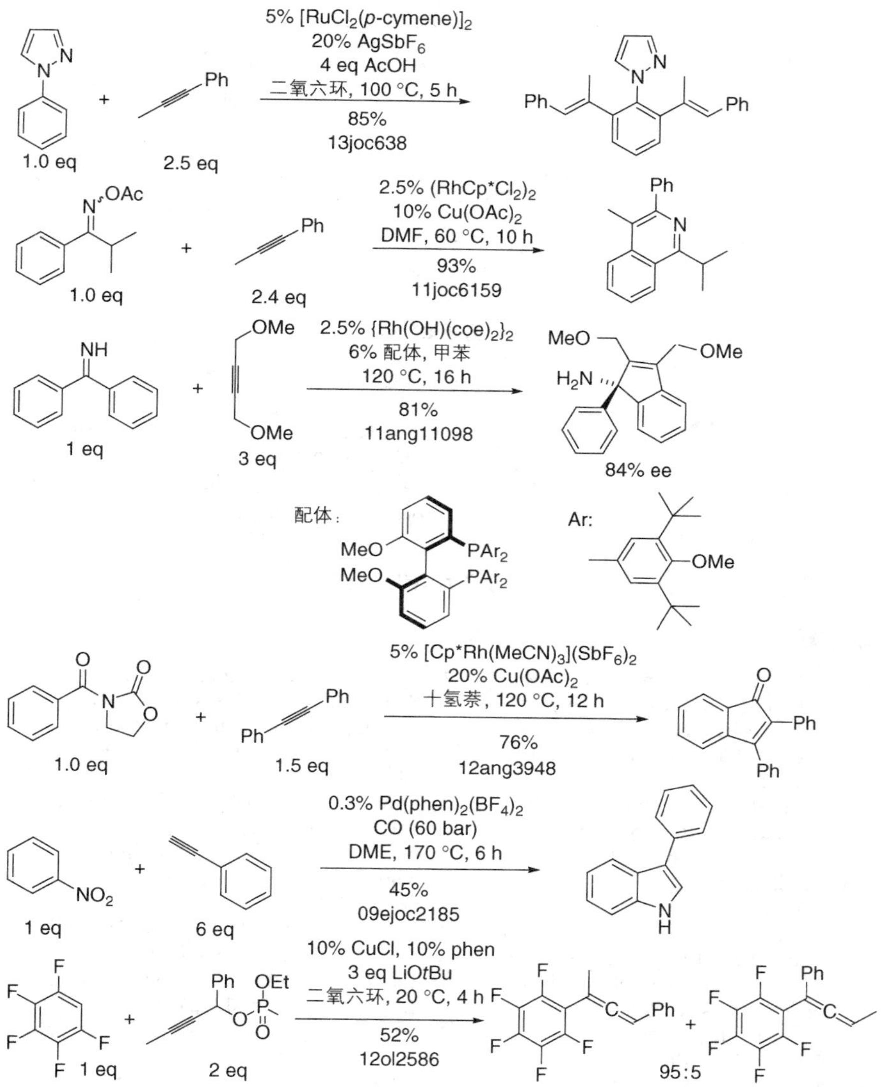

图式 2.13 芳烃与炔烃的烯基化反应中的副反应[56-61]（更多实例：[62,63]）

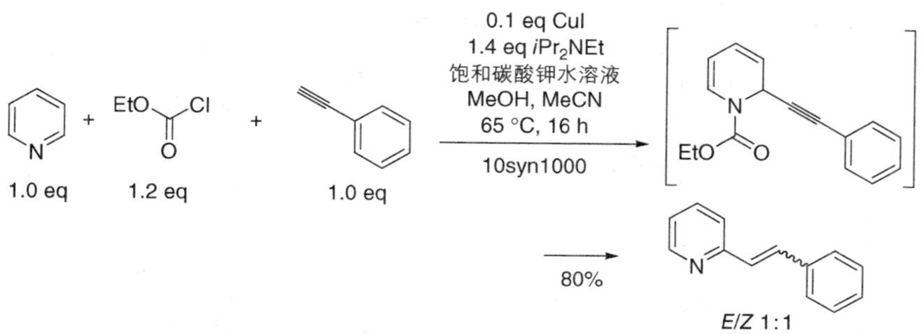

图式 2.14 吡啶与炔烃的烯化[64,65]（更多实例：[66]）

续图式 2.14

图式 2.15 芳烃的氧化炔化反应[67]

参考文献

1. Stang, P.J. and Anderson, A.G. (1978) Preparation and chemistry of vinyl triflates. 16. Mechanism of alkylation of aromatic substrates. *J. Am. Chem. Soc.*, **100**, 1520–1525.
2. Roberts, R.M. and Abdel-Baset, M.B. (1976) Friedel–Crafts alkylations with vinyl halides. Vinyl cations and spirobiindans. *J. Org. Chem.*, **41**, 1698–1701.
3. Sun, H., Hua, R., Chen, S., and Yin, Y. (2006) An efficient bismuth(III) chloride-catalyzed synthesis of 1,1-diarylalkenes via Friedel–Crafts reaction of acyl chloride or vinyl chloride with arenes. *Adv. Synth. Catal.*, **348**, 1919–1925.
4. Brand, J.P., Chevalley, C., Scopelliti, R., and Waser, J. (2012) Ethynyl benziodoxolones for the direct alkynylation of heterocycles: structural requirement, improved procedure for pyrroles, and insights into the mechanism. *Chem. Eur. J.*, **18**, 5655–5666.
5. Beletskaya, I.P. and Cheprakov, A.V. (2004) Copper in cross-coupling reactions; the post-Ullmann chemistry. *Coord. Chem. Rev.*, **248**, 2337–2364.
6. Hassan, J., Sévignon, M., Gozzi, C., Schulz, E., and Lemaire, M. (2002) Aryl–aryl bond formation one century after the discovery of the Ullmann reaction. *Chem. Rev.*, **102**, 1359–1469.
7. Sperotto, E., van Klink, G.P.M., van Koten, G., and de Vries, J.G. (2010) The mechanism of the modified Ullmann reaction. *Dalton Trans.*, **39**, 10338–10351.
8. Cohen, T. and Cristea, I. (1976) Kinetics and mechanism of the copper(I)-induced homogeneous Ullmann coupling of o-bromonitrobenzene. *J. Am. Chem. Soc.*, **98**, 748–753.
9. Do, H.-Q., Khan, R.M.K., and Daugulis, O. (2008) A general method for copper-catalyzed arylation of arene C–H bonds. *J. Am. Chem. Soc.*, **130**, 15185–15192.
10. Mousseau, J.J., Bull, J.A., and Charette, A.B. (2010) Copper-catalyzed direct alkenylation of N-iminopyridinium ylides. *Angew. Chem. Int. Ed.*, **49**, 1115–1118.
11. Daugulis, O., Zaitsev, V.G., Shabashov, D., Pham, Q.-N., and Lazareva, A. (2006) Regioselective functionalization of unreactive carbon–hydrogen bonds. *Synlett*, 3382–3388.
12. Zaitsev, V.G. and Daugulis, O. (2005) Catalytic coupling of haloolefins with anilides. *J. Am. Chem. Soc.*, **127**, 4156–4157.
13. Matsuura, Y., Tamura, M., Kochi, T., Sato, M., Chatani, N., and Kakiuchi, F. (2007) The Ru(cod)(cot)-catalyzed alkenylation of aromatic C–H bonds

with alkenyl acetates. *J. Am. Chem. Soc.*, **129**, 9858–9859.

14. Jin, W., Yu, Z., He, W., Ye, W., and Xiao, W.-J. (2009) Efficient Rh(I)-catalyzed direct arylation and alkenylation of arene C–H bonds via decarbonylation of benzoic and cinnamic anhydrides. *Org. Lett.*, **11**, 1317–1320.

15. Ueno, S., Kochi, T., Chatani, N., and Kakiuchi, F. (2009) Unique effect of coordination of an alkene moiety in products on ruthenium-catalyzed chemoselective C–H alkenylation. *Org. Lett.*, **11**, 855–858.

16. Beletskaya, I.P. and Cheprakov, A.V. (2000) The Heck reaction as a sharpening stone of palladium catalysis. *Chem. Rev.*, **100**, 3009–3066.

17. Le Bras, J. and Muzart, J. (2011) Intermolecular dehydrogenative Heck reactions. *Chem. Rev.*, **111**, 1170–1214.

18. Yokota, T., Tani, M., Sakaguchi, S., and Ishii, Y. (2003) Direct coupling of benzene with olefins catalyzed by Pd(OAc)$_2$ combined with heteropolyoxometalate under dioxygen. *J. Am. Chem. Soc.*, **125**, 1476–1477.

19. Kubota, A., Emmert, M.H., and Sanford, M.S. (2012) Pyridine ligands as promoters in Pd$^{II/0}$-catalyzed C–H olefination reactions. *Org. Lett.*, **14**, 1760–1763.

20. Yamada, T., Sakaguchi, S., and Ishii, Y. (2005) Oxidative coupling of benzenes with α,β-unsaturated aldehydes by the Pd(OAc)$_2$/molybdovanadophosphoric acid/O$_2$ system. *J. Org. Chem.*, **70**, 5471–5474.

21. Milstein, D., Weissman, H., and Song, X.P. (2002) Method for the production of aryl alkenes. WO Patent 02055455.

22. Vasseur, A., Harakat, D., Muzart, J., and Le Bras, J. (2013) Aerobic dehydrogenative Heck reactions of heterocycles with styrenes: a negative effect of metallic co-oxidants. *Adv. Synth. Catal.*, **355**, 59–67.

23. Shang, X., Xiong, Y., Zhang, Y., Zhang, L., and Liu, Z. (2012) Pd(II)-catalyzed direct olefination of arenes with allylic esters and ethers. *Synlett*, 259–262.

24. Cho, S.H., Hwang, S.J., and Chang, S. (2008) Palladium-catalyzed C–H functionalization of pyridine N-oxides: highly selective alkenylation and direct arylation with unactivated arenes. *J. Am. Chem. Soc.*, **130**, 9254–9256.

25. Park, S.H., Kim, J.Y., and Chang, S. (2011) Rhodium-catalyzed selective olefination of arene esters via C–H bond activation. *Org. Lett.*, **13**, 2372–2375.

26. Patureau, F.W., Besset, T., and Glorius, F. (2011) Rhodium-catalyzed oxidative olefination of C–H bonds in acetophenones and benzamides. *Angew. Chem. Int. Ed.*, **50**, 1064–1067.

27. Padala, K. and Jeganmohan, M. (2012) Highly regio- and stereoselective ruthenium(II)-catalyzed direct *ortho*-alkenylation of aromatic and heteroaromatic aldehydes with activated alkenes under open atmosphere. *Org. Lett.*, **14**, 1134–1137.

28. Mochida, S., Hirano, K., Satoh, T., and Miura, M. (2010) Synthesis of stilbene and distyrylbenzene derivatives through rhodium-catalyzed *ortho*-olefination and decarboxylation of benzoic acids. *Org. Lett.*, **12**, 5776–5779.

29. García-Rubia, A., Fernández-Ibañez, M.A., Gomez Arrayás, R., and Carretero, J. (2011) 2-Pyridyl sulfoxide: a versatile and removable directing group for the Pd^{II}-catalyzed direct C–H olefination of arenes. *Chem. Eur. J.*, **17**, 3567–3570.

30. Neumann, J.J., Rakshit, S., Dröge, T., Würtz, S., and Glorius, F. (2011) Exploring the oxidative cyclization of substituted N-aryl enamines: Pd-catalyzed formation of indoles from anilines. *Chem. Eur. J.*, **17**, 7298–7303.

31. Rakshit, S., Grohmann, C., Besset, T., and Glorius, F. (2011) Rh(III)-catalyzed directed C–H olefination using an oxidizing directing group: mild, efficient, and versatile. *J. Am. Chem. Soc.*, **133**, 2350–2353.

32. Boele, M.D.K., van Strijdonck, G.P.F., de Vries, A.H.M., Kamer, P.C.J., de Vries, J.G., and van Leeuwen, P.W.N.M. (2002) Selective Pd-catalyzed oxidative coupling of anilides with olefins through C–H bond activation at room temperature. *J. Am. Chem. Soc.*, **124**, 1586–1587.

33. Zhang, X., Fan, S., He, C.-Y., Wan, X., Min, Q.-Q., Yang, J., and Jiang, Z.-X. (2010) Pd(OAc)$_2$ catalyzed olefination of

highly electron-deficient perfluoroarenes. *J. Am. Chem. Soc.*, **132**, 4506–4507.
34. Padala, K. and Jeganmohan, M. (2011) Ruthenium-catalyzed *ortho*-alkenylation of aromatic ketones with alkenes by C–H bond activation. *Org. Lett.*, **13**, 6144–6147.
35. Zheng, L. and Wang, J. (2012) Direct oxidative coupling of arenes with olefins by Rh-catalyzed C–H activation in air: observation of a strong cooperation of the acid. *Chem. Eur. J.*, **18**, 9699–9704.
36. Liu, B., Fan, Y., Gao, Y., Sun, C., Xu, C., and Zhu, J. (2013) Rhodium(III)-catalyzed *N*-nitroso-directed C–H olefination of arenes. High-yield, versatile coupling under mild conditions. *J. Am. Chem. Soc.*, **135**, 468–473.
37. Wang, C., Chen, H., Wang, Z., Chen, J., and Huang, Y. (2012) Rhodium(III)-catalyzed C–H activation of arenes using a versatile and removable triazene directing group. *Angew. Chem. Int. Ed.*, **51**, 7242–7245.
38. Zhang, Y.-H., Shi, B.-F., and Yu, J.-Q. (2009) Pd(II)-catalyzed olefination of electron-deficient arenes using 2,6-dialkylpyridine ligands. *J. Am. Chem. Soc.*, **131**, 5072–5074.
39. Truong, T. and Daugulis, O. (2012) Directed functionalization of C–H bonds: now also *meta* selective. *Angew. Chem. Int. Ed.*, **51**, 11677–11679.
40. Shang, Y., Jie, X., Zhou, J., Hu, P., Huang, S., and Su, W. (2013) Pd-catalyzed C–H olefination of (hetero)arenes by using saturated ketones as an olefin source. *Angew. Chem. Int. Ed.*, **52**, 1299–1303.
41. Wrigglesworth, J.W., Cox, B., Lloyd-Jones, G.C., and Booker-Milburn, K.I. (2011) New heteroannulation reactions of *N*-alkoxybenzamides by Pd(II) catalyzed C–H activation. *Org. Lett.*, **13**, 5326–5329.
42. Wang, H. and Glorius, F. (2012) Mild rhodium(III)-catalyzed C–H activation and intermolecular annulation with allenes. *Angew. Chem. Int. Ed.*, **51**, 7318–7322.
43. Li, Z., Zhang, Y., and Liu, Z.-Q. (2012) Pd-catalyzed olefination of perfluoroarenes with allyl esters. *Org. Lett.*, **14**, 74–77.
44. Ackermann, L., Novák, P., Vicente, R., and Hofmann, N. (2009) Ruthenium-catalyzed regioselective direct alkylation of arenes with unactivated alkyl halides through C–H bond cleavage. *Angew. Chem. Int. Ed.*, **48**, 6045–6048.
45. Jiang, H., Feng, Z., Wang, A., Liu, X., and Chen, Z. (2010) Palladium-catalyzed alkenylation of 1,2,3-trizoles with terminal conjugated alkenes by direct C–H bond functionalization. *Eur. J. Org. Chem.*, **2010**, 1227–1230.
46. Vasseur, A., Harakat, D., Muzart, J., and Le Bras, J. (2012) ESI-MS studies of the dehydrogenative Heck reaction of furans with acrylates using benzoquinone as the reoxidant and DMSO as the solvent. *J. Org. Chem.*, **77**, 5751–5758.
47. Zhang, Y., Li, Z., and Liu, Z.-Q. (2012) Pd-catalyzed olefination of furans and thiophenes with allyl esters. *Org. Lett.*, **14**, 226–229.
48. Pindur, U. and Adam, R. (1990) Thermal 1,6-electrocyclization reactions of acceptor-substituted 2,3-divinyl-1*H*-indoles yielding functionalized carbazoles. *Helv. Chim. Acta*, **73**, 827–838.
49. Rodríguez-Escrich, C., Davis, R.L., Jiang, H., Stiller, J., Johansen, T.K., and Jørgensen, K.A. (2013) Breaking symmetry with symmetry: bifacial selectivity in the asymmetric cycloaddition of anthracene derivatives. *Chem. Eur. J.*, **19**, 2932–2936.
50. de la Hoz, A., Díaz-Ortiz, A., Fraile, J.M., Gómez, M.V., Mayoral, J.A., Moreno, A., Saiz, A., and Vázquez, E. (2001) Synergy between heterogeneous catalysis and microwave irradiation in an efficient one-pot synthesis of benzene derivatives via ring-opening of Diels–Alder cycloadducts of substituted furans. *Synlett*, 753–756.
51. Muller, G.W., Saindane, M., Ge, C., Kothare, M.A., Cameron, L.M., and Rogers, M.E. (2007) Process for the preparation of substituted 2-(2,6-dioxopiperidin-3-yl)isoindole-1,3-dione. WO Patent 2007136640.
52. Tsukada, N., Mitsuboshi, T., Setoguchi, H., and Inoue, Y. (2003) Stereoselective *cis*-addition of aromatic C–H bonds to

alkynes catalyzed by dinuclear palladium complexes. *J. Am. Chem. Soc.*, **125**, 12102–12103.

53. Cheng, K., Yao, B., Zhao, J., and Zhang, Y. (2008) RuCl$_3$-catalyzed alkenylation of aromatic C–H bonds with terminal alkynes. *Org. Lett.*, **10**, 5309–5312.

54. Ding, Z. and Yoshikai, N. (2012) Mild and efficient C2-alkenylation of indoles with alkynes catalyzed by a cobalt complex. *Angew. Chem. Int. Ed.*, **51**, 4698–4701.

55. Li, H.-J., Guillot, R., and Gandon, V. (2010) A gallium-catalyzed cycloisomerization/Friedel–Crafts tandem. *J. Org. Chem.*, **75**, 8435–8449.

56. Hashimoto, Y., Hirano, K., Satoh, T., Kakiuchi, F., and Miura, M. (2013) Regioselective C–H bond cleavage/alkyne insertion under ruthenium catalysis. *J. Org. Chem.*, **78**, 638–646.

57. Too, P.C., Chua, S.H., Wong, S.H., and Chiba, S. (2011) Synthesis of azaheterocycles from aryl ketone O-acetyl oximes and internal alkynes by Cu–Rh bimetallic relay catalysts. *J. Org. Chem.*, **76**, 6159–6168.

58. Tran, D.N. and Cramer, N. (2011) Enantioselective rhodium(I)-catalyzed [3 + 2] annulations of aromatic ketimines induced by directed C–H activations. *Angew. Chem. Int. Ed.*, **50**, 11098–11102.

59. Li, B.-J., Wang, H.-Y., Zhu, Q.-L., and Shi, Z.-J. (2012) Rhodium/copper-catalyzed annulation of benzimides with internal alkynes: indenone synthesis through sequential C–H and C–N cleavage. *Angew. Chem. Int. Ed.*, **51**, 3948–3952.

60. Ragaini, F., Ventriglia, F., Hagar, M., Fantauzzi, S., and Cenini, S. (2009) Synthesis of indoles by intermolecular cyclization of unfunctionalized nitroarenes and alkynes: one-step synthesis of the skeleton of fluvastatin. *Eur. J. Org. Chem.*, **2009**, 2185–2189.

61. Nakatani, A., Hirano, K., Satoh, T., and Miura, M. (2012) A concise access to (polyfluoroaryl)allenes by Cu-catalyzed direct coupling with propargyl phosphates. *Org. Lett.*, **14**, 2586–2589.

62. Chidipudi, S.R., Khan, I., and Lam, H.W. (2012) Functionalization of C_{sp3}–H and C_{sp2}–H bonds: synthesis of spiroindenes by enolate-directed ruthenium-catalyzed oxidative annulation of alkynes with 2-aryl-1,3-dicarbonyl compounds. *Angew. Chem. Int. Ed.*, **51**, 12115–12119.

63. Dong, W., Wang, L., Parthasarathy, K., Pan, F., and Bolm, C. (2013) Rhodium-catalyzed oxidative annulation of sulfoximines and alkynes as an approach to 1,2-benzothiazines. *Angew. Chem. Int. Ed.*, **52**, 11573–11576.

64. Beveridge, R.E. and Arndtsen, B.A. (2010) A direct, copper-catalyzed functionalization of pyridines with alkynes. *Synthesis*, 1000–1008.

65. Murakami, M. and Hori, S. (2003) Ruthenium-mediated regio- and stereoselective alkenylation of pyridine. *J. Am. Chem. Soc.*, **125**, 4720–4721.

66. Agawa, T. and Miller, S.I. (1961) Reaction of silver acetylide with acylpyridinium salts: N-benzoyl-2-phenylethynyl-1,2-dihydropyridine. *J. Am. Chem. Soc.*, **83**, 449–453.

67. de Haro, T. and Nevado, C. (2010) Gold-catalyzed ethynylation of arenes. *J. Am. Chem. Soc.*, **132**, 1512–1513.

3 芳烃的亲电芳基化反应

3.1 概述

大部分属于芳烃的芳基化反应类型如图式3.1所示。反应第一步是通过金属对C—H键或C—X键的插入实现芳烃的金属化,第二步或者通过形成二芳基金属中间体后进一步经还原消除,或者通过芳族亲核取代进行反应。在下面给出的许多实例中,也存在通过自由基中间体反应的情况。

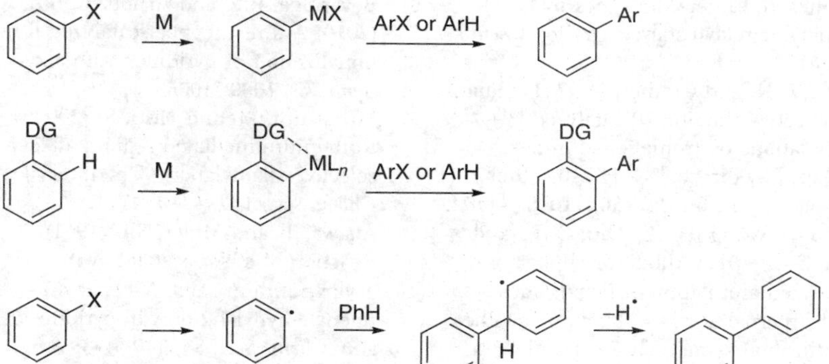

图式 3.1 联芳烃的形成机理(M: 金属; X: 氢或离去基团; DG: 导向基团)

直到最近,仅有几个无取代芳烃与卤代芳烃的芳基化反应实例被报道[1]。最早的例子是铜催化的卤代芳烃(主要是碘化物)与其他芳烃或杂芳烃的乌尔曼反应。通常,乌尔曼反应[2,3]将卤代芳烃转化为同二聚体、异二聚体的收率往往较低。乌尔曼反应的另一缺点是需要使用化学计量的铜催化剂和高反应温度,典型副反应包括卤素交换、还原脱卤和芳香亲核取代反应。

近几十年来,已经开发出一些更方便的、可以替代经典乌尔曼反应的方法[4]。这些改进包括仅使用催化量过渡金属、较低的反应温度和更好的官能团耐受性。

3.2 卤代芳烃的芳基化反应

3.2.1 通过阳离子中间体

当没有催化剂存在时,大多数卤代芳烃的亲电性还不足以对富电子芳烃进行芳基化反应。即使在强路易斯酸的存在下,芳基卤通常也不反应。仅亲电性较高的杂芳烃,如 2,4,6-三氯三嗪(氰脲酰氯),或者通过卤原子攫取或质子化能够形成稳定阳离子的芳烃(如 9,10-二氯蒽和噻吩)等在三氯化铝的存在下与其他芳烃发生芳基化反应(图式 3.2)。

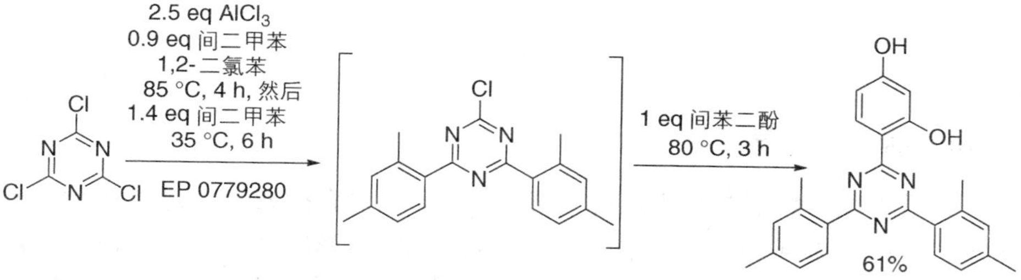

图式 3.2　氰脲酰氯的芳基化[5] (更多实例:[6])

由于在芳基阳离子中电荷不能被有效地离域化而导致通过完全攫取卤代芳烃中的卤不利于芳基阳离子的生成,因此,无过渡金属参与的卤代芳烃的芳基化反应多数是通过加成-消除机理进行的。在酸催化的芳基化反应中,卤代芳烃可被质子化得到更活泼的中间体。然而,这一中间体不必与卤素原子相连的碳原子反应(图式 3.3),也不会强行与最稳定的或最具代表性的共振结构反应,因为它是反应活性最低的一个共振结构。未质子化的卤代芳烃的质子化进一步加大了上述反应的复杂性(图式 3.3 中最后一个反应)。

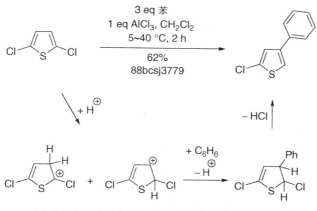

图式 3.3　酸引发的 2-氯噻吩的芳基化反应[7-9]

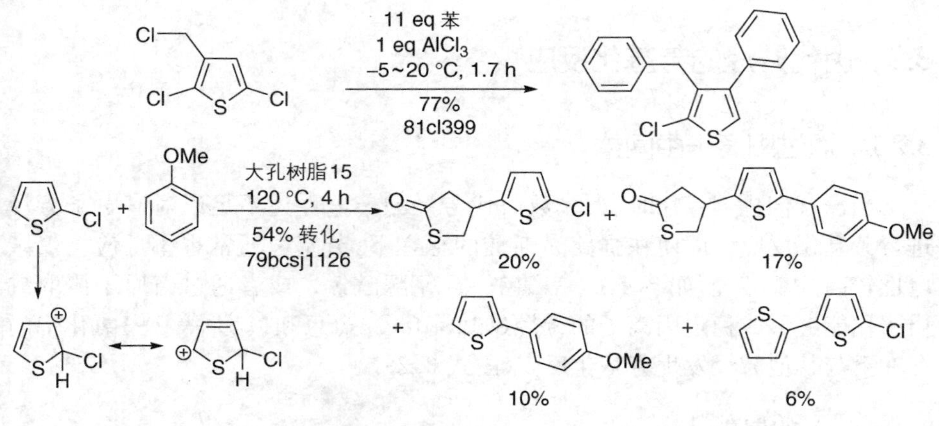

续图式 3.3

3.2.2 通过自由基

多数卤代芳烃的芳基化反应是经过金属-卤素交换或者通过形成芳基自由基进行的。芳基自由基可通过多种方式产生,例如用叔醇盐处理卤代芳烃(图式 3.4)。最佳结果通常在两种具有不同电子密度的芳烃间发生,并且用来"捕获自由基"的芳烃需要大大过量,以阻止同二聚体的形成。这种不需催化的反应类型和过渡金属催化的醇盐引发的芳基化反应均有实例报道(图式 3.4)。铜催化的卤代芳烃乌尔曼型芳基化反应也被认为是基于自由基机理[10]。

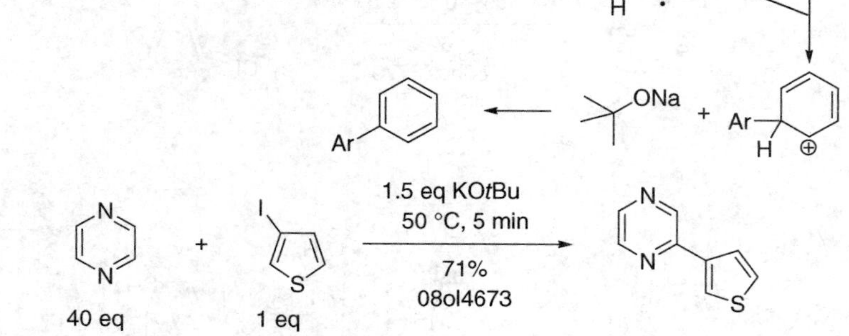

图式 3.4 强碱引发的、过渡金属催化的卤代芳烃的芳基化反应[11-15] (更多实例:[16])

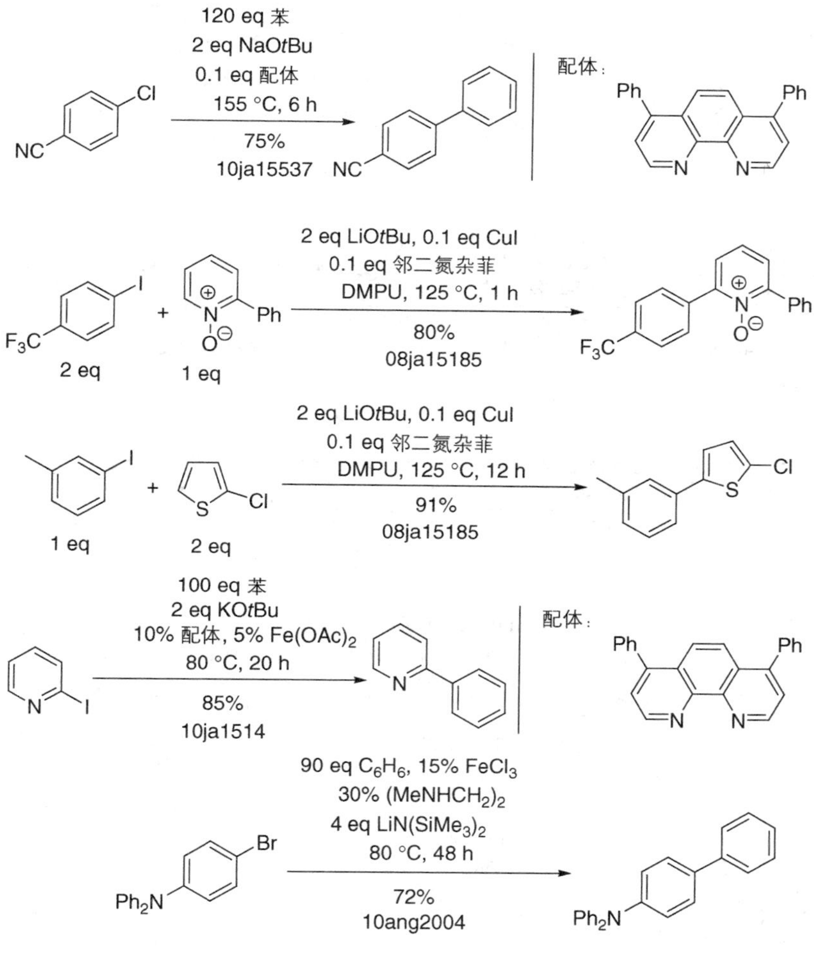

续图式 3.4

与带电荷的中间体相比,自由基对反应底物电子密度的变化不太敏感而导致反应的区域选择性通常较低。因此,芳基自由基进行芳基化反应经常得到不同位置异构体的混合物(图式 3.5)。

在显著过量强碱的存在下反应很容易得到意外的产物。弱酸性化合物去质子化常常会成为强还原剂并且会被弱氧化剂氧化。因此,反应应严格排除空气和其他潜在的氧化剂(酮类、多卤代烷或硝基化合物)。此外,强碱可以导致芳炔的形成,从而产生不同位置异构体的混合物。

芳基自由基还可以通过过氧化苯甲酰的热分解产生(图式 3.6)。由于酰基过氧化物也是酰化试剂,并且能够氧化很多官能团(胺、硫醚、亚砜、烯烃、酮、芳烃等),因此这个反应适用范围很小。此外,这些反应的结果因芳烃的取代模式变化而表现很大差别[21,22]。包括过氧二酰基在内的许多过氧化物都是易爆物种,并且能自发性地引爆。

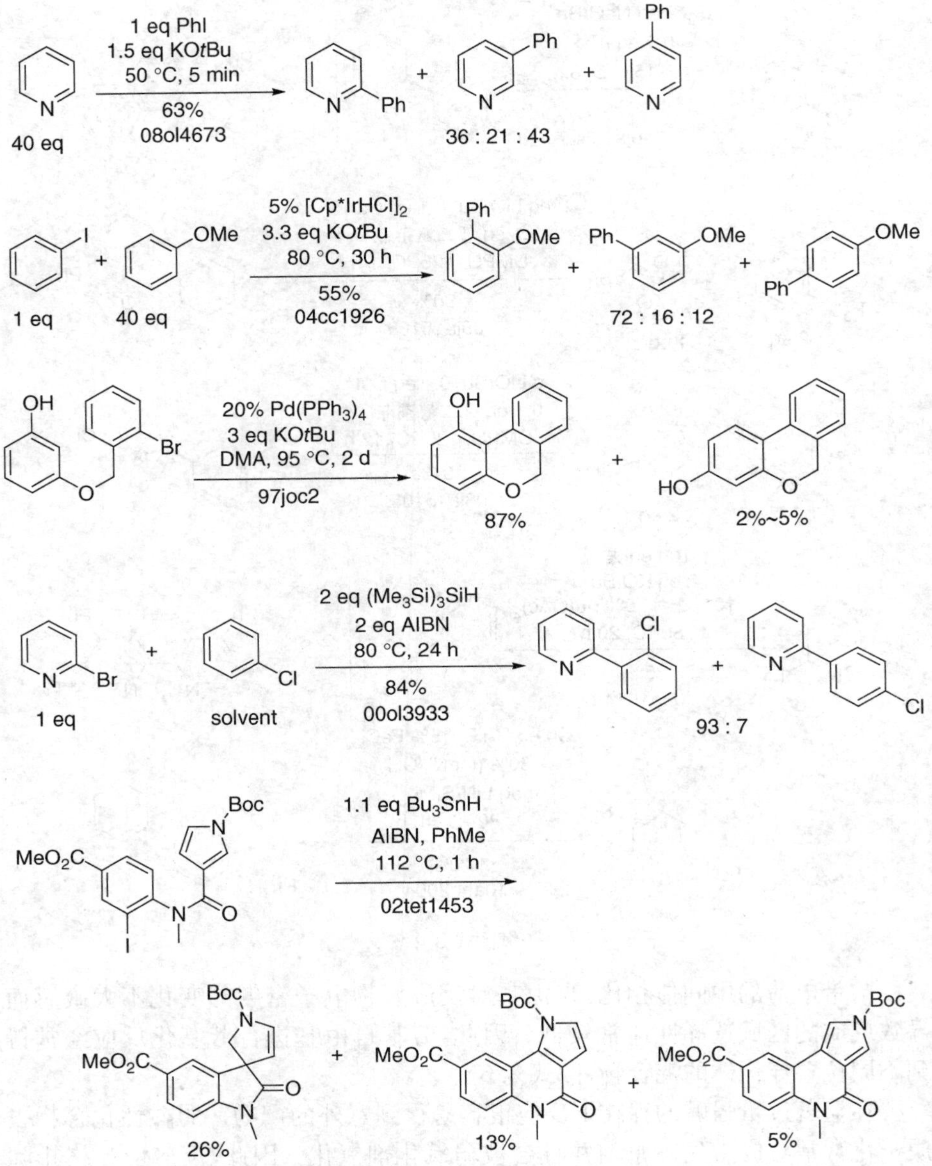

图式 3.5 在与芳基自由基进行芳基化反应中形成不同位置的异构体[11,17-20]

自由基是短寿命的中间体,为了获得满意的收率通常需要高浓度的自由基受体。与主反应竞争的反应包括自由基聚化(自偶联)、还原(即从溶剂或其他的氢供体获取氢原子)或者与氧气反应形成过氧自由基,自由基也可以诱发烯烃或炔烃的低聚。

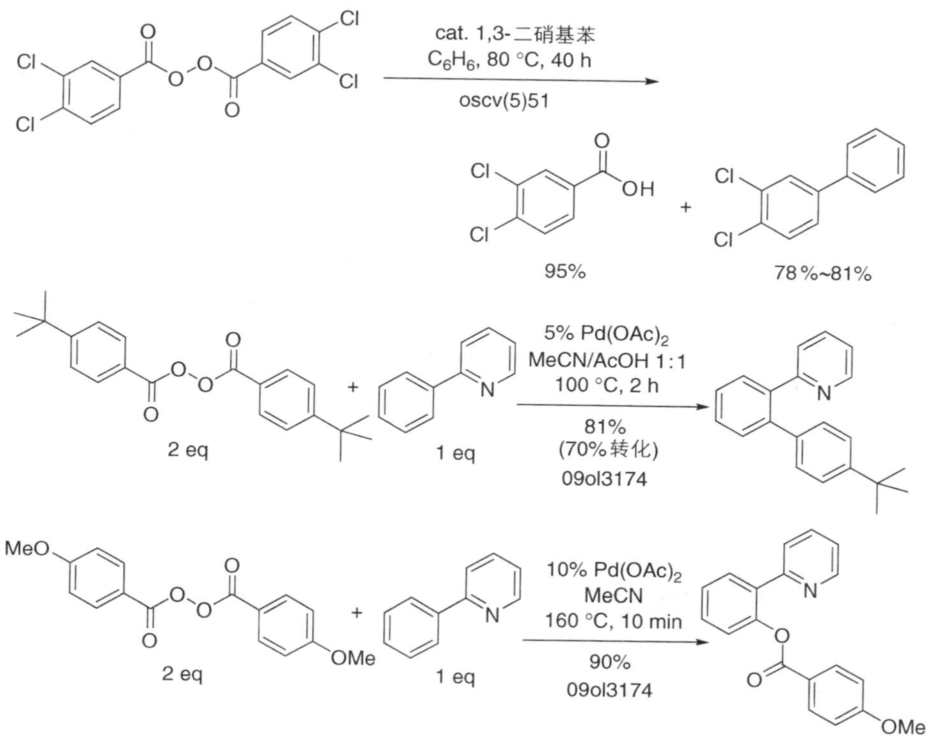

图式 3.6　由二酰基过氧化物产生芳基自由基及其反应[23,24]

3.2.3　通过过渡金属螯合物

配位官能团取代的芳烃能够在温和条件并且不需要强碱存在的情况下通过形成螯合物实现其金属化。通过这种方式形成的芳基钌和钯螯合物能够与芳基卤化物反应，经还原消除后得到联芳基产物并再生催化剂（图式 3.7）。

3.2.4　通过过渡金属催化

许多已报道的卤代芳烃的芳基化反应是假设通过未取代芳烃金属化而不是形成金属螯合物进行的。氧化剂虽非必需，但有时会在这些反应中添加。反应的区域选择性很低而且难以预测（图式 3.8），但控制反应区域选择性的原则（或反应缺乏区域选择性的原因）正在逐步得到认识[33]。此外，偶尔会产生自偶联的副产物。

过渡金属催化卤代芳烃的芳基化反应的一个典型副反应是卤代芳烃的自偶联，该反应可以在非常温和的条件下发生[46-48]（图式 3.9）。为防止自偶联，上面给出的反应实例中多数使其中一个反应物的用量显著过量。

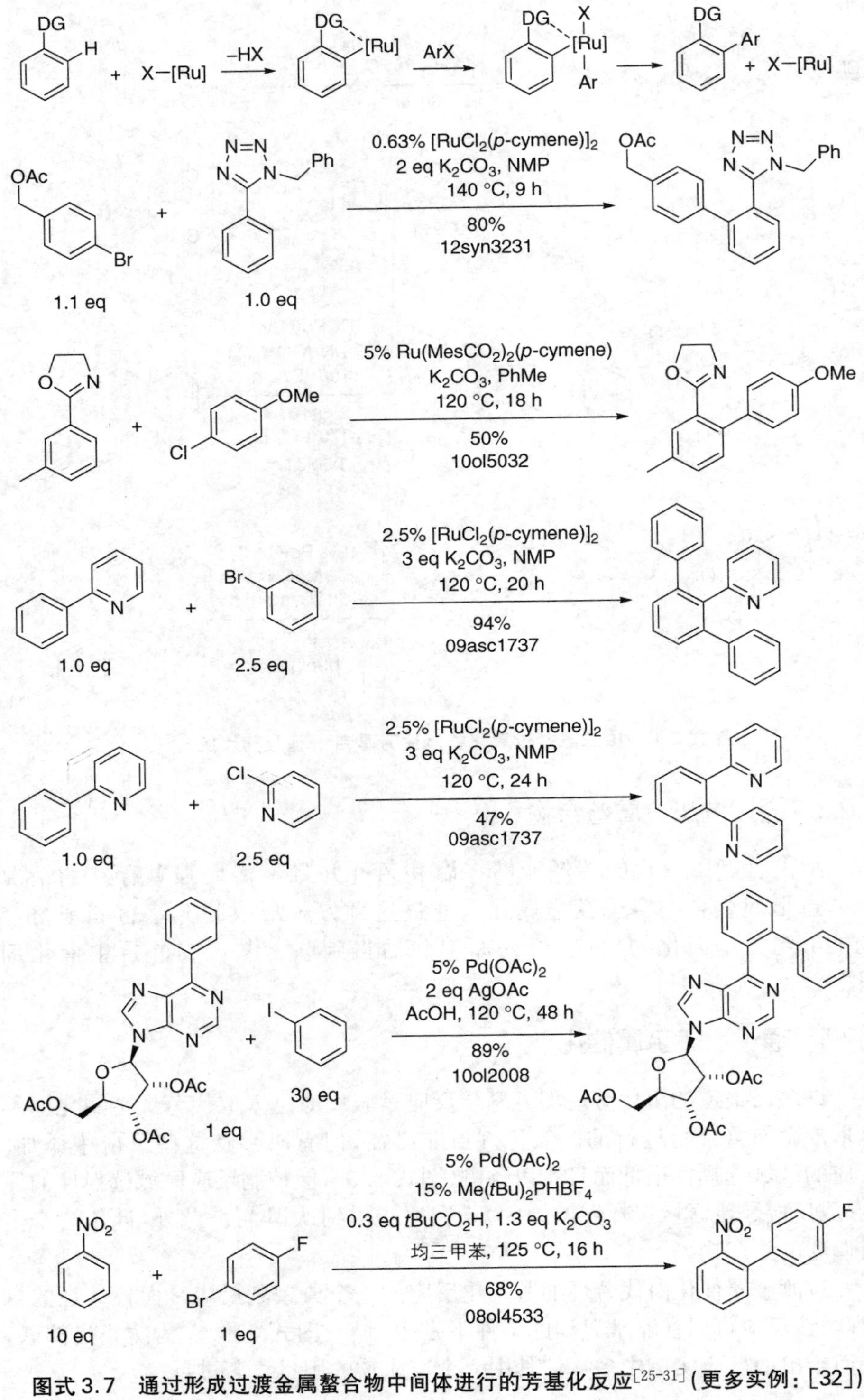

图式 3.7 通过形成过渡金属螯合物中间体进行的芳基化反应[25-31]（更多实例：[32]）

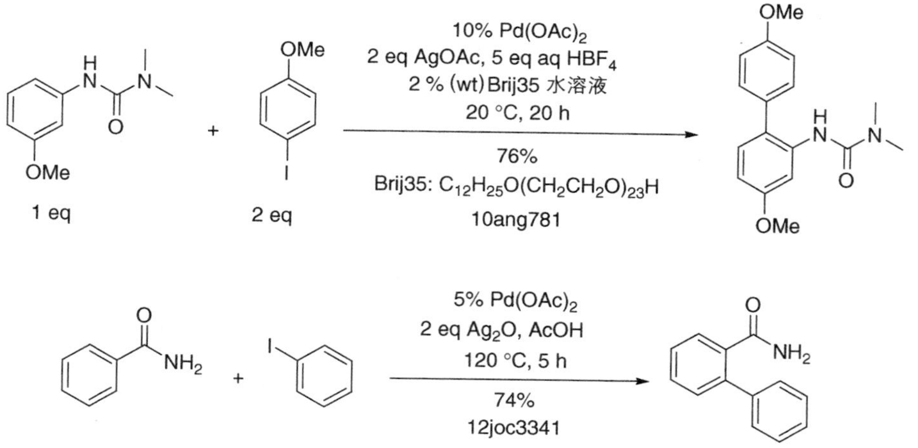

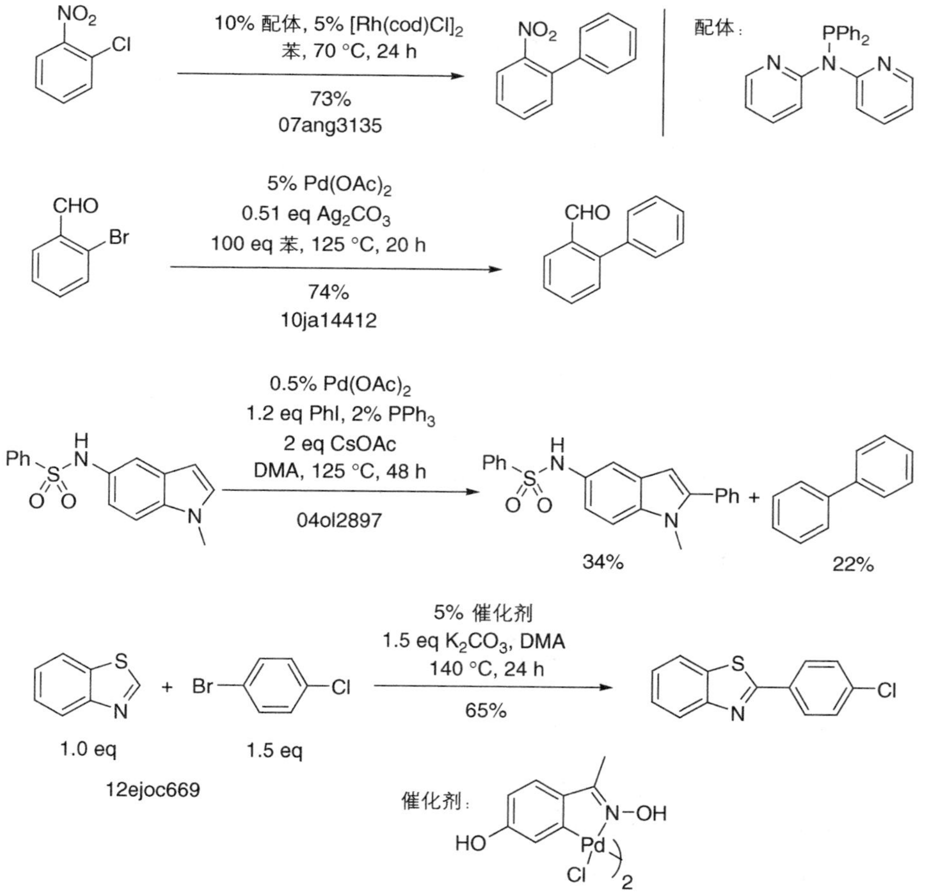

图式 3.8　未取代的芳烃与卤代芳烃或甲苯磺酸酯在过渡金属催化下芳基化反应[34-41]（更多实例：[33,42-45]）

续图式 3.8

图式 3.9 卤代芳烃的自偶联[49,50]

3.3 与重氮盐的芳基化反应

另一种产生芳基自由基的简单方法是用碱的水溶液处理芳基重氮盐[刚伯格-巴赫姆(Gomberg - Bachmann)反应]。由此形成的重氮酸酐(Ar—N=N—O—N=N—Ar)受热均裂得到芳基自由基与显著过量的另一种芳烃发生芳基化反应,从而得到非对称联芳基化合物(图式 3.10)。此外,芳基重氮盐也可通过光解产生芳基自由基[51]。

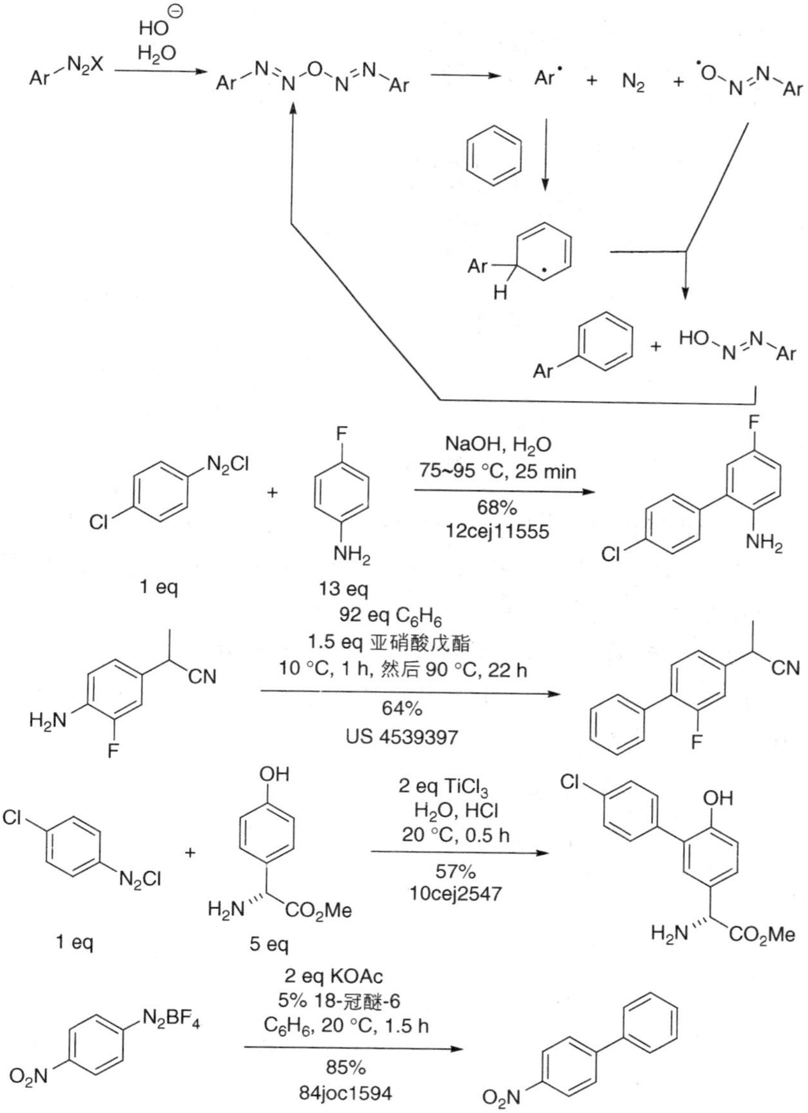

图式 3.10 由重氮盐产生芳基自由基的芳基化反应[52-58]（更多实例：[59]）

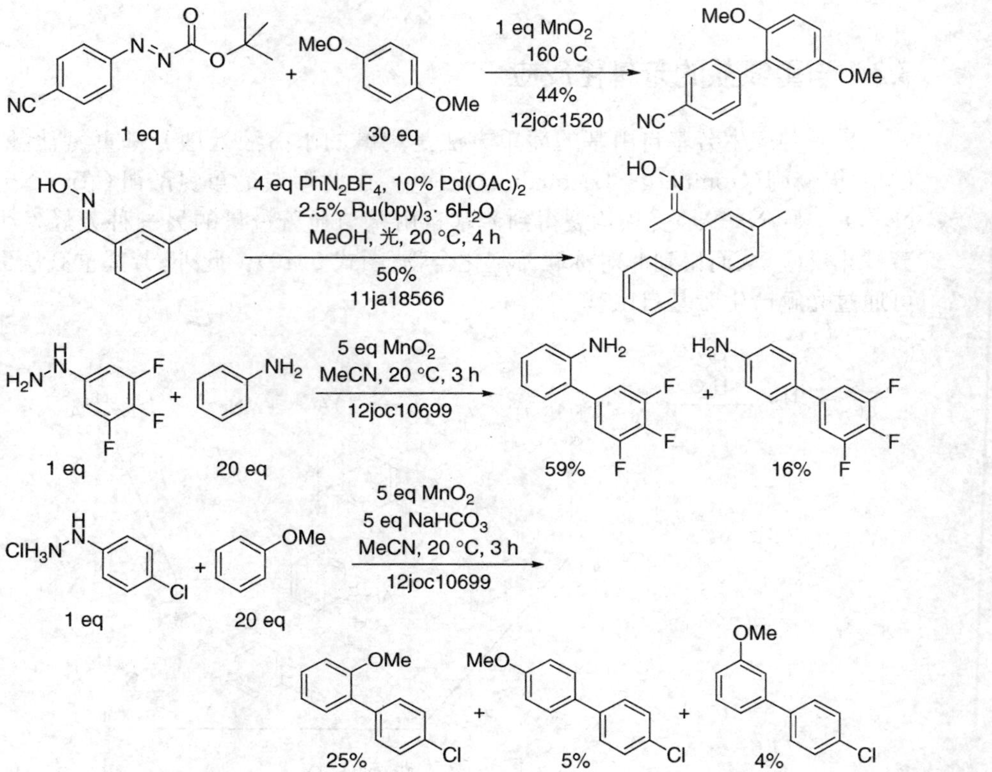

续图式 3.10

不是所有的芳胺或杂芳胺均能够被转化为重氮盐。缺电子的杂芳基重氮盐[例如,从 2-(或 4-)氨基吡啶或 2-氨基嘧啶制备的重氮盐]在氮原子上发生快速的芳香亲核取代,并常常水解成羟基芳烃。由 2-(伯烷基)苯胺制备的重氮盐对碱不稳定,容易环化成吲唑[60]。富电子的苯胺重氮化偶尔可能导致芳族亚硝化或硝化[61]。一些杂芳基自由基重排为更稳定的自由基的速度显著超过加成到芳烃上的速度(图式 3.11)。

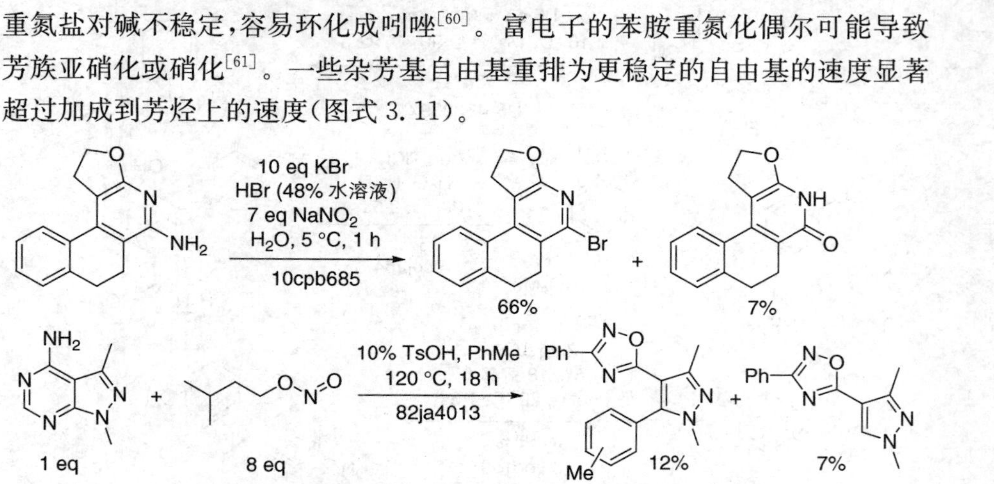

图式 3.11 2-吡啶和 4-嘧啶重氮盐分解途径[62,63]

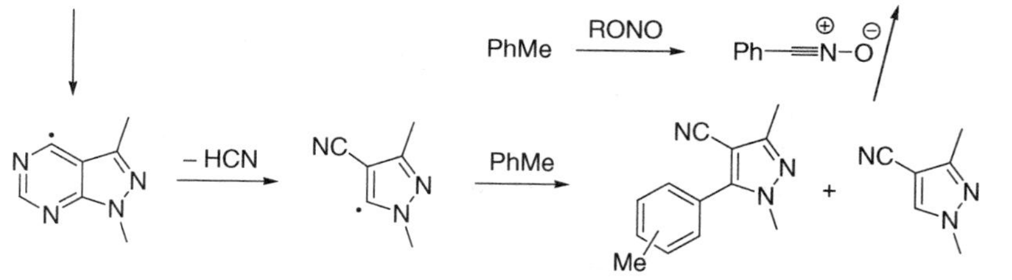

续图式 3.11

3.4 与其他官能团化芳烃的芳基化反应

苯甲酸和芳杂环羧酸是芳基自由基或金属化芳烃的前体,并且可用于芳族 C—H 键的芳基化反应(图式 3.12)。由苯甲酸盐氧化形成的苯甲酰自由基,其脱羧形成芳基自由基速度非常快,通常观察不到由苯甲酰自由基产生的副产物。

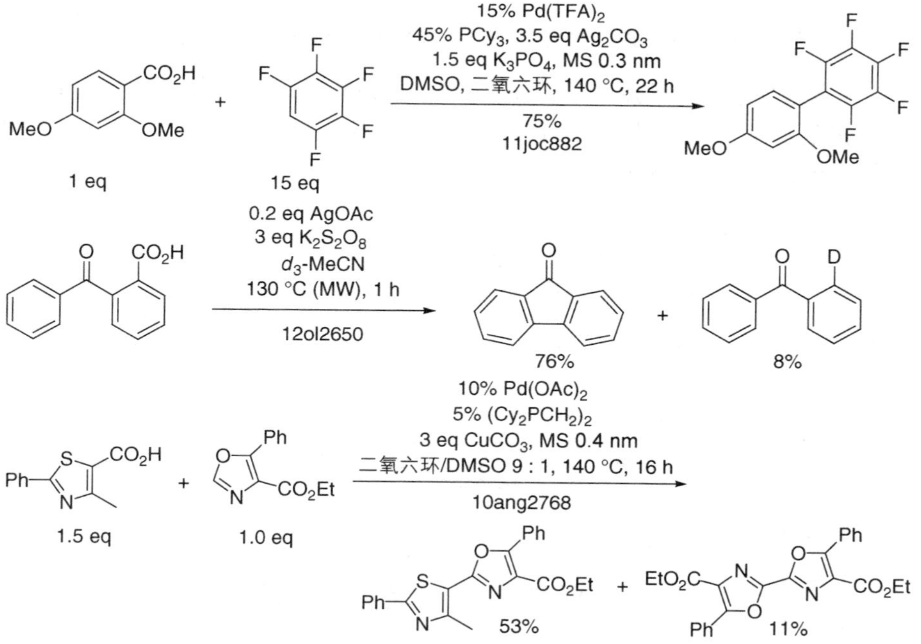

图式 3.12 芳香羧酸的芳基化反应[64-66] (更多实例:[67])

通过形成螯合对芳族 C—H 金属化生成的芳基钯络合物容易发生自偶联反应。实现不同芳烃的交叉偶联面临的挑战在于找到将零价钯转化为二价钯的氧化剂,且该氧化剂不与亲核偶联的物种或金属化芳烃反应。苯醌是一种这样的氧化剂,可用于铃木(Suzuki)偶联中。虽然苯醌也可以与芳基钯络合物发生赫克反应,但芳基钯络合物与硼酸偶联速度更快。用铁催化剂和以空气作为氧化剂可以实现与芳基硼酸进行类似的芳基化反应(图式 3.13)。

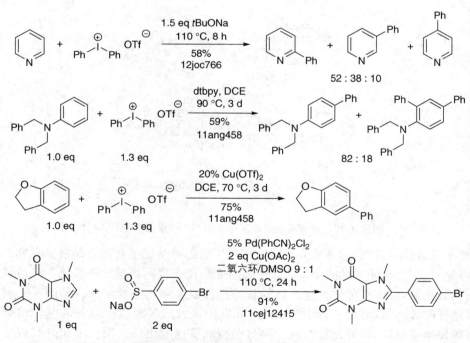

图式 3.13 钯和铁催化的芳香族 C—H 基团与硼酸的芳基化反应[68-70]

许多芳基亲电体被用于从未取代芳烃制备联芳基化合物,包括芳基碘鎓盐、磺酰氯和亚磺酸盐(图式 3.14)。仅有少许这类反应被报道,其适用范围和官能团耐受性鲜为人知。

图式 3.14 芳烃与芳烃碘鎓盐和亚磺酸盐的芳基化反应[71-73]

3.5 未取代芳烃的芳基化反应

芳烃氧化二聚为大型、对称分子合成提供了一个有吸引力的方法。这种类型反应的重要实例包括酚的偶联制备对称的 2,2′-(或 4,4′-)二羟基联苯 (例如,联萘酚的合成[74])和吡啶氧化二聚合成 2,2′-联吡啶,后者是除草剂敌草快(dequat)合成的关键中间体(图式 3.15)。

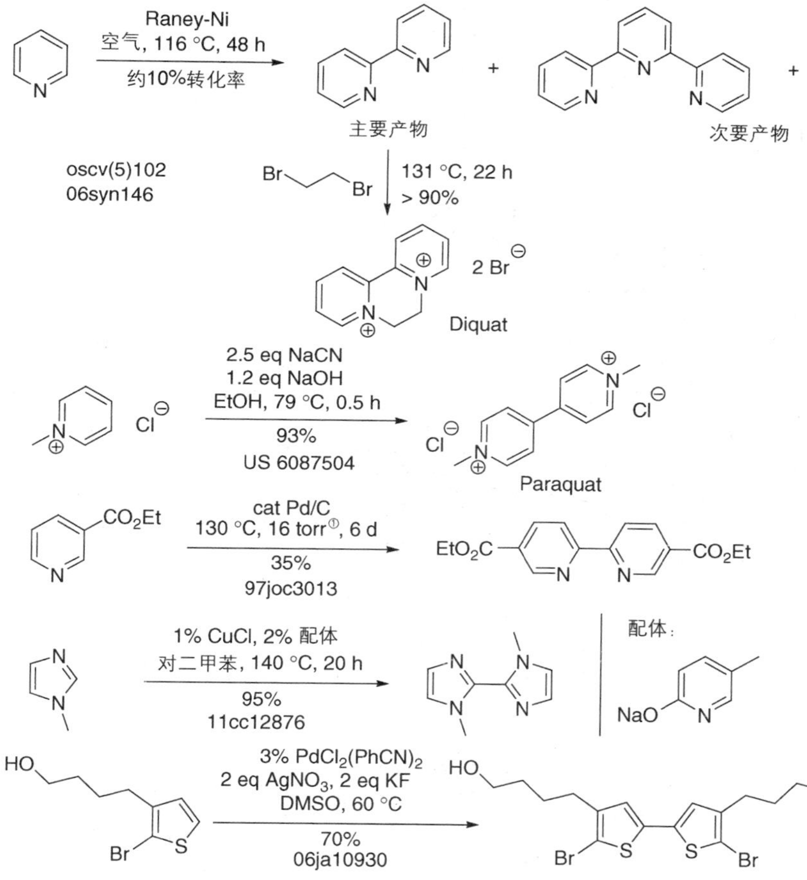

图式 3.15 芳烃氧化二聚[75-81]

两种不同的芳烃间的氧化偶联是比较困难的反应,容易产生自聚和异聚的位置异构体混合物[82]。不过,精心挑选偶联反应底物和反应条件有时能以高收率合成联芳基产物(图式 3.16)。常见的副反应包括官能团的氧化转化、氧化成醌以及羟基化、酰胺化、亚磺酰和芳烃的卤代。在氧气和铜盐的存在

---

① 1 torr=133.322 Pa。

下，N,N-二甲基甲酰胺（DMF）可被氧化成氰化物[83]，它可导致芳烃的氰化或过渡金属催化剂的中毒。

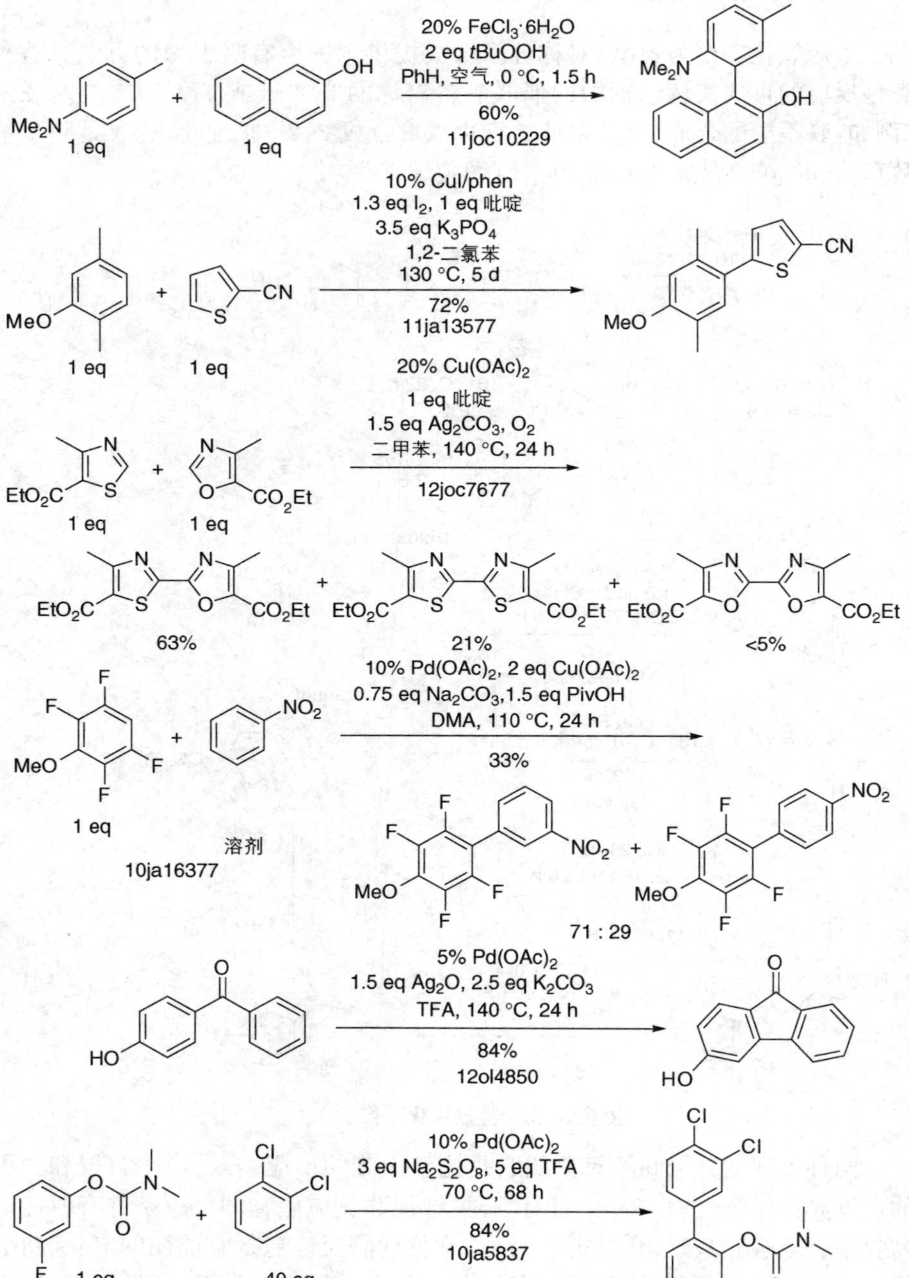

图式 3.16 芳烃的氧化异二聚[82,84-90]（更多例子：[91-96]）

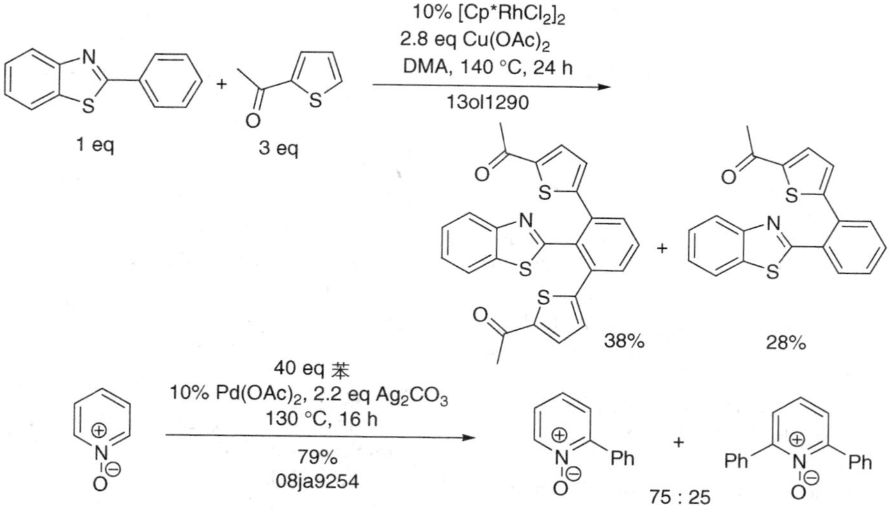

续图式 3.16

酸性芳烃可以被金属化和氧化二聚生成对称的联芳基产物（图式 3.17）。潜在的副反应包括金属化芳烃与氧化剂的反应和芳香亲核取代反应。

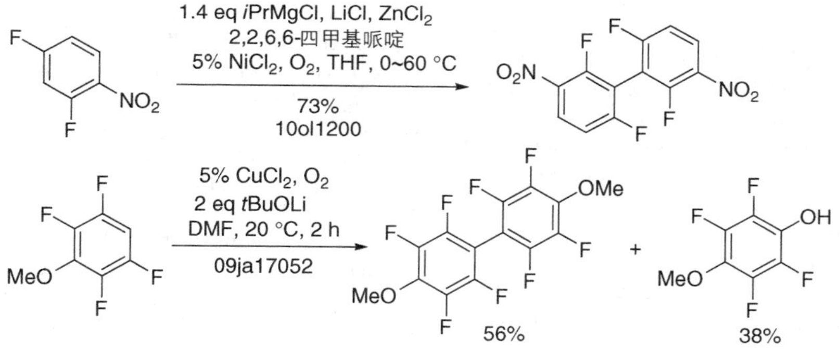

图式 3.17 金属化芳烃的氧化二聚[97,98]

氧化剂、易燃的有机溶剂和强碱的混合物会剧烈分解，对于大规模反应通常过于危险。为更好地操作图式 3.17 所示的反应，应使用不易燃溶剂并选用便于精确、可控加料的固体或液体氧化剂。

参考文献

1. Kuhl, N., Hopkinson, M.N., Wencel-Delord, J., and Glorius, F. (2012) Beyond directing groups: transition-metal-catalyzed C–H activation of simple arenes. *Angew. Chem. Int. Ed.*, **51**, 10236–10254.
2. Hassan, J., Sévignon, M., Gozzi, C., Schulz, E., and Lemaire, M. (2002) Aryl–aryl bond formation one century after the discovery of the Ullmann reaction. *Chem. Rev.*, **102**, 1359–1469.
3. Fanta, P.E. (1964) The Ullmann synthesis of biaryls, 1945–1963. *Chem. Rev.*, **64**, 613–632.
4. Lyons, T.W. and Sanford, M.S. (2010) Palladium-catalyzed ligand-directed C–H functionalization reactions. *Chem. Rev.*, **110**, 1147–1169.

5. Orban, I., Holer, M., and Kaufmann, A. (1997) Verfahren zur Herstellung von 2-(2,4-Dihydroxyphenyl)-4,6-bis-(2,4-dimethylphenyl)-s-triazin. EP Patent 0779280.
6. Gupta, R.B., Jakiela, D.J., Venimadhavan, S., and Cappadona, R. (2000) Process for preparing triazines using a combination of Lewis acids and reaction promoters. WO Patent 0029392.
7. Sone, T., Inoue, M., and Sato, K. (1988) A new and simple route to 3-arylthiophenes. *Bull. Chem. Soc. Jpn.*, **61**, 3779–3781.
8. Sone, T., Kawasaki, H., Nagasawa, S., Takahashi, N., Tate, K., and Sato, K. (1981) Unusual Friedel–Crafts reaction of 2,5-dichloro-3-(chloromethyl)thiophene with benzene and some alkylbenzenes. *Chem. Lett.*, 399–402.
9. Sone, T., Shiromaru, O., Igarashi, S., Kato, E., and Sawara, M. (1979) Acid-catalyzed oligomerization of thiophene nuclei. III. Reactions of 2-chlorothiophene with cation exchange resin and 100% orthophosphoric acid. Formation of the dimer type products containing a tetrahydro-2-thiophenone moiety. *Bull. Chem. Soc. Jpn.*, **52**, 1126–1130.
10. Jones, G.O., Liu, P., Houk, K.N., and Buchwald, S.L. (2010) Computational explorations of mechanisms and ligand-directed selectivities of copper-catalyzed Ullmann-type reactions. *J. Am. Chem. Soc.*, **132**, 6205–6213.
11. Yanagisawa, S., Ueda, K., Taniguchi, T., and Itami, K. (2008) Potassium *t*-butoxide alone can promote the biaryl coupling of electron-deficient nitrogen heterocycles and haloarenes. *Org. Lett.*, **10**, 4673–4676.
12. Shirakawa, E., Itoh, K., Higashino, T., and Hayashi, T. (2010) *tert*-Butoxide-mediated arylation of benzene with aryl halides in the presence of a catalytic 1,10-phenanthroline derivative. *J. Am. Chem. Soc.*, **132**, 15537–15539.
13. Do, H.-Q., Khan, R.M.K., and Daugulis, O. (2008) A general method for copper-catalyzed arylation of arene C–H bonds. *J. Am. Chem. Soc.*, **130**, 15185–15192.
14. Vallée, F., Mousseau, J.J., and Charette, A.B. (2010) Iron-catalyzed direct arylation through an aryl radical transfer pathway. *J. Am. Chem. Soc.*, **132**, 1514–1516.
15. Liu, W., Cao, H., and Lei, A. (2010) Iron-catalyzed direct arylation of unactivated arenes with aryl halides. *Angew. Chem. Int. Ed.*, **49**, 2004–2008.
16. Fagnou, K., Leclerc, J.-P., Campeau, L.-C., and Stuart, D.R. (2008) Use of *N*-oxide compounds in coupling reactions. US Patent 2008132698.
17. Fujita, K., Nonogawaa, M., and Yamaguchi, R. (2004) Direct arylation of aromatic C–H bonds catalyzed by Cp*Ir complexes. *Chem. Commun.*, 1926–1927.
18. Hennings, D.D., Iwasa, S., and Rawal, V.H. (1997) Anion-accelerated palladium-catalyzed intramolecular coupling of phenols with aryl halides. *J. Org. Chem.*, **62**, 2–3.
19. Martínez-Barrasa, V., García de Viedma, A., Burgos, C., and Alvarez-Builla, J. (2000) Synthesis of biaryls via intermolecular radical addition of heteroaryl and aryl bromides onto arenes. *Org. Lett.*, **2**, 3933–3935.
20. Escolano, C. and Jones, K. (2002) Aryl radical cyclisation onto pyrroles. *Tetrahedron*, **58**, 1453–1464.
21. Linhardt, R.J., Murr, B.L., Montgomery, E., Osby, J., and Sherbine, J. (1982) Mechanism for diacyl peroxide decomposition. *J. Org. Chem.*, **47**, 2242–2251.
22. Srinivas, S. and Taylor, K.G. (1990) Amine-induced reactions of diacyl peroxides. *J. Org. Chem.*, **55**, 1779–1786.
23. Hey, D.H. and Perkins, M.J. (1973) Arylbenzenes: 3,4-dichlorobiphenyl. *Org. Synth.*, Coll. Vol. **5**, 51–54.
24. Yu, W.-Y., Sit, W.N., Zhou, Z., and Chan, A.S.-C. (2009) Palladium-catalyzed decarboxylative arylation of C–H bonds by aryl acylperoxides. *Org. Lett.*, **11**, 3174–3177.
25. Seki, M. (2012) An efficient C–H arylation of a 5-phenyl-1*H*-tetrazole derivative: a practical synthesis of an angiotensin II receptor blocker. *Synthesis*, **44**, 3231–3237.
26. Ackermann, L., Vicente, R., Potukuchi, H.K., and Pirovano, V. (2010) Mechanistic insight into direct arylations with

ruthenium(II) carboxylate catalysts. *Org. Lett.*, **12**, 5032–5035.

27. Pozgan, F. and Dixneuf, P.H. (2009) Ruthenium(II) acetate catalyst for direct functionalisation of sp^2-C–H bonds with aryl chlorides and access to tris-heterocyclic molecules. *Adv. Synth. Catal.*, **351**, 1737–1743.

28. Guo, H.-M., Jiang, L.-L., Niu, H.-Y., Rao, W.-H., Liang, L., Mao, R.-Z., Li, D.-Y., and Qu, G.-R. (2011) Pd(II)-catalyzed *ortho* arylation of 6-arylpurines with aryl iodides via purine-directed C–H activation: a new strategy for modification of 6-arylpurine derivatives. *Org. Lett.*, **13**, 2008–2011.

29. Caron, L., Campeau, L.-C., and Fagnou, K. (2008) Palladium-catalyzed direct arylation of nitro-substituted aromatics with aryl halides. *Org. Lett.*, **10**, 4533–4536.

30. Nishikata, T., Abela, A.R., and Lipshutz, B.H. (2010) Room temperature C–H activation and cross-coupling of aryl ureas in water. *Angew. Chem. Int. Ed.*, **49**, 781–784.

31. Li, D.-D., Yuan, T.-T., and Wang, G.-W. (2012) Palladium-catalyzed *ortho*-arylation of benzamides via direct sp^2 C–H bond activation. *J. Org. Chem.*, **77**, 3341–3347.

32. Kametani, Y., Satoh, T., Miura, M., and Nomura, M. (2000) Regioselective arylation of benzanilides with aryl triflates or bromides under palladium catalysis. *Tetrahedron Lett.*, **41**, 2655–2658.

33. Lapointe, D., Markiewicz, T., Whipp, C.J., Toderian, A., and Fagnou, K. (2011) Predictable and site-selective functionalization of poly(hetero)arene compounds by palladium catalysis. *J. Org. Chem.*, **76**, 749–759.

34. Proch, S. and Kempe, R. (2007) An efficient bimetallic rhodium catalyst for the direct arylation of unactivated arenes. *Angew. Chem. Int. Ed.*, **46**, 3135–3138.

35. Mousseau, J.J., Vallée, F., Lorion, M.M., and Charette, A.B. (2010) Umpolung direct arylation reactions: facile process requiring only catalytic palladium and substoichiometric amount of silver salts. *J. Am. Chem. Soc.*, **132**, 14412–14414.

36. Lane, B.S. and Sames, D. (2004) Direct C–H bond arylation: selective palladium-catalyzed C2-arylation of N-substituted indoles. *Org. Lett.*, **6**, 2897–2900.

37. Zhang, G., Zhao, X., Yan, Y., and Ding, C. (2012) Direct arylation under catalysis of an oxime-derived palladacycle: search for a phosphane-free method. *Eur. J. Org. Chem.*, **2012**, 669–672.

38. Fan, S., Yang, J., and Zhang, X. (2011) Pd-catalyzed direct cross-coupling of electron-deficient polyfluoroarenes with heteroaromatic tosylates. *Org. Lett.*, **13**, 4374–4377.

39. Guo, P., Joo, J.M., Rakshit, S., and Sames, D. (2011) C–H arylation of pyridines: high regioselectivity as a consequence of the electronic character of C–H bonds and heteroarene ring. *J. Am. Chem. Soc.*, **133**, 16338–16341.

40. Ye, M., Gao, G.-L., Edmunds, A.J.F., Worthington, P.A., Morris, J.A., and Yu, J.-Q. (2011) Ligand-promoted C3-selective arylation of pyridines with Pd catalysts: gram-scale synthesis of (±)-preclamol. *J. Am. Chem. Soc.*, **133**, 19090–19093.

41. Effenberger, F., Agster, W., Fischer, P., Jogun, K.H., Stezowski, J.J., Daltrozzo, E., and Kollmannsberger-von Nell, G. (1983) Synthesis, structure, and spectral behavior of donor–acceptor substituted biphenyls. *J. Org. Chem.*, **48**, 4649–4658.

42. Lafrance, M. and Fagnou, K. (2006) Palladium-catalyzed benzene arylation: incorporation of catalytic pivalic acid as a proton shuttle and a key element in catalyst design. *J. Am. Chem. Soc.*, **128**, 16496–16497.

43. Lafrance, M., Shore, D., and Fagnou, K. (2006) Mild and general conditions for the cross-coupling of aryl halides with pentafluorobenzene and other perfluoroaromatics. *Org. Lett.*, **8**, 5097–5100.

44. Fall, Y., Reynaud, C., Doucet, H., and Santelli, M. (2009) Ligand-free-palladium-catalyzed direct 4-arylation of isoxazoles using aryl bromides. *Eur. J. Org. Chem.*, **2009**, 4041–4050.

45. Liégault, B., Petrov, I., Gorelsky, S.I., and Fagnou, K. (2010) Modulating reactivity and diverting selectivity in palladium-catalyzed heteroaromatic direct arylation through the use of a chloride activating/blocking group. *J. Org. Chem.*, **75**, 1047–1060.

46. Dhital, R.N., Kamonsatikul, C., Somsook, E., Bobuatong, K., Ehara, M., Karanjit, S., and Sakurai, H. (2012) Low-temperature carbon–chlorine bond activation by bimetallic gold/palladium alloy nanoclusters: an application to Ullmann coupling. *J. Am. Chem. Soc.*, **134**, 20250–20253.

47. Li, J.-H., Xie, Y.-X., and Yin, D.-L. (2003) New role of CO_2 as a selective agent in palladium-catalyzed reductive Ullmann coupling with zinc in water. *J. Org. Chem.*, **68**, 9867–9869.

48. Kashiwabara, T. and Tanaka, M. (2005) Pd-catalyzed desulfonylative homocoupling of arenesulfonyl chlorides in the presence of hexamethyldisilane forming biaryls. *Tetrahedron Lett.*, **46**, 7125–7128.

49. Penalva, V., Hassan, J., Lavenot, L., Gozzi, C., and Lemaire, M. (1998) Direct homocoupling of aryl halides catalyzed by palladium. *Tetrahedron Lett.*, **39**, 2559–2560.

50. Monopoli, A., Calò, V., Ciminale, F., Cotugno, P., Angelici, C., Cioffi, N., and Nacci, A. (2010) Glucose as a clean and renewable reductant in the Pd-nanoparticle-catalyzed reductive homocoupling of bromo- and chloroarenes in water. *J. Org. Chem.*, **75**, 3908–3911.

51. Hari, D.P., Schroll, P., and König, B. (2012) Metal-free, visible-light-mediated direct C–H arylation of heteroarenes with aryl diazonium salts. *J. Am. Chem. Soc.*, **134**, 2958–2961.

52. Pratsch, G., Wallaschkowski, T., and Heinrich, M.R. (2012) The Gomberg–Bachmann reaction for the arylation of anilines with aryl diazotates. *Chem. Eur. J.*, **18**, 11555–11559.

53. Ramachandran, V. (1985) (Alkoxydiazo)halobenzeneacetonitriles. US Patent 4539397.

54. Wetzel, A., Pratsch, G., Kolb, R., and Heinrich, M.R. (2010) Radical arylation of phenols, phenyl ethers, and furans. *Chem. Eur. J.*, **16**, 2547–2556.

55. Beadle, J.R., Korzeniowski, S.H., Rosenberg, D.E., Garcia-Slanga, B.J., and Gokel, G.W. (1984) Phase-transfer-catalyzed Gomberg–Bachmann synthesis of unsymmetrical biarenes: a survey of catalysts and substrates. *J. Org. Chem.*, **49**, 1594–1603.

56. Jasch, H., Höfling, S.B., and Heinrich, M.R. (2012) Nucleophilic substitutions and radical reactions of phenylazocarboxylates. *J. Org. Chem.*, **77**, 1520–1532.

57. Kalyani, D., McMurtrey, K.B., Neufeldt, S.R., and Sanford, M.S. (2011) Room-temperature C–H arylation: merger of Pd-catalyzed C–H functionalization and visible-light photocatalysis. *J. Am. Chem. Soc.*, **133**, 18566–18569.

58. Jasch, H., Scheumann, J., and Heinrich, M.R. (2012) Regioselective radical arylation of anilines with arylhydrazines. *J. Org. Chem.*, **77**, 10699–10706.

59. Wetzel, A., Ehrhardt, V., and Heinrich, M.R. (2008) Synthesis of amino- and hydroxybiphenyls by radical chain reaction of arenediazonium salts. *Angew. Chem. Int. Ed.*, **47**, 9130–9133.

60. Crestey, F., Collot, V., Stiebing, S., and Rault, S. (2006) A new and efficient synthesis of 2-azatryptophans. *Tetrahedron*, **62**, 7772–7775.

61. Ferrarini, P.L., Mori, C., Badawneh, M., Manera, C., Martinelli, A., Miceli, M., Romagnoli, F., and Saccomanni, G. (1997) Unusual nitration of substituted 7-amino-1,8-naphthyridine in the synthesis of compounds with antiplatelet activity. *J. Heterocycl. Chem.*, **34**, 1501–1510.

62. Okuda, K., Deguchi, H., Kashino, S., Hirota, T., and Sasaki, K. (2010) Polycyclic N-heterocyclic compounds. Part 64: Synthesis of 5-amino-1,2,6,7-tetrahydrobenzo[*f*]furo[2,3-*c*] isoquinolines and related compounds. Evaluation of their bronchodilator activity and effects on lipoprotein lipase mRNA expression. *Chem. Pharm. Bull.*, **58**, 685–689.

63. Press, J.B., Eudy, N.H., Lovell, F.M., Morton, G.O., and Siegel, M.M. (1982) A remarkable fragmentation reaction of a pyrimidyl radical. *J. Am. Chem. Soc.*, **104**, 4013–4014.

64. Zhao, H., Wei, Y., Xu, J., Kan, J., Su, W., and Hong, M. (2011) Pd/PR_3-catalyzed cross-coupling of aromatic carboxylic acids with electron-deficient polyfluoroarenes via combination of

decarboxylation with sp² C–H cleavage. *J. Org. Chem.*, **76**, 882–893.

65. Seo, S., Slater, M., and Greaney, M.F. (2012) Decarboxylative C–H arylation of benzoic acids under radical conditions. *Org. Lett.*, **14**, 2650–2653.
66. Zhang, F. and Greaney, M.F. (2010) Decarboxylative C–H cross-coupling of azoles. *Angew. Chem. Int. Ed.*, **49**, 2768–2771.
67. Xie, K., Yang, Z., Zhou, X., Li, X., Wang, S., Tan, Z., An, X., and Guo, C.-C. (2010) Pd-catalyzed decarboxylative arylation of thiazole, benzoxazole, and polyfluorobenzene with substituted benzoic acids. *Org. Lett.*, **12**, 1564–1567.
68. Nishikata, T., Abela, A.R., Huang, S., and Lipshutz, B.H. (2010) Cationic palladium(II) catalysis: C–H activation/Suzuki–Miyaura couplings at room temperature. *J. Am. Chem. Soc.*, **132**, 4978–4979.
69. Wen, J., Qin, S., Ma, L.-F., Dong, L., Zhang, J., Liu, S.-S., Duan, Y.-S., Chen, S.-Y., Hu, C.-W., and Yu, X.-Q. (2010) Iron-mediated direct Suzuki–Miyaura reaction: a new method for the *ortho*-arylation of pyrrole and pyridine. *Org. Lett.*, **12**, 2694–2697.
70. Singh, P.P., Aithagani, S.K., Yadav, M., Singh, V.P., and Vishwakarma, R.A. (2013) Iron-catalyzed cross-coupling of electron-deficient heterocycles and quinone with organoboron species via innate C–H functionalization: application in total synthesis of pyrazine alkaloid botryllazine A. *J. Org. Chem.*, **78**, 2639–2648.
71. Wen, J., Zhang, R.-Y., Chen, S.-Y., Zhang, J., and Yu, X.-Q. (2012) Direct arylation of arene and *N*-heteroarenes with diaryliodonium salts without the use of transition metal catalyst. *J. Org. Chem.*, **77**, 766–771.
72. Ciana, C.-L., Phipps, R.J., Brandt, J.R., Meyer, F.-M., and Gaunt, M.J. (2011) A highly *para*-selective copper(II)-catalyzed direct arylation of aniline and phenol derivatives. *Angew. Chem. Int. Ed.*, **50**, 458–462.
73. Liu, B., Guo, Q., Cheng, Y., Lan, J., and You, J. (2011) Palladium-catalyzed desulfitative C–H arylation of heteroarenes with sodium sulfinates. *Chem. Eur. J.*, **17**, 13415–13419.
74. Matsushita, M., Kamata, K., Yamaguchi, K., and Mizuno, N. (2005) Heterogeneously catalyzed aerobic oxidative biaryl coupling of 2-naphthols and substituted phenols in water. *J. Am. Chem. Soc.*, **127**, 6632–6640.
75. Sasse, W.H.F. (1973) 2,2′-Dipyridine. *Org. Synth.*, Coll. Vol. **5**, 102–107.
76. Coe, B.J., Curati, N.R.M., and Fitzgerald, E.C. (2006) Unusually facile syntheses of diquat (6,7-dihydrodipyrido[1,2-*a*:2′,1′-*c*]pyrazinediium) and related cations. *Synthesis*, 146–150.
77. Xiao, Y., Chu, L., Sanakis, Y., and Liu, P. (2009) Revisiting the IspH catalytic system in the deoxyxylulose phosphate pathway: achieving high activity. *J. Am. Chem. Soc.*, **131**, 9931–9933.
78. Tsou, C.-P., Lin, S.-C., Huang, T.-K., and Chen, C.-Y. (2000) Process for the preparation of 1,1′-dialkyl-4,4′-bipyridinium salt compounds. US Patent 6087504.
79. Newkome, G.R., Gross, J., and Patri, A.K. (1997) Synthesis of unsymmetrical 5,5′-disubstituted 2,2′-bipyridines. *J. Org. Chem.*, **62**, 3013–3014.
80. Zhu, M., Fujita, K.-I., and Yamaguchi, R. (2011) Efficient synthesis of biazoles by aerobic oxidative homocoupling of azoles catalyzed by a copper(I)/2-pyridonate catalytic system. *Chem. Commun.*, **47**, 12876–12878.
81. Takahashi, M., Masui, K., Sekiguchi, H., Kobayashi, N., Mori, A., Funahashi, M., and Tamaoki, N. (2006) Palladium-catalyzed C–H homocoupling of bromothiophene derivatives and synthetic application to well-defined oligothiophenes. *J. Am. Chem. Soc.*, **128**, 10930–10933.
82. Do, H.-Q. and Daugulis, O. (2011) A general method for copper-catalyzed arene cross-dimerization. *J. Am. Chem. Soc.*, **133**, 13577–13586.
83. Kim, J., Choi, J., Shin, K., and Chang, S. (2012) Copper-mediated sequential cyanation of aryl C–B and arene C–H bonds using ammonium iodide and DMF. *J. Am. Chem. Soc.*, **134**, 2528–2531.
84. Chandrasekharam, M., Chiranjeevi, B., Gupta, K.S.V., and Sridhar, B. (2011)

84. Iron-catalyzed regioselective direct oxidative aryl–aryl cross-coupling. *J. Org. Chem.*, **76**, 10229–10235.
85. Qin, X., Feng, B., Dong, J., Li, X., Xue, Y., Lan, J., and You, J. (2012) Copper(II)-catalyzed dehydrogenative cross-coupling between two azoles. *J. Org. Chem.*, **77**, 7677–7683.
86. Wei, Y. and Su, W. (2010) Pd(OAc)$_2$-catalyzed oxidative C–H/C–H cross-coupling of electron-deficient polyfluoroarenes with simple arenes. *J. Am. Chem. Soc.*, **132**, 16377–16379.
87. Li, H., Zhu, R.-Y., Shi, W.-J., He, K.-H., and Shi, Z.-J. (2012) Synthesis of fluorenone derivatives through Pd-catalyzed dehydrogenative cyclization. *Org. Lett.*, **14**, 4850–4853.
88. Zhao, X., Yeung, C.S., and Dong, V.M. (2010) Palladium-catalyzed *ortho*-arylation of *O*-phenylcarbamates with simple arenes and sodium persulfate. *J. Am. Chem. Soc.*, **132**, 5837–5844.
89. Reddy, V.P., Qiu, R., Iwasaki, T., and Kambe, N. (2013) Rhodium-catalyzed intermolecular oxidative cross-coupling of (hetero)arenes with chalcogenophenes. *Org. Lett.*, **15**, 1290–1293.
90. Cho, S.H., Hwang, S.J., and Chang, S. (2008) Palladium-catalyzed C–H functionalization of pyridine *N*-oxides: highly selective alkenylation and direct arylation with unactivated arenes. *J. Am. Chem. Soc.*, **130**, 9254–9256.
91. Han, W., Mayer, P., and Ofial, A.R. (2011) Palladium-catalyzed dehydrogenative cross-couplings of benzazoles with azoles. *Angew. Chem. Int. Ed.*, **50**, 2178–2182.
92. Wencel-Delord, J., Nimphius, C., Wang, H., and Glorius, F. (2012) Rhodium(III) and hexabromobenzene—a catalyst system for the cross-dehydrogenative coupling of simple arenes and heterocycles with arenes bearing directing groups. *Angew. Chem. Int. Ed.*, **51**, 13001–13005.
93. Fan, S., Chen, Z., and Zhang, X. (2012) Copper-catalyzed dehydrogenative cross-coupling of benzothiazoles with thiazoles and polyfluoroarene. *Org. Lett.*, **14**, 4950–4953.
94. Chen, F., Feng, Z., He, C.-Y., Wang, H.-Y., Guo, Y.-L., and Zhang, X. (2012) Pd-catalyzed dehydrogenative cross-coupling of polyfluoroarenes with heteroatom-substituted enones. *Org. Lett.*, **14**, 1176–1179.
95. He, C.-Y., Fan, S., and Zhang, X. (2010) Pd-catalyzed oxidative cross-coupling of perfluoroarenes with aromatic heterocycles. *J. Am. Chem. Soc.*, **132**, 12850–12852.
96. Mukhopadhyay, S., Rothenberg, G., Gitis, D., and Sasson, Y. (2000) Tandem one-pot palladium-catalyzed reductive and oxidative coupling of benzene and chlorobenzene. *J. Org. Chem.*, **65**, 3107–3110.
97. Truong, T., Alvarado, J., Tran, L.D., and Daugulis, O. (2010) Nickel, manganese, cobalt, and iron-catalyzed deprotonative arene dimerization. *Org. Lett.*, **12**, 1200–1203.
98. Do, H.-Q. and Daugulis, O. (2009) An aromatic Glaser–Hay reaction. *J. Am. Chem. Soc.*, **131**, 17052–17053.

4 芳烃的亲电酰化反应

4.1 概述

芳烃和杂芳烃与亲电酰化剂的酰化反应,特别是由路易斯酸引发的反应,以其发现者名字命名为傅-克酰化反应(图式 4.1)。该反应为获取大多数各种不同类型的芳族酮提供了快捷的通道,并且是最重要的 C—C 键形成反应之一。由于产物会使 1 当量的路易斯酸失活,故在大多数情况下反应需要化学计量的三氯化铝。如果芳烃有足够强的亲核性,也可使用其他化学计量的路易斯酸(四氯化钛、四氯化锡或氯化铁)、三氟乙酸酐[1],或者催化量的质子酸(氢氟酸、盐酸、磷酸、甲磺酸或三氟甲磺酸)、路易斯酸(氯化锌、氧化铁[2]或三卤化硼),或者固体酸[3,4]来引发傅-克酰化反应。由于氧化铝成为三氯化铝时会形成氯化氢,故三氯化铝和铝的混合物可用于代替 1 当量三氯化铝。在微波条件下,锌粉[5]和铝粉[6]也可以用作催化剂。傅-克酰化反应还可以在纯的氢氟酸或三氟甲基磺酸中进行。在大规模的工业过程中废物处理和回收利用是一个重要问题,因此采用催化过程或可收回酸(如氢氟酸或三氟醋酸)具有重要的价值。

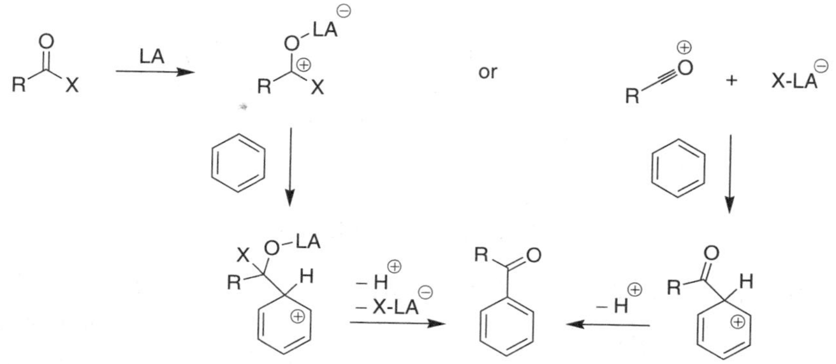

图式 4.1 傅-克酰化反应机理(X: 离去基团,LA: 路易斯酸,R: 烷基、芳基)

分子内傅-克酰化反应通常比分子间反应容易进行。给电子基团导向酰基进入它们的邻位和对位。

富电子的芳烃,如烷氧基芳烃、烷基芳烃,或富电子杂芳烃(吡咯、吲哚、噻吩等)最容易发生傅-克酰化反应,并且不需要化学计量的酸。事实上,在大量强酸的存在下富电子芳烃常常发生分解或低聚,因此必须使用弱酸(参见图式4.33)。

在大多数傅-克酰化反应中,酰卤被用作酰化剂。但对于富电子芳烃而言,弱酰化试剂,如酸酐、羧酸酯[7]、酸、腈或酰胺[8]等就具有足够反应活性(图式4.2)。无论如何,富电子性能低的芳烃需要强酰化试剂,如酰基卤、烯酮或羧酸磺酸混酐。

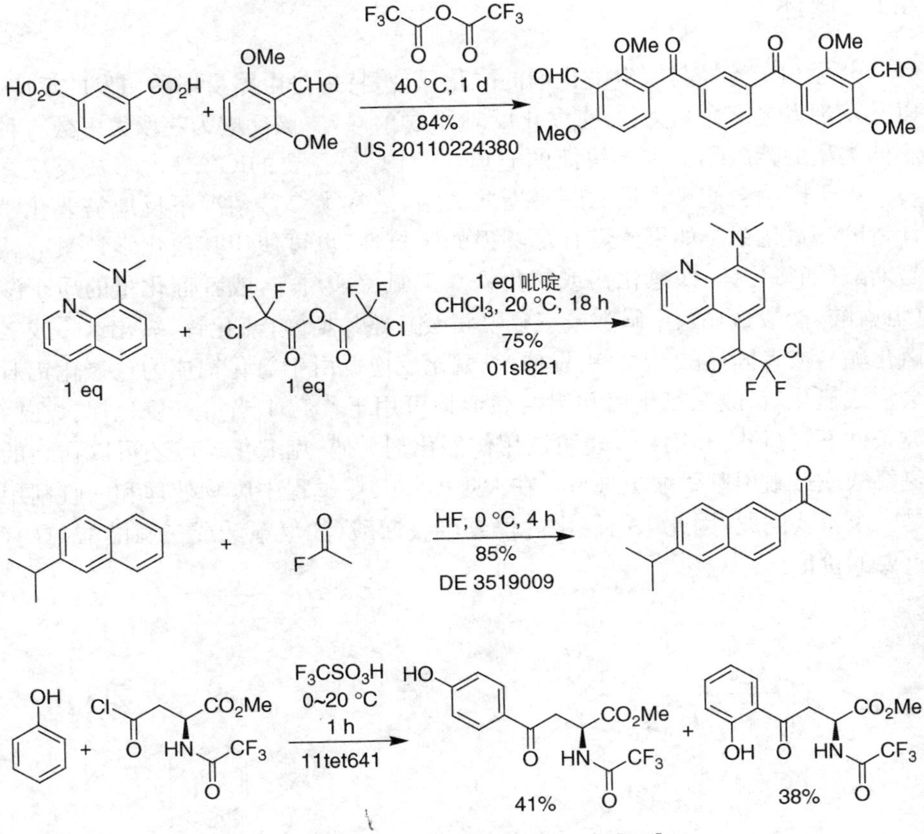

图式 4.2 傅-克酰化反应的实例[9-12]

假如需高温反应,溶剂的选择可能比较困难。最合适的(即非亲核性的)溶剂(例如:四氯化碳、三氯甲烷、二氯甲烷、1,2-二氯乙烷或醚)在强烈的傅-克酰化反应条件下也不稳定。如果芳烃本身不能用作溶剂,可选择硝基苯、硝基甲烷或环丁砜[13]。在离子液体中的傅-克酰化反应也有报道[14,15],但并非所有的离子液体都足够稳定,咪唑鎓盐对酰化试剂特别活泼[16]。

在傅-克反应条件下,酮可以作为亲电体,过量芳烃能够导致芳烃的羟基烷基化,并且形成烯烃(傅-克烷基化反应,图式 1.37 和图式 4.3)。

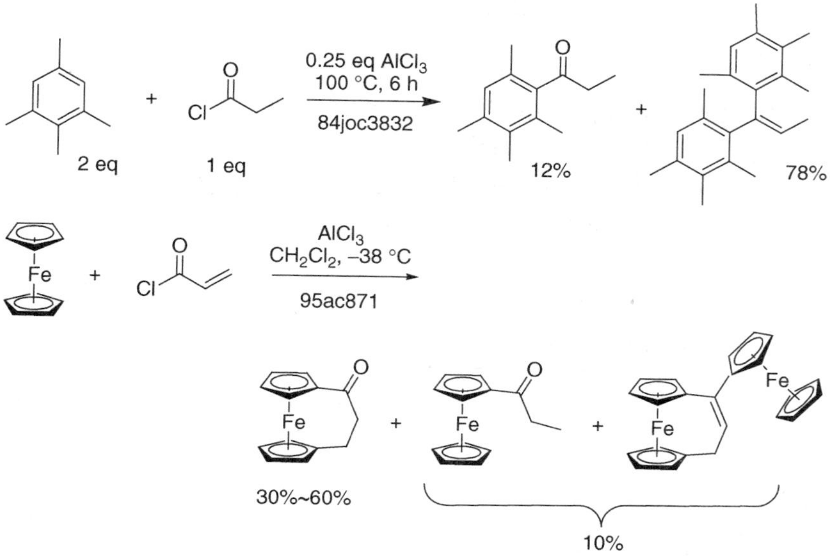

图式 4.3 烯基化是傅-克酰化反应中的副反应[17,18]

芳基酮也可以通过未取代芳烃与 1,1,1-三卤代烷烃或其合成等价物的烷基化反应及水解进行制备(图式 4.4)。潜在的副反应包括芳烃的卤代以及反应产物作为亲电体或卤代试剂引发的反应。

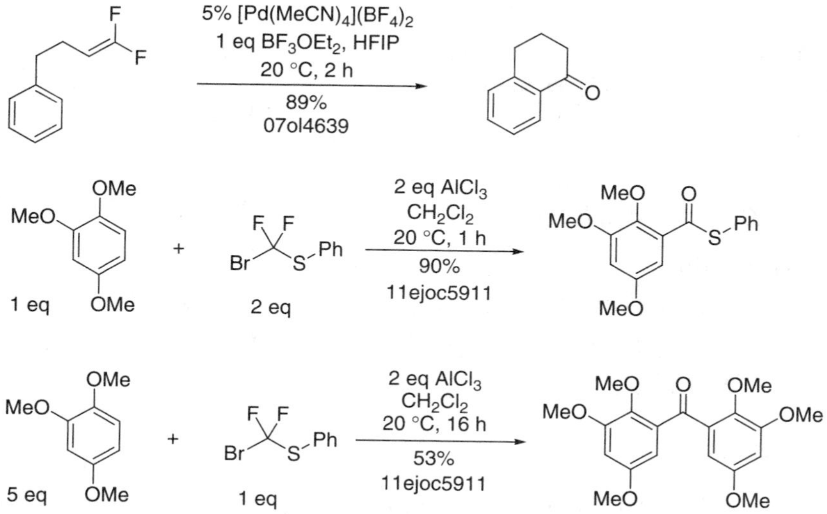

图式 4.4 与烯基和烷基卤代物的傅-克酰化反应[19,20]

4.2 芳烃的问题

4.2.1 芳烃的去烷基化/异构化反应

许多芳烃不以预期方式进行傅-克酰化反应。酸处理富电子烷基芳烃能够发生异构化或脱烷基化。事实上，与质子、金属或者三烷基甲硅基一样，仲或叔烷基也可以作为芳族亲电取代中的离去基团(本位取代)。

具有苄基 C—H 基团的富电子烷基苯是潜在的氢给体，可还原碳正离子，而自身通过傅-克烷基化反应发生聚合(图式 4.5)。

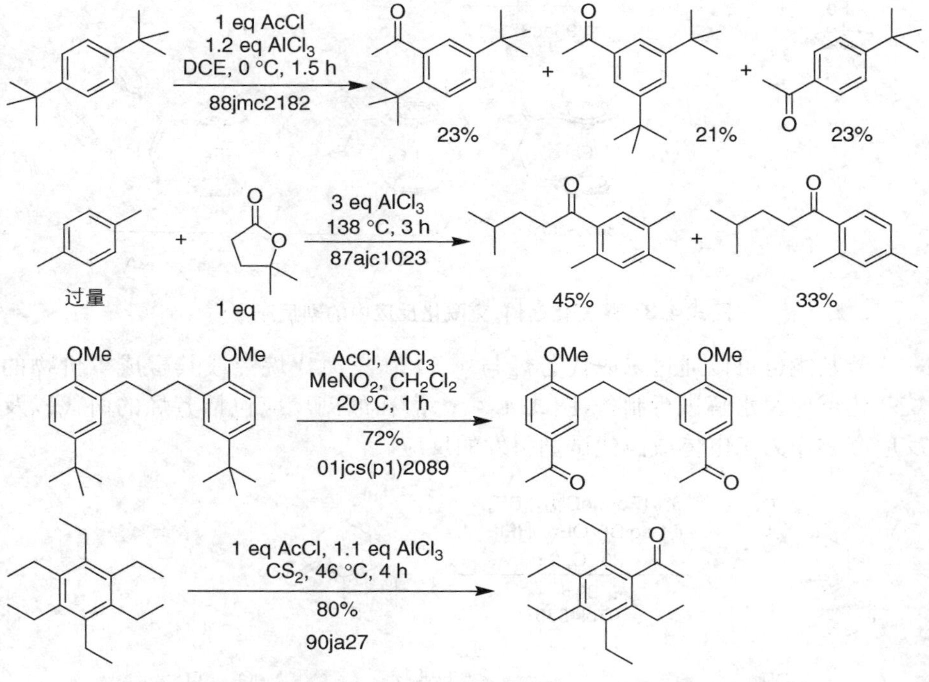

图式 4.5 傅-克酰化反应中烷基迁移、还原和原位取代的实例[21-24]

烷氧基取代的苯在傅-克酰化反应中能够发生去烷基化得到 2-酰基酚(图式 4.6)。在多烷氧基芳烃中，大部分 2-烷氧基会脱去烷基。形成螯合物是该反应的主要驱动力。根据所使用的酸的不同，脱烷基反应中形成的烷基正离子也可以对芳烃进行烷基化反应。

4.2.2 苯乙烯

烯烃和炔烃也可以被酰卤酰化。富电子烯烃(烯醇醚和烯胺)与强亲电性酰卤(例如：三卤代乙酰氯化物和光气)反应不需要另外加催化剂[28]。但对亲

78

核性较低的烯烃往往需要催化剂。在苯乙烯和炔基苯中,烯基或炔基可以与芳烃发生竞争性酰化反应。根据起始原料和精确的反应条件,可得到烯酮或2-卤代乙基酮。烯烃或炔烃的低聚是一种常见的副反应(图式4.7)。

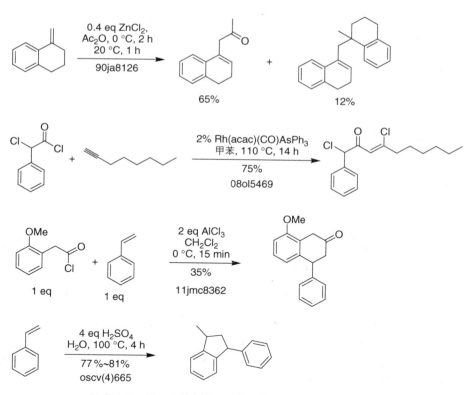

图式4.7 烯烃和炔烃的酰化[29-32] (更多实例: [33,34])

4.2.3 苯胺、苯酚和苯硫酚

苯胺和苯酚的杂原子在傅-克酰化反应中首先被酰化,但所得到的酰苯胺和苯酚酯仍然会以预期的邻/对位选择性进行C-酰化。为了避免苯胺的氮原子酰化,可以使用腈和三卤化硼的混合物作为酰化剂来代替酰卤或酸酐。酰苯胺类可以在钯催化下选择性地邻位酰化[35](图式4.8)。

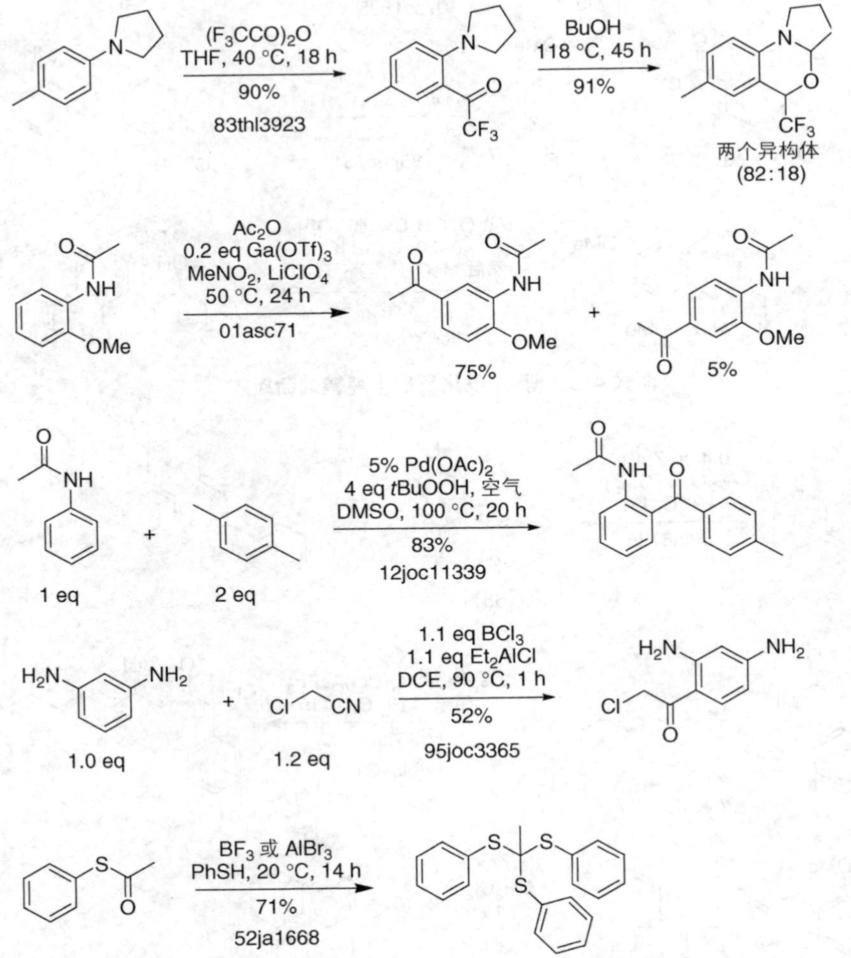

图式4.8 苯胺和苯硫酚衍生物与亲电试剂的反应[36-40](更多实例:[41])

用酸处理酰苯胺类和苯酚酯能将酰基从杂原子上转移到碳上[弗里斯(Fries)重排]。苯硫酚酯不发生弗里斯重排,而是得到原酸酯[42]。

带有碱性官能团(胺、吡啶、咪唑、酮、酰胺、脲等)的芳烃在傅-克酰化反应中将会额外消耗1当量的酸,并有可能形成不溶性的盐。由于傅-克酰化反应通常在非极性溶剂中进行,盐会沉淀出来并妨碍体系的有效搅拌和原料混合。

解决这个问题的一个方法是使用少量的 N,N-二甲基甲酰胺(DMF)[43]或环丁砜作为溶剂；或不使用溶剂，直接将原料与过量的三氯化铝一起熔化反应[44]。

包括 N,N-二烷基苯胺在内的叔胺，有时能在极其温和的反应条件下被酰化试剂去烷基化[45,46]（图式 4.9）。特别容易脱去的烷基是甲基、苄基及三苯甲基。

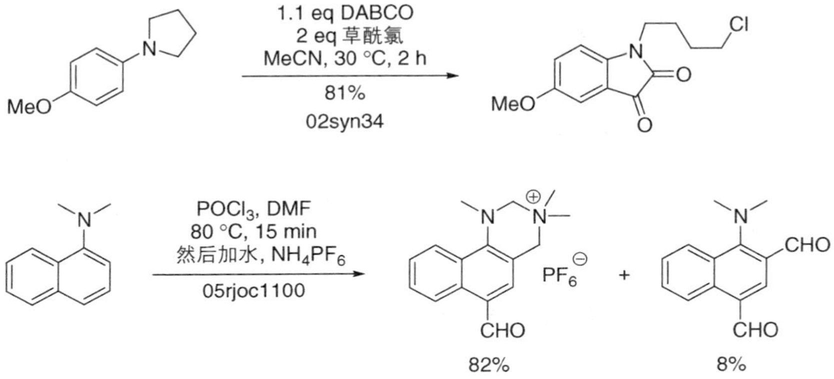

图式 4.9 酰化剂脱去叔胺的烷基[47,48]

4.2.4 缺电子芳烃

缺电子芳烃不容易被酰化。仅有几个磺酰苯、苯甲腈、其他苯甲酸衍生物或者缺电子杂芳烃(吡啶、嘧啶、三唑或吡唑)的分子间傅-克酰化反应被报道。在大多数实例中，这些底物都被给电子取代基活化。硝基芳烃和芳基酮可以在剧烈条件下发生傅-克酰化（图式 4.10 和图式 4.11）。

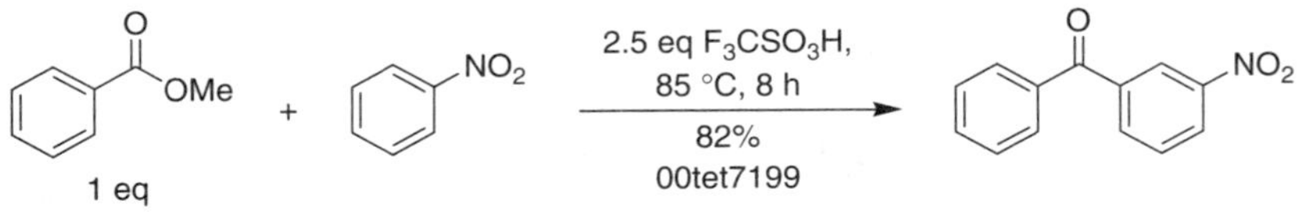

图式 4.10 硝基苯的苯甲酰化反应[49]

由于酰基能够显著降低芳烃与亲电试剂的反应活性，故傅-克酰化反应通常在第一次酰化后中止。但是，在较高的温度下芳烃有时可以发生两次酰化反应，有时甚至发生在第一个酰基的邻或对位（图式 4.11）。

假如缺电子芳烃和杂芳烃的亲电酰化反应太困难，有时可以通过其金属化[52-57]或酰基自由基[58,59]进行酰化反应（图式 4.12）。如图式 4.12 的最后两个实例所示，由于存在太多潜在的副反应，如芳基羟化反应，故氧化酰化反应条件很难优化[60]。

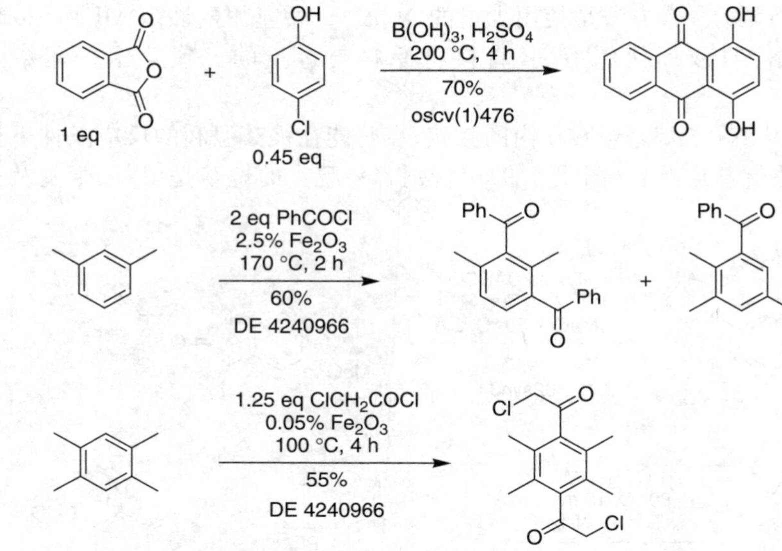

图式 4.11　酰基苯的傅-克酰化反应[50,51]

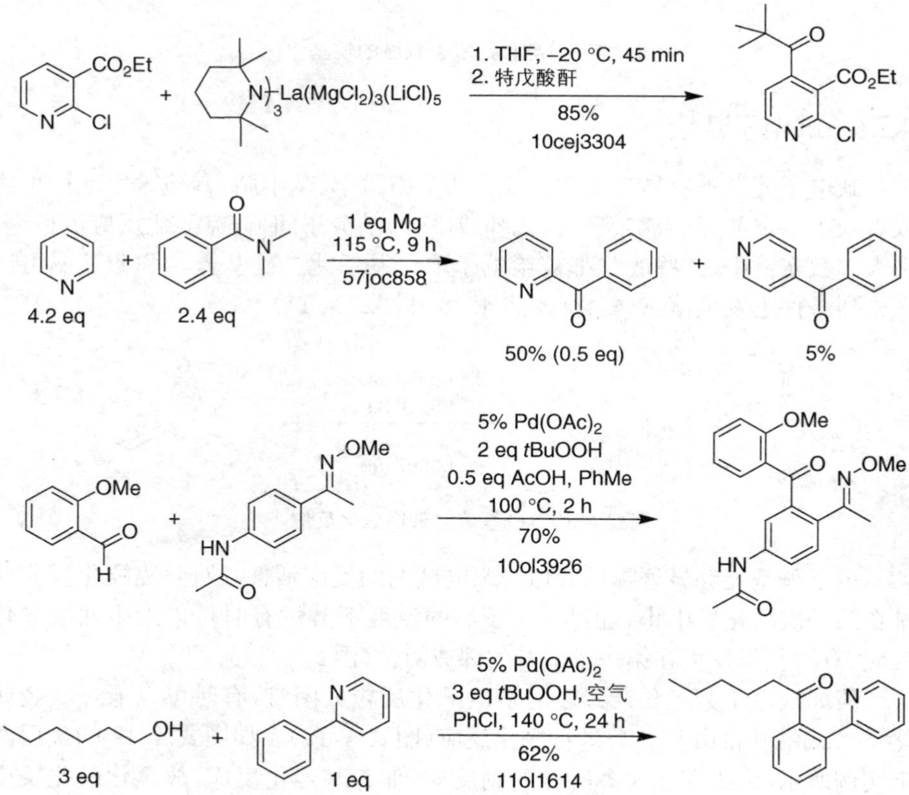

图式 4.12　缺电子芳烃通过金属化的酰化反应[52,59,61,62]（更多实例：[63,64]）

4.2.5 唑类

通常,吡咯、咪唑和吲哚可以在非常温和的反应条件下进行傅-克酰化反应(图式4.13),但其区域选择性难以预测。此反应原则上存在形成烯酮和烯酮二聚等竞争反应,因此图式4.13中最后一个咪唑酰化实例的高反应收率是不同寻常的。

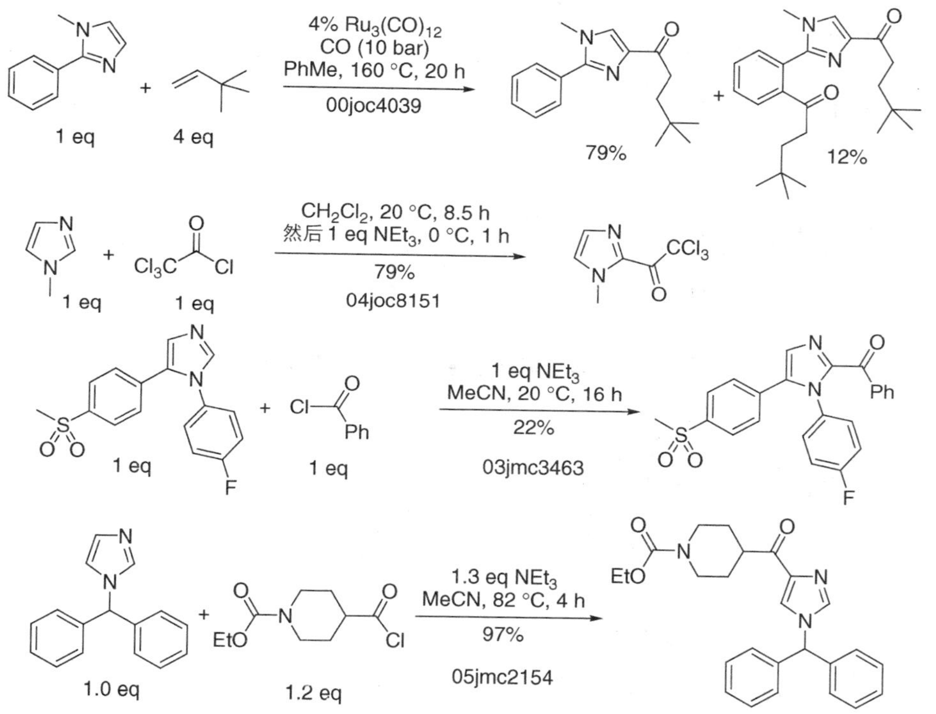

图式4.13 咪唑的傅-克酰化反应[65-68](更多实例:[69])

在唑类 C-酰化反应中,有许多意外反应会导致意想不到的结果。当用强亲电试剂处理某些唑类和吖嗪时,可能通过形成卡宾中间体导致二聚(图式4.14),它将会给傅-克酰化和其他亲电取代反应(烷基化、卤化、硝化等)带来问题。尽管咪唑具有较高的热稳定性,但在西奥特-鲍曼(Schotten-Baumann)酰化条件下很不稳定,通常裂解为1,2-双(酰胺)乙烯类化合物[班伯格尔(Bamberger)裂解[70]]。强酸可以降低杂芳环的亲核性并阻止卡宾的形成,从而避免它们的二聚或低聚。

吲哚傅-克酰化反应的区域选择性依赖于精准的取代模式和反应条件,并且很难预测。甚至酰化试剂的类型和溶剂都可以影响酰化的区域选择性[72]。与吡咯不同,无取代的吲哚通常酰化在3位[73]。但是,偶尔也会观察到苄位酰化和二聚(图式4.15)。

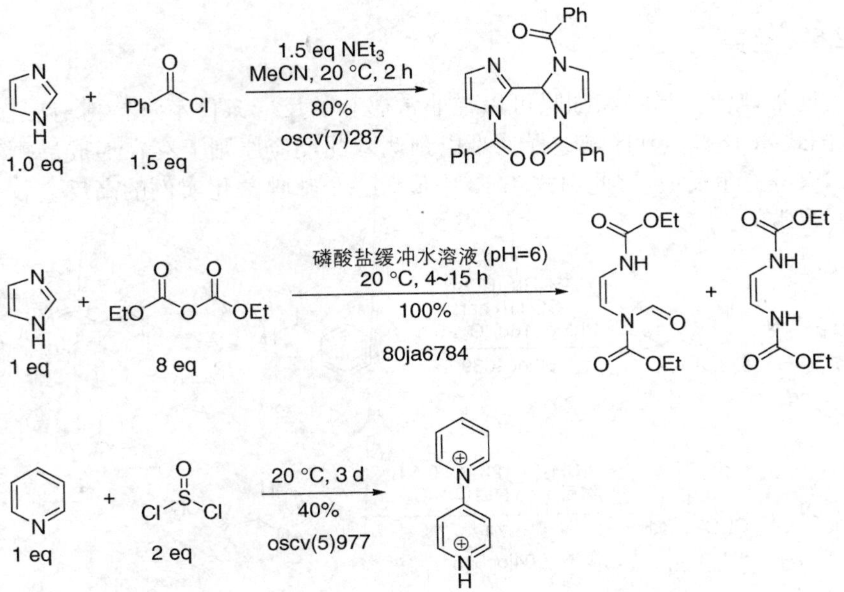

图式 4.14 用亲电试剂处理杂芳烃时的二聚和裂解[16,70,71]

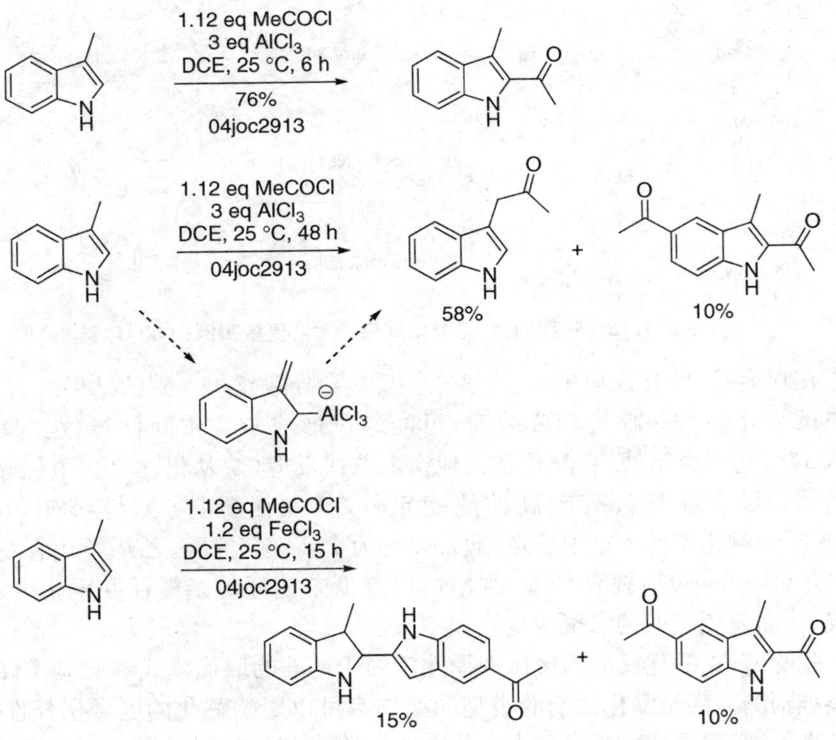

图式 4.15 吲哚和苯并噁唑的酰化反应[72,74-76]

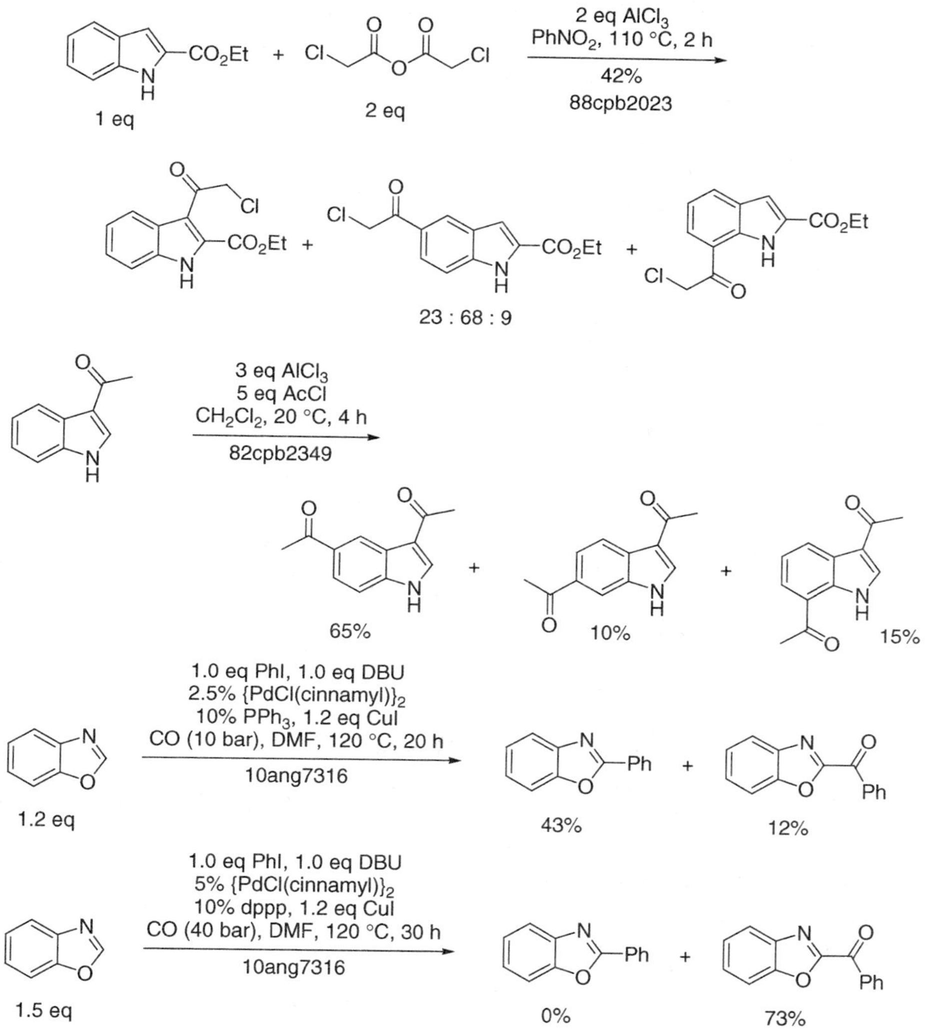

续图式 4.15

4.3 亲电试剂的问题

4.3.1 酰卤的问题

一些酰卤在用酸处理时会分解,导致意外产物的形成。如果相应的酰基阳离子脱羰基能得到一个稳定的碳正离子,脱羰基和烷基化反应就可能得以实现(图式 4.16)。其他易脱羰基的酰卤化物包括草酰氯、其他 α-羰基酰卤化物、α-氨基或 α-羟基羧酸衍生的酰化剂[77-80]。只有富电子芳烃(苯甲醚、噻吩等)能够与这种不稳定的酰卤发生傅-克酰化,得到预期的酮。酰基卤的脱羰基反应能被痕量钯[81-83]或铑[84,85]催化,因此为防止这种副反应发生,必须使用清洁的反应器和搅拌棒(或螯合试剂)。

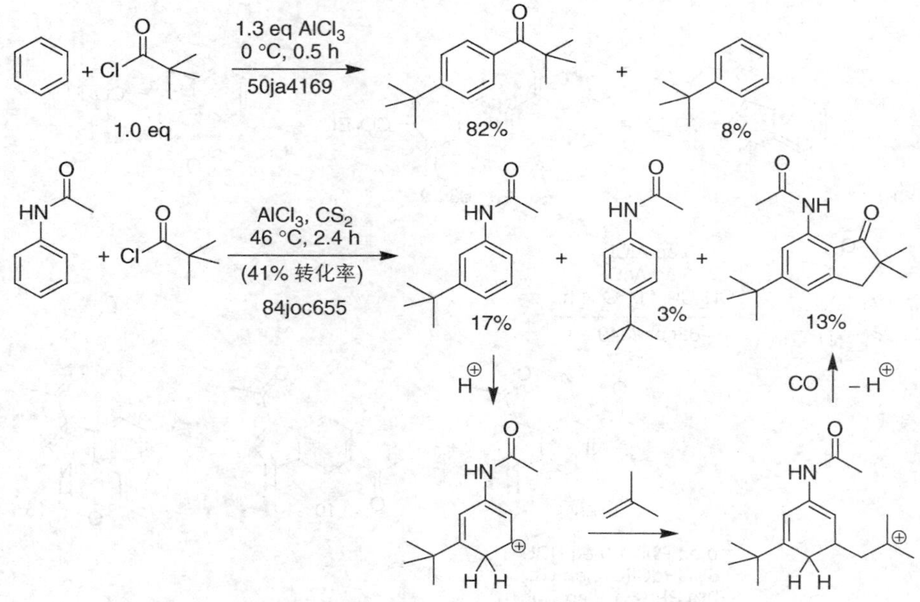

图式 4.16 特戊酰氯的傅-克酰基化反应[86,87]

酸引发的张力碳环或杂环的重排、裂解是傅-克酰化反应中另外的潜在副反应。芳基硫醚、环丙烷[88,89]及许多其他酸敏感的化合物在强酸性反应条件下可以发生重排反应(图式 4.17)。

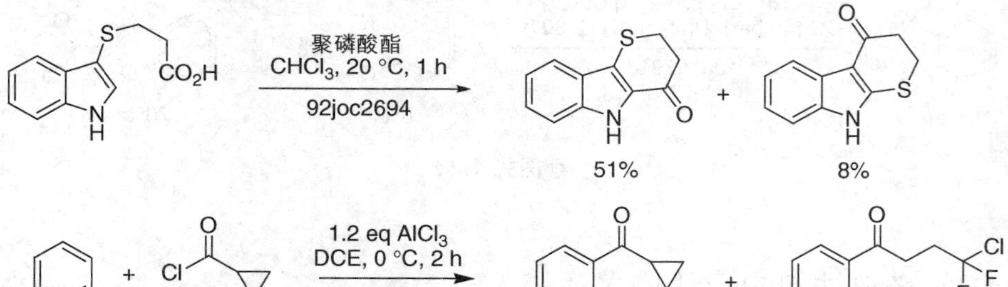

图式 4.17 酸引发的酰卤转化反应[90-92]

丙烯酸衍生物既可作为酰化试剂又可作为烷基化试剂。丙烯酰氯通常首先进行酰化反应，生成的芳基乙烯酮进一步环化成为茚满酮的反应也被观察到，特别是α-取代的丙烯酸酯（图式1.28）。进一步的副反应是氯或其他亲核试剂对产物中高亲电性乙烯基的加成反应（图式4.18）。

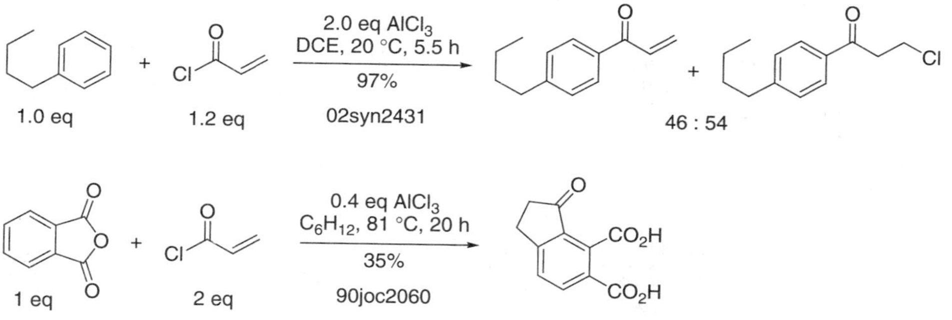

图式4.18 丙烯酰氯的傅-克酰化反应[93,94]

三氯乙酰氯、光气或三氟乙酸酐等强亲电试剂能作为氧化剂。当使用三乙胺作为碱与这些试剂一起使用时，经常形成酰化烯胺副产品[95,96]（图式4.19）。叔胺也能被酰卤或磺酰卤进行N-酰化和去烷基化。这就是以叔胺作碱、在无水条件下的酰化反应中往往得到复杂混合物的一些原因。在制备酰胺或磺酰胺时，无机碱水溶液（西奥特-鲍曼条件）要远优于无水条件。作为替代方法，三烷基胺可用吡啶或（便宜的）3-甲基吡啶代替。2-烷基吡啶不适于用作酰化反应的碱，因为它容易在（酸性的）2-烷基上发生酰化反应[97]。

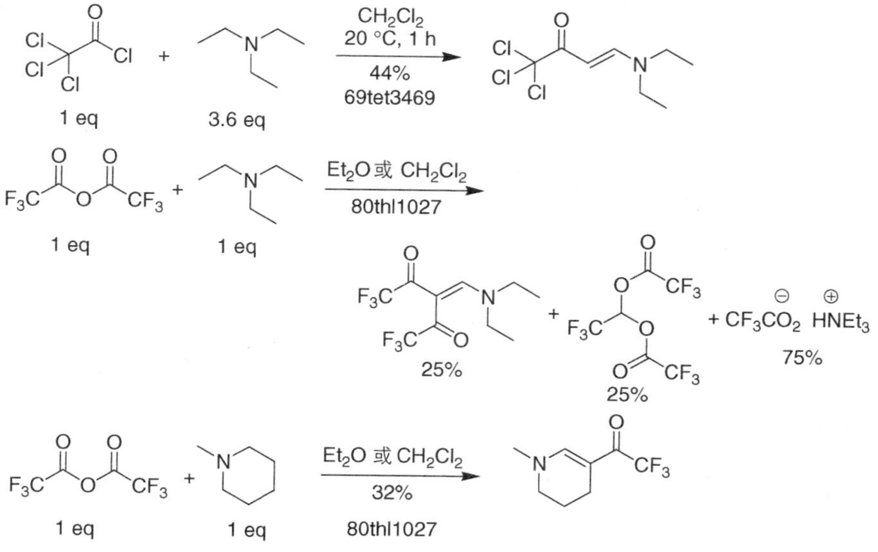

图式4.19 三烷基胺被酰化试剂氧化[98,99]

4.3.2 羧酸酯类和内酯

内酯和非环状羧酸酯属于双亲电试剂，它们可以作为烷基化或酰基化试剂与亲核试剂进行反应。通常简单的烷基酯对芳烃烷基化，但通过优化反应条件可以实现傅-克酰化反应(图式 4.20)。

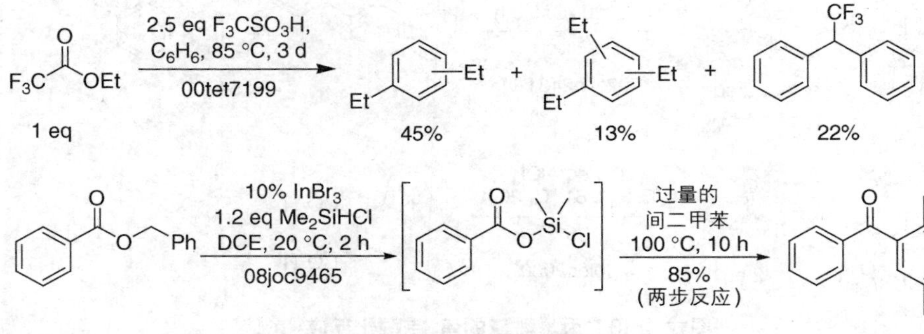

图式 4.20　与酯的傅-克酰化和烷基化[7,49]

在三氯化铝存在下，丁内酯和戊内酯首先作为烷基化试剂与芳烃反应得到 ω-芳基烷基酸。如果反应继续进行，可能进一步发生酰化得到环状或非环状酮。如果多聚磷酸作为催化剂，芳烃首先发生酰化反应(图式 4.21)。

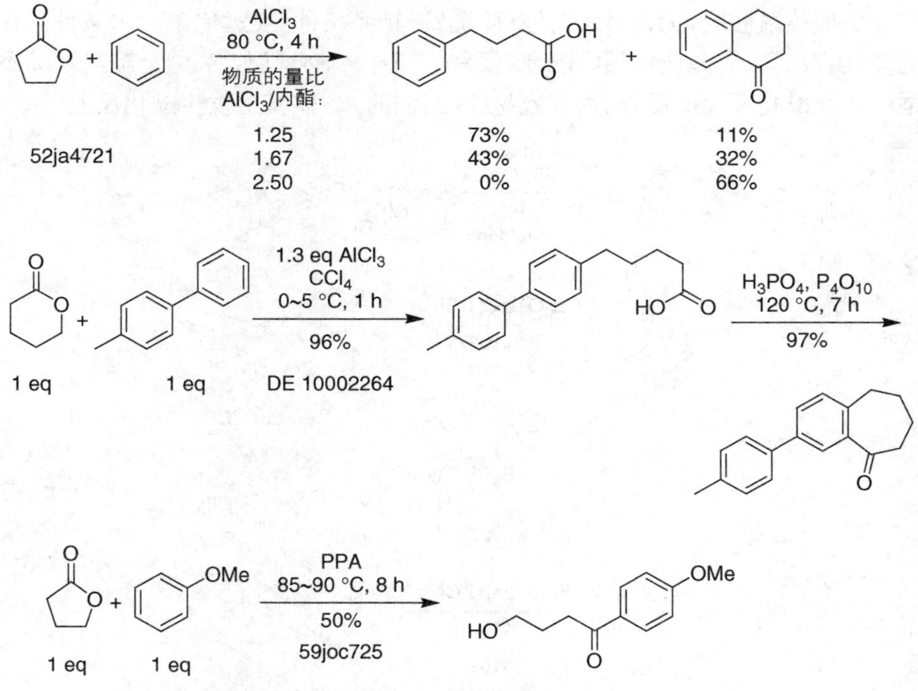

图式 4.21　与丁内酯和戊内酯的傅-克酰化和烷基化反应[100-102]

如果反应条件太剧烈，内酯在与芳烃反应之前断裂，所得碳正离子通过氢迁移重排成更稳定的碳正离子或者消除成为烯烃，从而导致形成意外的异构体(图式 4.22)。

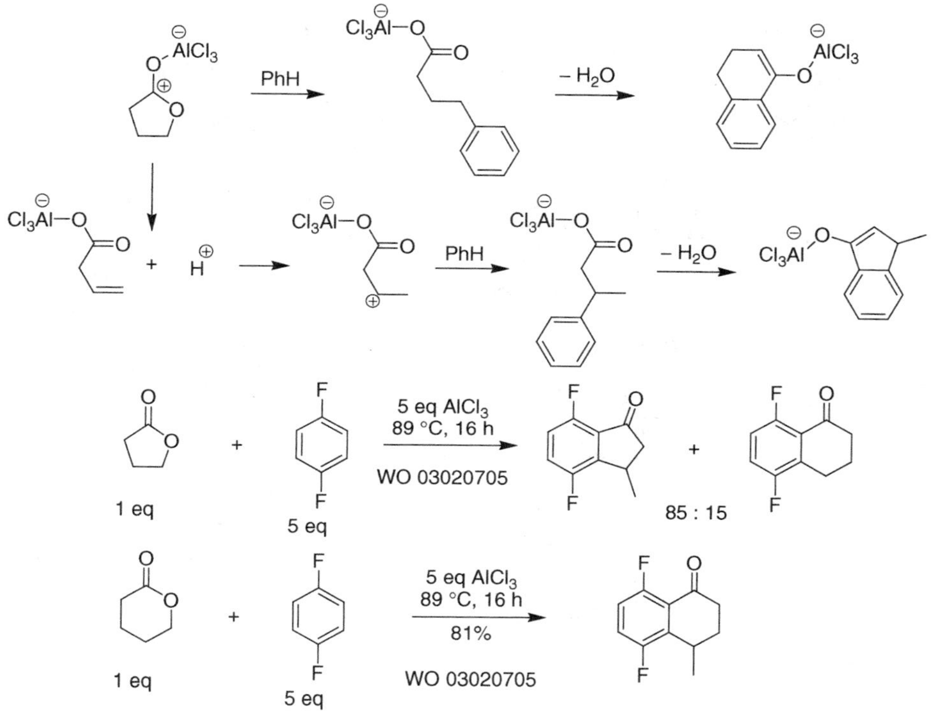

图式 4.22 内酯的傅-克酰化反应[103]

硬的有机金属试剂，例如格氏试剂或有机锂试剂，首先与内酯的羰基反应，得到 ω-羟基烷基酮(图式 4.23)。而软金属有机化合物，例如铜试剂，常常被内酯烷基化[104]。

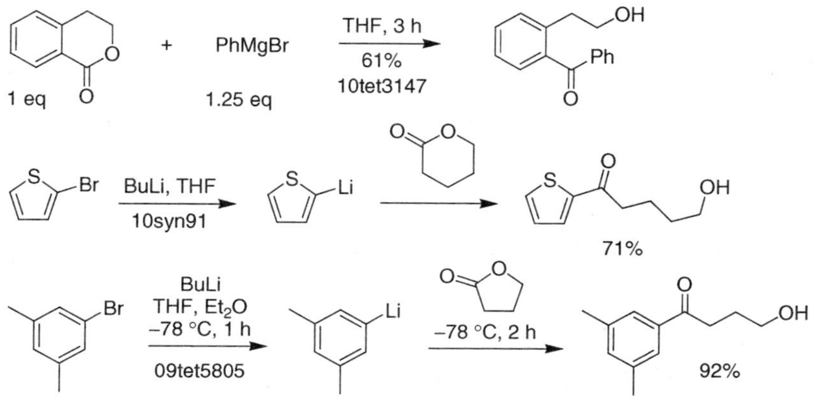

图式 4.23 内酯与有机镁和有机锂试剂的反应[105-107]

丙内酯对亲核试剂的反应性正好相反：在强路易斯酸的存在下，丙内酯酰化芳烃。与之前所述的与硬的有机金属化合物反应情况相同[108,109]，与软的有机金属化合物优先进行烷基化反应[108,110]（图式 4.24）。

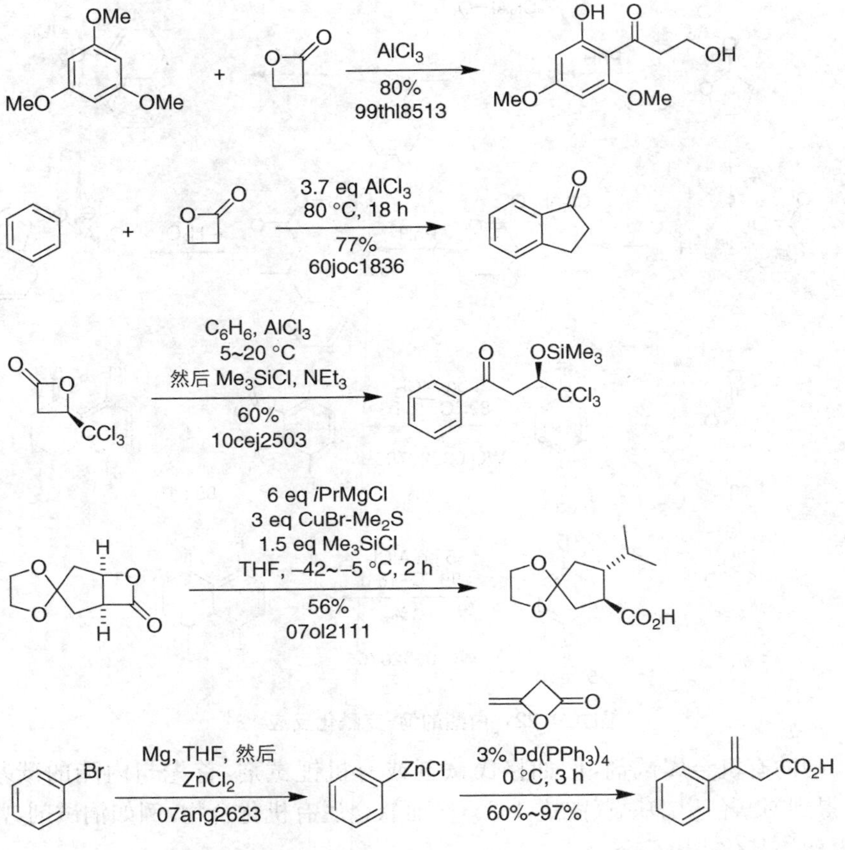

图式 4.24　丙醇酸内酯与各种亲核试剂的反应[111-115]

4.3.3　碳酸衍生物

二氧化碳可以在强碱或强酸存在下与芳烃反应。这个反应的一个重要的例子是由苯酚钠羧化制备水杨酸[100 bar, 190℃；科尔贝-施密特（Kolbe-Schmitt）合成]。不幸的是，这一有价值的反应适用范围有限，仅仅对一些酚盐能够获得高收率[116]，而不适用于烷氧基取代的苯、苯胺或苯硫酚，甚至咔唑二锂盐也不容易与二氧化碳反应[117]。

许多能够使活化程度较低的芳烃与二氧化碳直接羧基化的反应条件（但是更昂贵）已经被发展（图式 4.25），如图式 4.25 给出的第一个实例，可以通过芳基铝中间体反应。

在三氯化铝和芳烃存在的情况下,二硫化碳是不活泼的,并且经常被用作傅-克反应的溶剂。然而,二硫化碳却能够使富电子的芳烃二硫代羰基化(图式 4.25)。例如,在三溴化铝存在下,甲苯可转化为二硫代-4-甲苯甲酸(20℃反应 24 h,收率为 24%[118])。

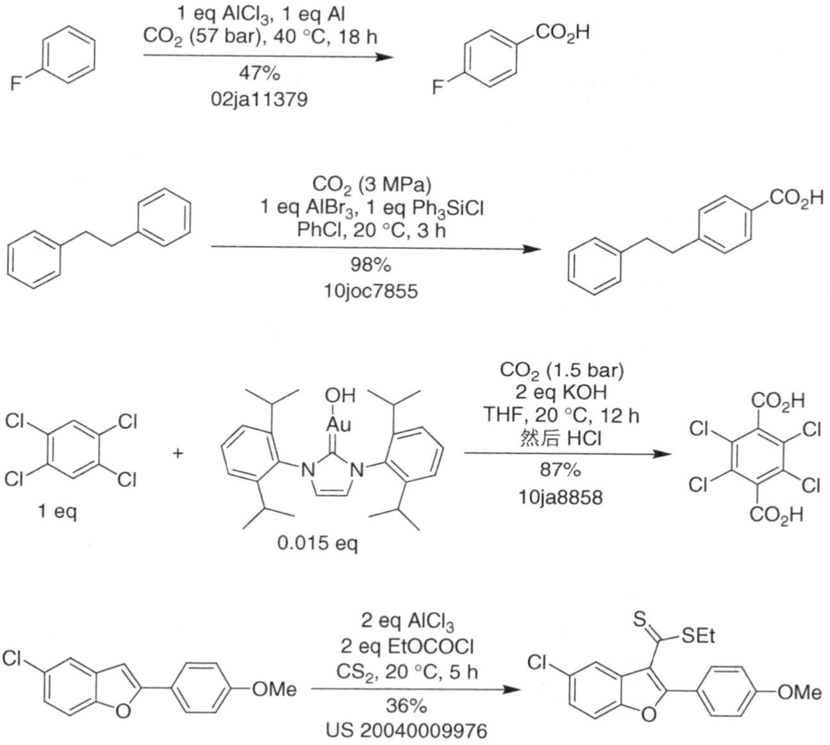

图式 4.25 芳烃与二氧化碳[118-120]和二硫化碳[121]的羧基化反应

另外一种替代不活泼二氧化碳的反应是钯、铑或者钌催化的一氧化碳或甲酸的羰基化反应(图式 4.26)。溴代苯是钯催化羰基化反应最常见的原料,但是也可以用未取代的芳烃代替[53,55,122](图式 4.26)。其他替代方法包括乙酰化后进行卤仿反应,或锂化后用二氧化碳处理。在图式 4.26 的最后一个反应中,在羧酸和氧化剂的存在下,钯催化反应也可以使芳烃羟基化。在醇的存在下,也有可能发生烷氧基化反应[123]。

光气或三光气可用于制备苯甲酰氯[126],常见的副产物是二苯甲酮。草酰氯是光气的合成等价体,在三氯化铝的存在下脱羰基,可用来高收率地制备苯甲酰氯(图式 4.27)。

卤化氰(Hal-CN)既可用作亲电氰化试剂,又可用作卤化剂。在没有强路易斯酸存在时,溴化氰特别适合用作卤化试剂(图式 4.28,氰化见[131])。

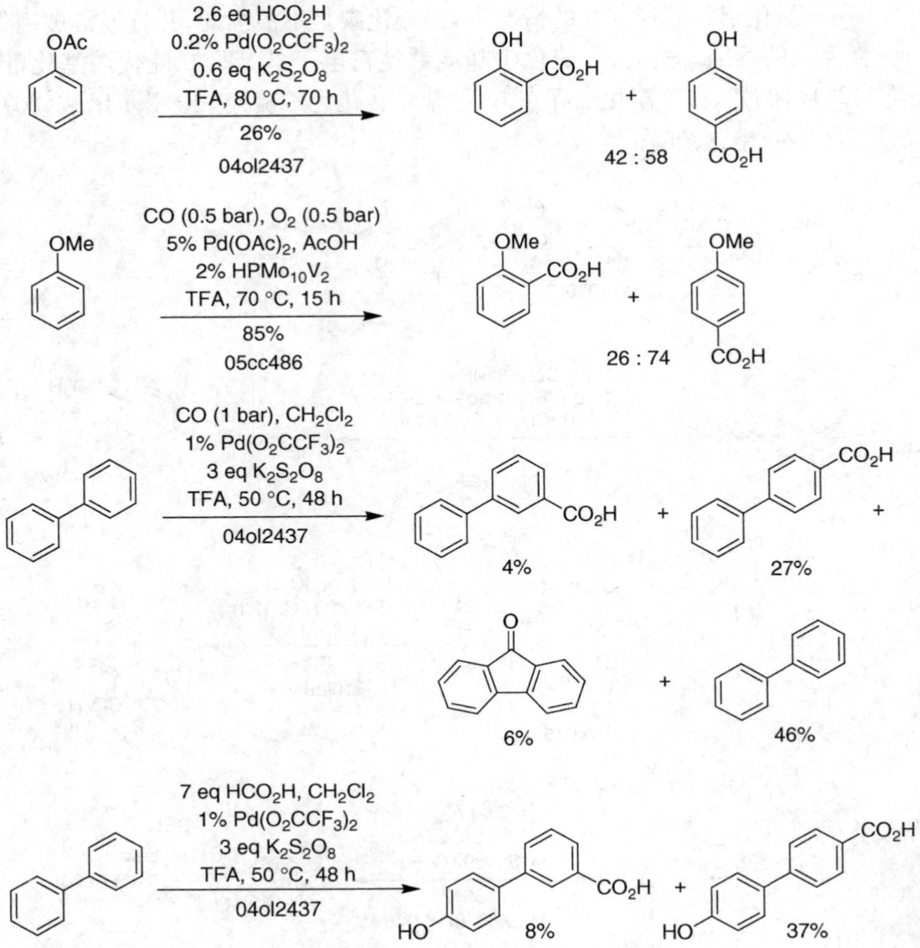

图式 4.26　钯催化芳烃的羧化[124,125]

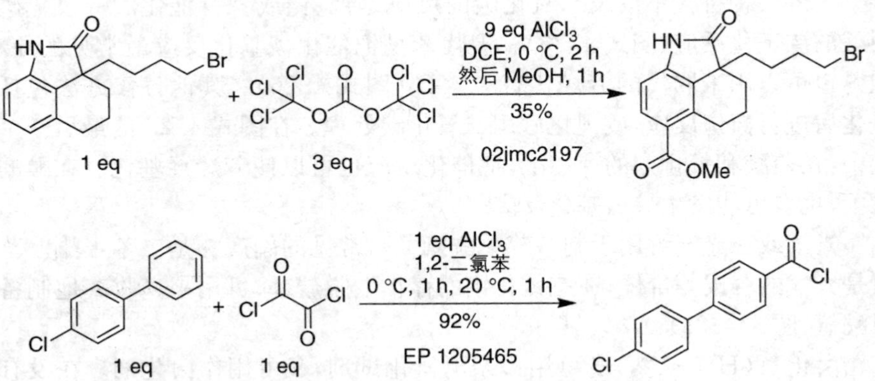

图式 4.27　三光气和草酰氯与芳烃的反应[127-129]（更多实例：[130]）

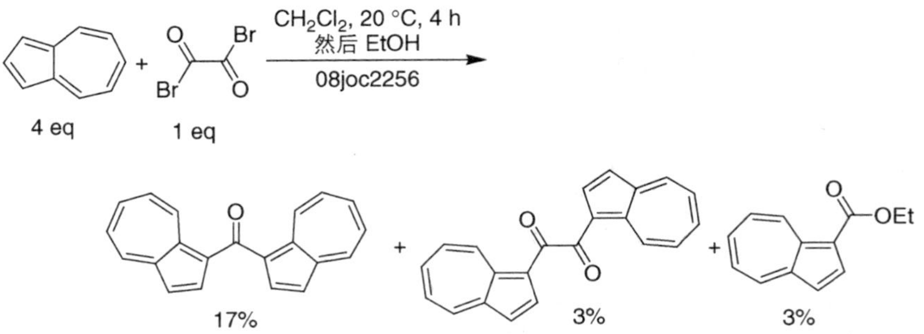

续图式 4.27

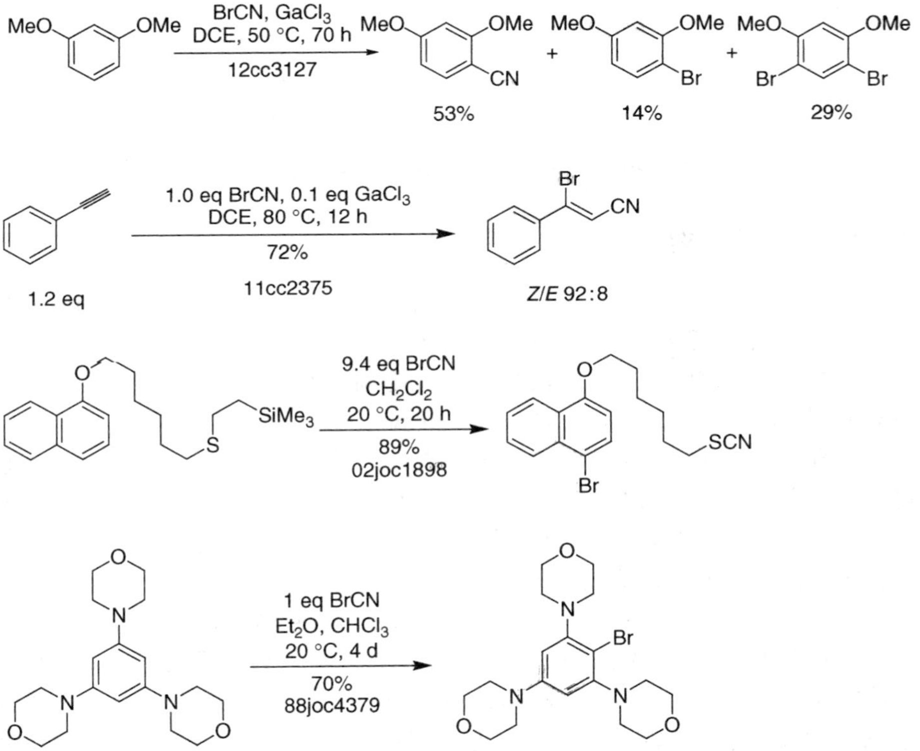

图式 4.28 与溴化氰的氰化和溴化反应[132-135]

也有几个使用碳酸的衍生物，如氨基甲酰氯（R_2NCOCl）、芳基氯甲酸酯（ArOCOCl）或异氰酸酯的傅-克反应被报道（图式 4.29）。这些反应与富电子芳烃的反应效果最佳，甚至可与氨基甲酸酯或脲反应[136]。脂肪族氯甲酸酯（ROCOCl）在酸性条件下不能被用作 C-酰化试剂，它们主要是对芳烃进行烷基化反应[137]。在高温和三氯化铝的存在下，芳基氯甲酸酯能被转化成为芳基氯（图式 8.65）。

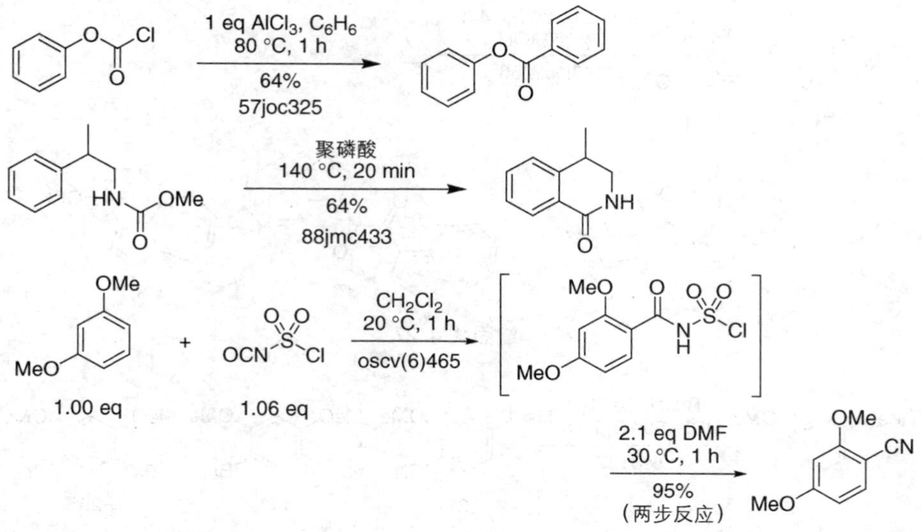

图式 4.29 与碳酸衍生物的傅-克羧基化反应[138-140]

4.3.4 甲酸衍生物

甲酰基卤化物分解成一氧化碳和卤化氢，对于合成应用而言，它们的寿命太短[141]。只有甲酰氟(沸点 29℃)能够稳定用于傅-克酰化反应中，而且已经被用于制备苯甲醛[142]。苯甲醛也可以在三氯化铝的存在下，用 HCl/CO 处理芳烃来制备[盖特曼-科赫(Gattermann-Koch)反应[143]]。

混合酸酐酰化反应通常发生在弱亲电性的酰基上，因此，乙酸甲酸混酐是进行乙酰化而非甲酰化反应。没有应用甲酸酯在傅-克酰化反应中的报道[144]，可能是由于甲酸酯在酸性(和碱性)条件下很容易脱羧基。不过，一些催化剂可以促进甲酸与苯或其他芳烃的甲酰化反应[145]。富电子芳基甲酯能以较好的收率发生弗里斯重排反应[146]，无论如何，威尔斯迈尔(Vilsmeier)反应(DMF/POCl$_3$)能以最好的收率实现富电子芳烃的亲电甲酰化。其他可用的甲酰化方法包括瑞梅尔-梯曼(Reimer-Tiemann)反应(酚与氯仿和碱进行甲酰化[147])，以及和乌洛托品[达夫(Duff)反应[148]]、甲醛、氢氰酸或二氯甲基甲醚进行甲酰化(图式 4.30)。

瑞梅尔-梯曼反应的甲酰化中间体是二氯卡宾[153]，通常只有酚和吡咯可以被甲酰化。一个与之互补的反应是三氯甲基负离子对强缺电子芳烃的亲核取代反应产生二氯芳烃，它们是苯甲醛的合成等价体[154]。

未络合的卡宾反应活性很高，因此选择性较差，故瑞梅尔-梯曼反应的收率一般较低。典型副产物来自于卡宾对有取代基位置的进攻、环丙烷的形成及二氯卡宾或三氯甲基负离子与其他官能团的反应(图式 4.31)。在优化条件后，其中一些副产物的收率可以得到控制，从而具备制备意义[155]。

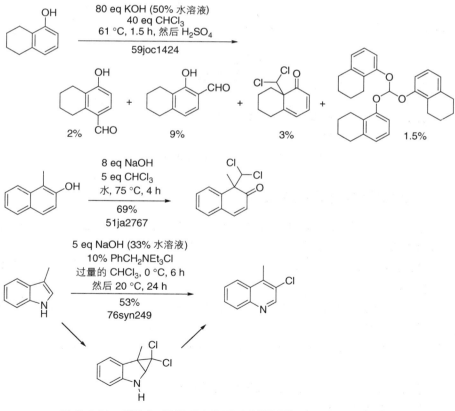

图式 4.30 亲电芳基甲酰化的实例[149-152]

图式 4.31 瑞梅尔-梯曼反应的副反应[156-159]（更多实例：[160]）

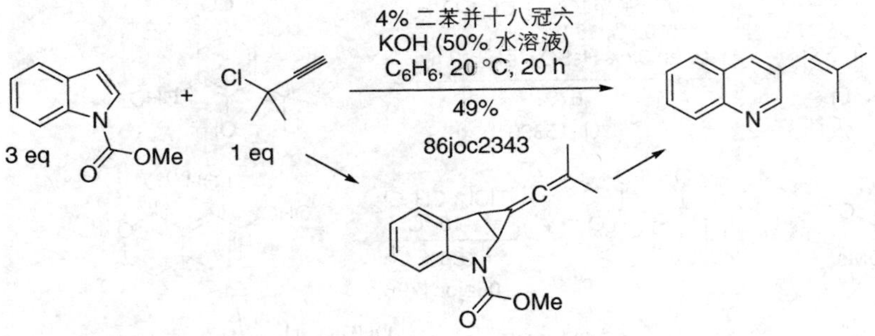

续图式 4.31

亲电甲酰化反应仅有富电子芳烃能够很好地进行,缺电性芳烃和杂芳烃最好通过金属化或自由基的方式进行甲酰化。无论如何,后者的区域选择性往往较差(图式 4.32)。

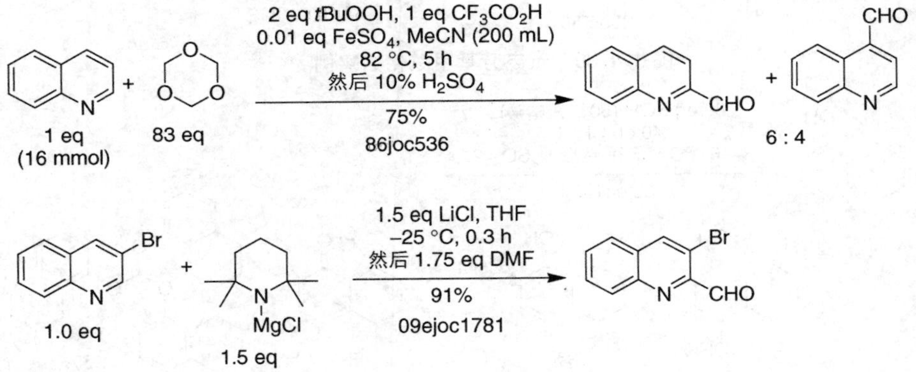

图式 4.32 缺电子芳烃的甲酰化[56,161]

4.3.5 混合羧酸酐和其他多重亲电试剂

对于具有一个以上相似反应活性的亲电官能团的亲电试剂而言,存在的问题难以避免。一类重要的多重亲电试剂是混合羧酸酐。不符合常规的是,在酸性条件下混合酸酐与芳烃的反应主要发生在亲电性较低的羰基上(图式 4.33)。

例如,芳烃与混合乙酸甲酸酐进行酰化反应产生的是苯乙酮而非苯甲醛,可能的原因是亲电性低的酰基更容易形成酰基阳离子。为了实现更好的化学选择性,混合酸酐其中的一个酰基必须是位阻受屏蔽或电性高度失活的。

与混合羧酸酐不同,混合羧酸碳酸酐的反应与所预期的一致,反应发生在亲电性更好的酰基上(图式 4.34)。

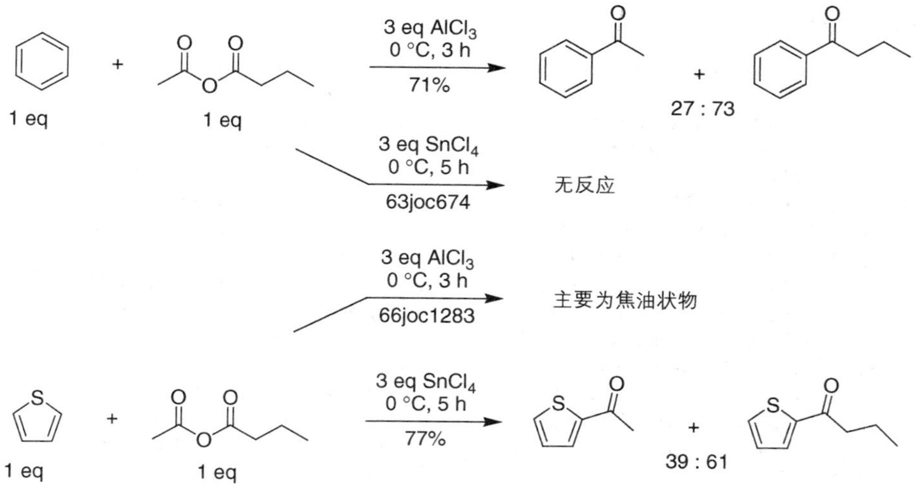

图式 4.33 傅-克酰化反应中混合酸酐的选择性[162,163]

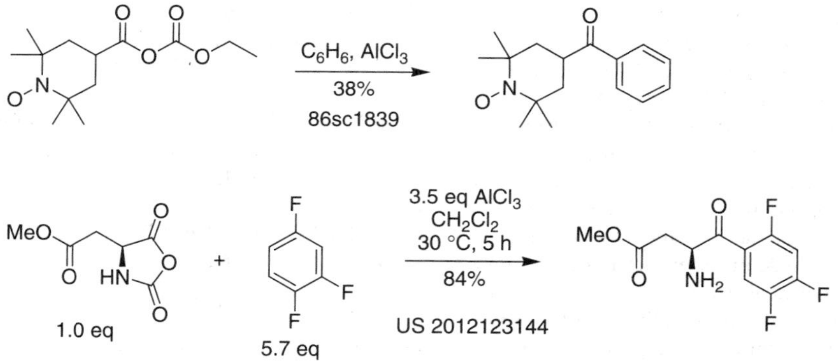

图式 4.34 混合羧酸碳酸酐的傅-克酰化反应[164,165]

酸引发的缩醛和相关化合物形成碳正离子的速度常常比形成酰基阳离子更快。因此当化合物中同时带有活泼的酰基和缩醛结构的合成等价体时,烷基化反应速度远超酰基化(图式 4.35)。但是在碱性条件下,酰基活性往往更高[166]。

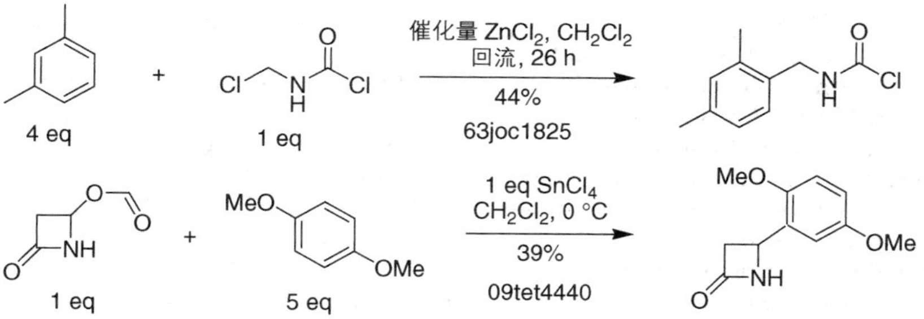

图式 4.35 醛衍生物的傅-克烷基化[167,168]

参考文献

1. Smyth, T.P. and Corby, B.W. (1997) Industrially viable alternative to the Friedel–Crafts acylation reaction: tamoxifen case study. *Org. Process Res. Dev.*, **1**, 264–267.
2. Sharghi, H., Jokar, M., Doroodmand, M.M., and Khalifeh, R. (2010) Catalytic Friedel–Crafts acylation and benzoylation of aromatic compounds using activated hematite as a novel heterogeneous catalyst. *Adv. Synth. Catal.*, **352**, 3031–3044.
3. Sartori, G. and Maggi, R. (2011) Use of solid catalysts in Friedel–Crafts acylation reactions. *Chem. Rev.*, **111**, PR181–PR214.
4. Fürstner, A., Voigtländer, D., Schrader, W., Giebel, D., and Reetz, M.T. (2001) A "hard/soft" mismatch enables catalytic Friedel–Crafts acylations. *Org. Lett.*, **3**, 417–420.
5. Paul, S., Nanda, P., Gupta, R., and Loupy, A. (2003) Zinc mediated Friedel–Crafts acylation in solvent-free conditions under microwave irradiation. *Synthesis*, **2003**, 2877–2881.
6. Gopalakrishnan, M., Sureshkumar, P., Kanagarajan, V., and Thanusu, J. (2005) Aluminum metal powder (atomized) catalyzed Friedel–Crafts acylation in solvent-free conditions: a facile and rapid synthesis of aryl ketones under microwave irradiation. *Catal. Commun.*, **6**, 753–756.
7. Nishimoto, Y., Babu, S.A., Yasuda, M., and Baba, A. (2008) Esters as acylating reagent in a Friedel–Crafts reaction: indium tribromide catalyzed acylation of arenes using dimethylchlorosilane. *J. Org. Chem.*, **73**, 9465–9468.
8. Raja, E.K., DeSchepper, D.J., Lill, S.O.N., and Klumpp, D.A. (2012) Friedel–Crafts acylation with amides. *J. Org. Chem.*, **77**, 5788–5793.
9. Seo, A., Imai, H., Iwase, N., Takata, T., Koyama, Y., and Yonekawa, M. (2011) Crosslinker, crosslinked polymer material, and production method of the crosslinked polymer material. US Patent 2011/0224380.
10. Médebielle, M., Keirouz, R., Okada, E., and Ashidab, T. (2001) Tetrakis(dimethylamino)ethylene (TDAE) mediated addition of heterocyclic difluoromethyl anions to heteroaryl aldehydes. A facile synthetic method for new *gem*-difluorinated alcohols derived from 4-bromo-1-naphthylamine and 8-quinolylamine. *Synlett*, 821–823.
11. Steinbach, R. and Ruppert, I. (1986) Verfahren zur selektiven Acylierung aromatischer Verbindungen. DE Patent 3519009.
12. Murashigea, R., Hayashia, Y., Ohmoria, S., Toriia, A., Aizua, Y., Mutoa, Y., Muraic, Y., Odaa, Y., and Hashimoto, M. (2011) Comparisons of *O*-acylation and Friedel–Crafts acylation of phenols and acyl chlorides and Fries rearrangement of phenyl esters in trifluoromethanesulfonic acid: effective synthesis of optically active homotyrosines. *Tetrahedron*, **67**, 641–649.
13. Tilstam, U. (2012) Sulfolane: a versatile dipolar aprotic solvent. *Org. Process Res. Dev.*, **16**, 1273–1278.
14. Boon, J.A., Levisky, J.A., Pflug, J.L., and Wilkes, J.S. (1986) Friedel–Crafts reactions in ambient-temperature molten salts. *J. Org. Chem.*, **51**, 480–483.
15. Hardacre, C., Nancarrow, P., Rooney, D.W., and Thompson, J.M. (2008) Friedel–Crafts benzoylation of anisole in ionic liquids: catalysis, separation, and recycle studies. *Org. Process Res. Dev.*, **12**, 1156–1163.
16. Bastiaansen, L.A.M. and van Lier, P.M. (1990) Imidazole-2-carboxaldehyde. *Org. Synth.*, Coll. Vol. **7**, 287–290.
17. El-Khawaga, A.M. and Roberts, R.M. (1984) Friedel–Crafts reactions of tetramethylphenyl ketones with tetramethylbenzenes. *J. Org. Chem.*, **49**, 3832–3834.
18. Trnka, T.M. and Kerber, R.C. (1995) An unexpected by-product in the Friedel–Crafts acylation of ferrocene. *Acta Crystallogr.*, **C51**, 871–873.
19. Yokota, M., Fujita, D., and Ichikawa, J. (2007) Activation of 1,1-difluoro-1-alkenes with a transition-metal

19. complex: palladium(II)-catalyzed Friedel–Crafts-type cyclization of 4,4-(difluorohomoallyl)arenes. *Org. Lett.*, **9**, 4639–4642.

20. Kuhakarn, C., Surapanich, N., Kamtonwong, S., Pohmakotr, M., and Reutrakul, V. (2011) Friedel–Crafts-type alkylation with bromodifluoro(phenylsulfanyl)methane through α-fluorocarbocations: syntheses of thioesters, benzophenones and xanthones. *Eur. J. Org. Chem.*, **2011**, 5911–5918.

21. Kagechika, H., Kawachi, E., Hashimoto, Y., Himi, T., and Shudo, K. (1988) Retinobenzoic acids. 1. Structure–activity relationships of aromatic amides with retinoidal activity. *J. Med. Chem.*, **31**, 2182–2192.

22. Rae, I.D. and Woolcock, M.L. (1987) Hydride transfers during Friedel–Crafts reactions of 5,5-dimethyl-4,5-dihydrofuran-2(3*H*)-one. *Aust. J. Chem.*, **40**, 1023–1029.

23. Yamato, T., Fujita, K., and Tsuzuki, H. (2001) Medium-sized cyclophanes. Part 58. Synthesis and conformational studies of [2.*n*]metacyclophan-1-enes and [*n*.1]metacyclophanes. *J. Chem. Soc., Perkin Trans. 1*, 2089–2097.

24. Downton, P.A., Mailvaganam, B., Frampton, C.S., Sayer, B,G., and McGlinchey, M.J. (1990) Unequivocal proof of slowed chromium tricarbonyl rotation in a sterically crowded arene complex: an X-ray crystallographic and variable-temperature high-field NMR study of $(C_6Et_5COCH_3)Cr(CO)_3$. *J. Am. Chem. Soc.*, **112**, 21–32.

25. Li, W.-M., Lai, H.-Q., Ge, Z.-H., Ding, C.-R., and Zhou, Y. (2007) Solvent affected facile synthesis of hydroxynaphthyl ketones: Lewis acids promoted Friedel–Crafts and demethylation reaction. *Synth. Commun.*, **37**, 1595–1601.

26. Strupczewski, J.T., Allen, R.C., Gardner, B.A., Schmid, B.L., Stache, U., Glamkowski, E.J., Jones, M.C., Ellis, D.B., Huger, F.P., and Dunn, R.W. (1985) Synthesis and neuroleptic activity of 3-(1-substituted-4-piperidinyl)-1,2-benzisoxazoles. *J. Med. Chem.*, **28**, 761–769.

27. Kim, K.S., Sack, J.S., Tokarski, J.S., Qian, L., Chao, S.T., Leith, L., Kelly, Y.F., Misra, R.N., Hunt, J.T., Kimball, S.D., Humphreys, W.G., Wautlet, B.S., Mulheron, J.G., and Webster, K.R. (2000) Thio- and oxoflavopiridols, cyclin-dependent kinase 1-selective inhibitors: synthesis and biological effects. *J. Med. Chem.*, **43**, 4126–4134.

28. Moriguchi, T., Endo, T., and Takata, T. (1995) Addition–elimination reaction in the trifluoroacetylation of electron-rich olefins. *J. Org. Chem.*, **60**, 3523–3528.

29. Song, Z. and Beak, P. (1990) Investigation of the mechanisms of ene reactions of carbonyl enophiles by intermolecular and intramolecular hydrogen–deuterium isotope effects: partitioning of reaction intermediates. *J. Am. Chem. Soc.*, **112**, 8126–8134.

30. Kashiwabara, T., Fuse, K., Hua, R., and Tanaka, M. (2008) Rhodium-complex-catalyzed addition reactions of chloroacetyl chlorides to alkynes. *Org. Lett.*, **10**, 5469–5472.

31. Bedini, A., Lucarini, S., Spadoni, G., Tarzia, G., Scaglione, F., Dugnani, S., Pannacci, M., Lucini, V., Carmi, C., Pala, D., Rivara, S., and Mor, M. (2011) Toward the definition of stereochemical requirements for MT2-selective antagonists and partial agonists by studying 4-phenyl-2-propionamidotetralin derivatives. *J. Med. Chem.*, **54**, 8362–8372.

32. Rosen, M.J. (1963) 1-Methyl-3-phenylindane. *Org. Synth.*, Coll. Vol. **4**, 665–667.

33. Zhou, H., Zeng, C., Ren, L., Liao, W., and Huang, X. (2006) $GaCl_3$-catalyzed chloroacylation of alkynes: a simple, convenient and efficient method to β-chlorovinyl ketones. *Synlett*, 3504–3506.

34. Iwai, T., Fujihara, T., Terao, J., and Tsuji, Y. (2012) Iridium-catalyzed addition of aroyl chlorides and aliphatic acid chlorides to terminal alkynes. *J. Am. Chem. Soc.*, **134**, 1268–1274.

35. Fang, P., Li, M., and Ge, H. (2010) Room temperature palladium-catalyzed decarboxylative *ortho*-acylation of acetanilides with α-oxocarboxylic acids. *J. Am. Chem. Soc.*, **132**, 11898–11899.

36. Verboom, W., van Dijk, B.G., and Reinhoudt, D.N. (1983) Novel applications of the 't-amino effect' in heterocyclic chemistry; synthesis of 5H-pyrrolo- and 1H,6H-pyrido[1,2-a][3,1]benzoxazines. *Tetrahedron Lett.*, **24**, 3923–3926.
37. Kobayashi, S., Komoto, I., and Matsuo, J. (2001) Catalytic Friedel–Crafts acylation of aniline derivatives. *Adv. Synth. Catal.*, **343**, 71–74.
38. Yin, Z. and Sun, P. (2012) Palladium-catalyzed direct *ortho*-acylation through an oxidative coupling of acetanilides with toluene derivatives. *J. Org. Chem.*, **77**, 11339–11344.
39. Merlic, C.A., Motamed, S., and Quinn, B. (1995) Structure determination and synthesis of fluoro nissl green: an RNA-binding fluorochrome. *J. Org. Chem.*, **60**, 3365–3369.
40. Tarbell, D.S. and Herz, A.H. (1953) An attempted Fries reaction with thiolesters. The formation of trithioorthoesters. *J. Am. Chem. Soc.*, **75**, 1668–1672.
41. Staskun, B. (1964) Nuclear acylation of arylamines. *J. Org. Chem.*, **29**, 2856–2860.
42. Aslam, M., Davenport, K.G., and Stansbury, W.F. (1991) Anhydrous hydrogen fluoride catalyzed Friedel–Crafts reactions of thioaromatic compounds. *J. Org. Chem.*, **56**, 5955–5958.
43. Yous, S., Poupaert, J.H., Lesieur, I., Depreux, P., and Lesieur, D. (1994) AlCl$_3$–DMF reagent in the Friedel–Crafts reaction. Application to the acylation reaction of 2(3H)-benzothiazolones. *J. Org. Chem.*, **59**, 1574–1576.
44. Yin, L., Hu, Q., and Hartmann, R.W. (2013) Tetrahydropyrroloquinolinone type dual inhibitors of aromatase/aldosterone synthase as a novel strategy for breast cancer patients with elevated cardiovascular risks. *J. Med. Chem.*, **56**, 460–470.
45. Lemoucheux, L., Rouden, J., Ibazizene, M., Sobrio, F., and Lasne, M.-.C. (2003) Debenzylation of tertiary amines using phosgene or triphosgene: an efficient and rapid procedure for the preparation of carbamoyl chlorides and unsymmetrical ureas. Application in carbon-11 chemistry. *J. Org. Chem.*, **68**, 7289–7297.
46. Dillard, R.D., Carr, F.P., McCullough, D., Haisch, K.D., Rinkema, L.E., and Fleisch, J.H. (1987) Leukotriene receptor antagonists. 2. The [[(tetrazol-5-ylaryl)oxy]methyl]acetophenone derivatives. *J. Med. Chem.*, **30**, 911–918.
47. Cheng, Y., Zhan, Y.-H., and Meth-Cohn, O. (2002) A simple route to N-ω-chloroalkylisatins from cyclic t-anilines, oxalyl chloride and DABCO. *Synthesis*, 34–38.
48. Semenova, O.N., Kudryavtseva, Y.A., Ermolenko, I.G., and Patsenker, L.D. (2005) Behavior of dimethylaminonaphthalenes in the Vilsmeier–Haak reaction. *Russ. J. Org. Chem.*, **41**, 1100–1101.
49. Hwang, J.P., Prakash, G.K.S., and Olah, G.A. (2000) Trifluoromethanesulfonic acid catalyzed novel Friedel–Crafts acylation of aromatics with methyl benzoate. *Tetrahedron*, **56**, 7199–7203.
50. Bigelow, L.A. and Reynolds, H.H. (1941) Quinizarin. *Org. Synth.*, Coll. Vol. **1**, 476–478.
51. Siegel, W. and Schroeder, J. (1994) Verfahren zur Herstellung mehrfach acylierter Aromaten. DE Patent 4240966.
52. Xiao, F., Shuai, Q., Zhao, F., Baslé, O., Deng, G., and Li, C.-J. (2011) Palladium-catalyzed oxidative sp^2 C–H bond acylation with alcohols. *Org. Lett.*, **13**, 1614–1617.
53. Asaumi, T., Chatani, N., Matsuo, T., Kakiuchi, F., and Murai, S. (2003) Ruthenium-catalyzed C–H/CO/olefin coupling reaction of N-arylpyrazoles. Extraordinary reactivity of N-arylpyrazoles toward carbonylation at C–H bonds. *J. Org. Chem.*, **68**, 7538–7540.
54. Snégaroff, K., Nguyen, T.T., Marquise, N., Halauko, Y.S., Harford, P.J., Roisnel, T., Matulis, V.E., Ivashkevich, O.A., Chevallier, F., Wheatley, A.E.H., Gros, P.C., and

Mongin, F. (2011) Deprotonative metalation of chloro- and bromopyridines using amido-based bimetallic species and regioselectivity-computed CH acidity relationships. *Chem. Eur. J.*, **17**, 13284–13297.

55. Kochi, T., Urano, S., Seki, H., Mizushima, E., Sato, M., and Kakiuchi, F. (2009) Ruthenium-catalyzed amino- and alkoxycarbonylations with carbamoyl chlorides and alkyl chloroformates via aromatic C–H bond cleavage. *J. Am. Chem. Soc.*, **131**, 2792–2793.

56. Rohbogner, C.J., Wunderlich, S.H., Clososki, G.C., and Knochel, P. (2009) New mixed Li/Mg and Li/Mg/Zn amides for the chemoselective metallation of arenes and heteroarenes. *Eur. J. Org. Chem.*, **2009**, 1781–1795.

57. Kaminski, T., Gros, P., and Fort, Y. (2003) Side-chain retention during lithiation of 4-picoline and 3,4-lutidine: easy access to molecular diversity in pyridine series. *Eur. J. Org. Chem.*, **2003**, 3855–3860.

58. Fontana, F., Minisci, F., Barbosa, M.C.N., and Vismara, E. (1991) Homolytic acylation of protonated pyridines and pyrazines with α-keto acids: the problem of monoacylation. *J. Org. Chem.*, **56**, 2866–2869.

59. Chan, C.-W., Zhou, Z., Chan, A.S.C., and Yu, W.-Y. (2010) Pd-catalyzed ortho–C–H acylation/cross coupling of aryl ketone O-methyl oximes with aldehydes using *tert*-butyl hydroperoxide as oxidant. *Org. Lett.*, **12**, 3926–3929.

60. Zheng, X., Song, B., and Xu, B. (2010) Palladium-catalyzed regioselective C–H bond *ortho*-acetoxylation of arylpyrimidines. *Eur. J. Org. Chem.*, **2010**, 4376–4380.

61. Wunderlich, S.H. and Knochel, P. (2010) Atom-economical preparation of aryl- and heteroaryl-lanthanum reagents by directed *ortho*-metalation by using TMP$_3$[La]. *Chem. Eur. J.*, **16**, 3304–3307.

62. Bachman, G.B. and Schisla, R.M. (1957) The direct *C*-acylation of pyridine. *J. Org. Chem.*, **22**, 858.

63. Frank, R.L. and Smith, P.V. (1955) 4-Ethylpyridine. *Org. Synth.*, Coll. Vol. 3, 410–412.

64. Zhou, W., Li, H., and Wang, L. (2012) Direct carbo-acylation reactions of 2-arylpyridines with α-diketones via Pd-catalyzed C–H activation and selective C(sp2)–C(sp2) cleavage. *Org. Lett.*, **14**, 4594–4597.

65. Chatani, N., Fukuyama, T., Tatamidani, H., Kakiuchi, F., and Murai, S. (2000) Acylation of five-membered *N*-heteroaromatic compounds by ruthenium carbonyl-catalyzed direct carbonylation at a C–H bond. *J. Org. Chem.*, **65**, 4039–4047.

66. Jaramillo, D., Liu, Q., Aldrich-Wright, J., and Tor, Y. (2004) Synthesis of *N*-methylpyrrole and *N*-methylimidazole amino acids suitable for solid-phase synthesis. *J. Org. Chem.*, **69**, 8151–8153.

67. Almansa, C., Alfón, J., de Arriba, A.F., Cavalcanti, F.L., Escamilla, I., Gómez, L.A., Miralles, A., Soliva, R., Bartrolí, J., Carceller, E., Merlos, M., and García-Rafanell, J. (2003) Synthesis and structure–activity relationship of a new series of COX-2 selective inhibitors: 1,5-diarylimidazoles. *J. Med. Chem.*, **46**, 3463–3475.

68. Janssens, F., Leenaerts, J., Diels, G., De Boeck, B., Megens, A., Langlois, X., van Rossem, K., Beetens, J., and Borgers, M. (2005) Norpiperidine imidazoazepines as a new class of potent, selective, and nonsedative H$_1$ antihistamines. *J. Med. Chem.*, **48**, 2154–2166.

69. Heckel, A. and Dervan, P.B. (2003) U-pin polyamide motif for recognition of the DNA minor groove. *Chem. Eur. J.*, **9**, 3353–3366.

70. Grace, M.E., Hosemore, M.J., Semmel, M.L., and Pratt, R.F. (1980) Kinetics and mechanism of the Bamberger cleavage of imidazole and of histidine derivatives by diethyl pyrocarbonate in aqueous solution. *J. Am. Chem. Soc.*, **102**, 6784–6789.

71. Evans, R.F. and Brown, H.C. (1973) 4-Pyridinesulfonic acid. *Org. Synth.*, Coll. Vol. **5**, 977–981.

72. Murakami, Y., Tani, M., Tanaka, K., and Yokoyama, Y. (1988) Synthetic studies on indoles and related compounds. XV. An unusual acylation of ethyl indole-2-carboxylate in the Friedel–Crafts acylation. *Chem. Pharm. Bull.*, **36**, 2023–2035.

73. Ottoni, O., Neder, A.de.V.F., Dias, A.K.B., Cruz, R.P.A., and Aquino, L.B. (2001) Acylation of indole under Friedel–Crafts conditions – an improved method to obtain 3-acylindoles regioselectively. *Org. Lett.*, **3**, 1005–1007.

74. Pal, M., Dakarapu, R., and Padakanti, S. (2004) A direct access to 3-(2-oxoalkyl)indoles via aluminum chloride induced C–C bond formation. *J. Org. Chem.*, **69**, 2913–2916.

75. Hino, T., Torisawa, Y., and Nakagawa, M. (1982) The acetylation of 3-acylindoles. *Chem. Pharm. Bull.*, **30**, 2349–2356.

76. Wu, X.-F., Anbarasan, P., Neumann, H., and Beller, M. (2010) Palladium-catalyzed carbonylative C–H activation of heteroarenes. *Angew. Chem. Int. Ed.*, **49**, 7316–7319.

77. Yonezawa, N., Hino, T., Matsuda, K., Matsuki, T., Narushima, D., Kobayashi, M., and Ikeda, T. (2000) Specific and chemoselective multi-α-arylation reaction of benzoylformic acid with or without decarbonylation in P_2O_5–MsOH and related acidic media. *J. Org. Chem.*, **65**, 941–944.

78. Seong, M.R., Lee, H.J., and Kim, J.N. (1998) Decarbonylative diarylation reaction of *N*-tosylated α-amino acids. *Tetrahedron Lett.*, **39**, 6219–6222.

79. Yonezawa, N., Hino, T., Tokita, Y., Matsuda, K., and Ikeda, T. (1997) Structural requirements for decarbonylative α,α-diarylation reaction of 2-methoxyalkanoic acids in phosphorus pentoxide-methanesulfonic acid mixture yielding 1,1-diarylalkane homologs. *Tetrahedron*, **53**, 14287–14296.

80. Palmer, M.H. and McVie, G.J. (1966) The elimination of carbon monoxide from acid chlorides. A new method for chloromethyl ether formation. *Tetrahedron Lett.*, **7**, 6405–6408.

81. Burhardt, M.N., Taaning, R., Nielsen, N.C., and Skrydstrup, T. (2012) Isotope-labeling of the fibril binding compound FSB via a Pd-catalyzed double alkoxycarbonylation. *J. Org. Chem.*, **77**, 5357–5363.

82. Verbicky, J.W., Dellacoletta, B.A., and Williams, L. (1982) Palladium catalyzed decarbonylation of aromatic acyl chlorides. *Tetrahedron Lett.*, **23**, 371–372.

83. Tsuji, J., Ohno, K., and Kajimoto, T. (1965) Organic syntheses by means of noble metal compounds XX. Decarbonylation of acyl chloride and aldehyde catalyzed by palladium and its relationship with the Rosenmund reduction. *Tetrahedron Lett.*, **6**, 4565–4568.

84. Ohno, K. and Tsuji, J. (1968) Organic syntheses by means of noble metal compounds. XXXV. Novel decarbonylation reactions of aldehydes and acyl halides using rhodium complexes. *J. Am. Chem. Soc.*, **90**, 99–107.

85. Blum, J., Oppenheimer, E., and Bergmann, E.D. (1967) Decarbonylation of aromatic carbonyl compounds catalyzed by rhodium complexes. *J. Am. Chem. Soc.*, **89**, 2338–2341.

86. Pearson, D.E. (1950) Pivalophenones from Friedel–Crafts and Grignard reactions. *J. Am. Chem. Soc.*, **72**, 4169–4170.

87. Maslak, P., Fanwick, P.E., and Guthrie, R.D. (1984) Aluminum chloride catalyzed reaction of acetanilide with pivalyl chloride. *J. Org. Chem.*, **49**, 655–659.

88. Iwama, T., Matsumoto, H., and Kataoka, T. (1997) Acid-promoted isomerisation of 1-acceptor-1-sulfenyl-substituted 2-vinylcyclopropanes with C^1–C^2 bond fission and novel 1,5-sulfenyl rearrangement. *J. Chem. Soc., Perkin Trans. 1*, 835–843.

89. Hogeveen, H., Roobeek, C.F., and Volger, H.C. (1972) Inversion of configuration in a fused cyclopropane ring opening by hydrochloric acid. *Tetrahedron Lett.*, **13**, 221–226.

90. Hamel, P., Girard, Y., and Atkinson, J.G. (1992) Acid-catalyzed isomerization of 3-indolyl sulfides to

91. Dolbier, W.R., Cornett, E., Martinez, H., and Xu, W. (2011) Friedel–Crafts reactions of 2,2-difluorocyclopropanecarbonyl chloride: unexpected ring-opening chemistry. *J. Org. Chem.*, **76**, 3450–3456.

2-indolyl sulfides: first synthesis of 3-unsubstituted 2-(arylthio)indoles. Evidence for a complex intermolecular process. *J. Org. Chem.*, **57**, 2694–2699.

92. Sakito, Y. and Suzukamo, G. (1986) Lewis acid catalyzed rearrangement of vinylcyclopropanecarbonyl chloride to cyclopentenecarbonyl chloride. *Chem. Lett.*, 621–624.

93. Ward, R.S., Davies, J., Hodges, G., and Roberts, D.W. (2002) Synthesis of quaternary alkylammonium sulfobetaines. *Synthesis*, 2431–2439.

94. Gaviña, F., Costero, A.M., and González, A.M. (1990) Reactive annulenones: a comparative study. *J. Org. Chem.*, **55**, 2060–2063.

95. Itaya, T. and Kanai, T. (1997) New reactions of phosgene with tertiary amines. *Heterocycles*, **46**, 101–104.

96. Talley, J.J. (1981) Oxidation of tertiary amines by hexachloroacetone. *Tetrahedron Lett.*, **22**, 823–826.

97. Kawasi, M., Teshima, M., Saito, S., and Tani, S. (1998) Trifluoroacetylation of methylpyridines and other methylazines: a convenient access to trifluoroacetonylazines. *Heterocycles*, **48**, 2103–2109.

98. Fraser, R.R. and Swingle, R.B. (1969) The oxidation of triethylamine by trichloroacetyl chloride. *Tetrahedron*, **25**, 3469–3475.

99. Schreiber, S.L. (1980) Hydrogen transfer from tertiary amines to trifluoroacetic anhydride. *Tetrahedron Lett.*, **21**, 1027–1030.

100. Truce, W.E. and Olson, C.E. (1952) The aluminum chloride-catalyzed condensation of γ-butyrolactone with benzene. *J. Am. Chem. Soc.*, **74**, 4721.

101. Meudt, A., Scherer, S., Nörenberg, A., and Vogt, W. (2001) Verfahren zur Herstellung von 3-*p*-Tolyl-8,9-dihydro-7*H*-benzocyclohepten-6-carbonsäure. DE Patent 10002264.

102. House, H.O. and McCaully, R.J. (1959) Polyphosphoric acid-catalyzed reaction of anisole with γ-butyrolactone. *J. Org. Chem.*, **24**, 725–726.

103. Sahlberg, C., Antonov, D., Wallenberg, H., and Noreen, R. (2003) Urea and thiourea derivatives as non-nucleoside reverse transcriptase inhibitors. WO Patent 03020705.

104. Davissont, V.J. and Poulter, C.D. (1993) Farnesyl-diphosphate synthase. Interplay between substrate topology, stereochemistry, and regiochemistry in electrophilic alkylations. *J. Am. Chem. Soc.*, **115**, 1245–1260.

105. Pirrung, M.C., Roy, B.G., and Gadamsetty, S. (2010) Structure–reactivity relationships in (2-hydroxyethyl)benzophenone photoremovable protecting groups. *Tetrahedron*, **66**, 3147–3151.

106. Miao, L., DiMaggio, S.C., and Trudell, M.L. (2010) Hydroxyarylketones via ring-opening of lactones with aryllithium reagents: an expedient synthesis of (±)-anabasamine. *Synthesis*, 91–97.

107. Andrews, D.M., Arnould, J.-C., Boutron, P., Délouvrie, B., Delvare, C., Foote, K.M., Hamon, A., Harris, C.S., Lambert-van der Brempt, C., Lamorlette, M., and Matusiak, Z.M. (2009) Fischer synthesis of isomeric thienopyrrole LHRH antagonists. *Tetrahedron*, **65**, 5805–5816.

108. Nelson, S.G., Wan, Z., and Stan, M.A. (2002) S_N2 ring opening of β-lactones: an alternative to catalytic asymmetric conjugate additions. *J. Org. Chem.*, **67**, 4680–4683.

109. Shen, K.-H., Kuo, C.-W., and Yao, C.-F. (2007) An efficient Grignard-type procedure for the preparation of *gem*-diallylated compounds. *Tetrahedron Lett.*, **48**, 6348–6351.

110. Smith, N.D., Wohlrab, A.M., and Goodman, M. (2005) Enantiocontrolled synthesis of α-methyl amino acids via Bn_2N-α-methylserine-β-lactone. *Org. Lett.*, **7**, 255–258.

111. Kraus, G.A. and Wang, X. (1999) A direct synthesis of aflatoxin M_2. *Tetrahedron Lett.*, **40**, 8513–8514.

112. Rinehart, K.L. and Gustafson, D.H. (1960) α-Indanone from β-propiolactone. *J. Org. Chem.*, **25**, 1836.

113. Tiseni, P.S. and Peters, R. (2010) Catalytic asymmetric formation of δ-lactones from unsaturated acyl halides. *Chem. Eur. J.*, **16**, 2503–2517.

114. Zhang, W., Matla, A.S., and Romo, D. (2007) Alkyl C–O ring cleavage of bicyclic β-lactones with Normant reagents: synthesis of a Merck IND intermediate. *Org. Lett.*, **9**, 2111–2114.

115. Sun, X., Zhou, L., Wang, C.-J., and Zhang, X. (2007) Rh-catalyzed highly enantioselective synthesis of 3-arylbutanoic acids. *Angew. Chem. Int. Ed.*, **46**, 2623–2626.

116. Baine, O., Adamson, G.F., Barton, J.W., Fitch, J.L., Swayampati, D.R., and Jeskey, H. (1954) A study of the Kolbe–Schmitt reaction. II. The carbonation of phenols. *J. Org. Chem.*, **19**, 510–514.

117. Katritzky, A.R., Rewcastle, G.W., and Vazquez de Miguel, L.M. (1988) Improved syntheses of substituted carbazoles and benzocarbazoles via lithiation of the (dialkylamino)methyl (aminal) derivatives. *J. Org. Chem.*, **53**, 794–799.

118. Nemoto, K., Yoshida, H., Egusa, N., Morohashi, N., and Hattori, T. (2010) Direct carboxylation of arenes and halobenzenes with CO_2 by the combined use of $AlBr_3$ and R_3SiCl. *J. Org. Chem.*, **75**, 7855–7862.

119. Olah, G.A., Török, B., Joschek, J.P., Bucsi, I., Esteves, P.M., Rasul, G., and Prakash, G.K.S. (2002) Efficient chemoselective carboxylation of aromatics to arylcarboxylic acids with a superelectrophilically activated carbon dioxide–Al_2Cl_6/Al system. *J. Am. Chem. Soc.*, **124**, 11379–11391.

120. Boogaerts, I.I.F. and Nolan, S.P. (2010) Carboxylation of C–H bonds using N-heterocyclic carbene gold(I) complexes. *J. Am. Chem. Soc.*, **132**, 8858–8859.

121. Takeuchi, K., Jirousek, M.R., Paal, M., Ruhter, G., and Schotten, T. (2004) Hypoglycemic imidazoline compounds. US Patent 2004009976.

122. Zhang, H., Shi, R., Gan, P., Liu, C., Ding, A., Wang, Q., and Lei, A. (2012) Palladium-catalyzed oxidative double C–H functionalization/carbonylation for the synthesis of xanthones. *Angew. Chem. Int. Ed.*, **51**, 5204–5207.

123. Wang, G.-W. and Yuan, T.-T. (2010) Palladium-catalyzed alkoxylation of N-methoxybenzamides via direct sp^2 C–H bond activation. *J. Org. Chem.*, **75**, 476–479.

124. Shibahara, F., Kinoshita, S., and Nozaki, K. (2004) Palladium(II)-catalyzed sequential hydroxylation–carboxylation of biphenyl using formic acid as a carbonyl source. *Org. Lett.*, **6**, 2437–2439.

125. Ohashi, S., Sakaguchi, S., and Ishii, Y. (2005) Carboxylation of anisole derivatives with CO and O_2 catalyzed by $Pd(OAc)_2$ and molybdovanadophosphates. *Chem. Commun.*, 486–488.

126. Neubert, M.E. and Fishel, D.L. (1990) Preparation of 4-alkyl and 4-halobenzoyl chlorides: 4-pentylbenzoyl chloride. *Org. Synth.*, Coll. Vol. **7**, 420–424.

127. Kikuchi, C., Ando, T., Watanabe, T., Nagaso, H., Okuno, M., Hiranuma, T., and Koyama, M. (2002) 2a-[4-(Tetrahydropyridoindol-2-yl)butyl]tetrahydrobenzindole derivatives: new selective antagonists of the 5-hydroxytryptamine[7] receptor. *J. Med. Chem.*, **45**, 2197–2206.

128. Rodefeld, L., Klausener, A., and Ullrich, F.-W. (2002) Verfahren zur Herstellung von gegebenfalls substituierten Biphenylcarbonsäurechloriden. EP Patent 1205465.

129. Ito, S., Okujima, T., Kikuchi, S., Shoji, T., Morita, N., Asao, T., Ikoma, T., Tero-Kubota, S., Kawakami, J., and Tajiri, A. (2008) Synthesis and intramolecular pericyclization of 1-azulenyl thioketones. *J. Org. Chem.*, **73**, 2256–2263.

130. Sokol, P.E. (1973) Mesitoic acid. *Org. Synth.*, Coll. Vol. **5**, 706–708.

131. Gore, P.H., Kamounah, F.S., and Miri, A.Y. (1979) Friedel–Crafts cyanation of some reactive aromatic hydrocarbons. *Tetrahedron*, **35**, 2927–2929.

132. Okamoto, K., Watanabe, M., Murai, M., Hatano, R., and Ohe, K. (2012) Practical synthesis of aromatic nitriles via

gallium-catalysed electrophilic cyanation of aromatic C–H bonds. *Chem. Commun.*, **48**, 3127–3129.

133. Murai, M., Hatano, R., Kitabata, S., and Ohe, K. (2011) Gallium(III)-catalysed bromocyanation of alkynes: regio- and stereoselective synthesis of β-bromo-α,β-unsaturated nitriles. *Chem. Commun.*, **47**, 2375–2377.

134. Chambert, S., Thomasson, F., and Décout, J.-L. (2002) 2-Trimethylsilylethyl sulfides in the von Braun cyanogen bromide reaction: selective preparation of thiocyanates and application to nucleoside chemistry. *J. Org. Chem.*, **67**, 1898–1904.

135. Effenberger, F., Mack, K.-E., Niess, R., Reisinger, F., Steinbach, A., Stohrer, W.-D., Stezowski, J.J., Rommel, I., and Maier, A. (1988) Aminobenzenes. 19. Dimeric σ-complexes: intermediates in the oxidative dimerization of aromatics. *J. Org. Chem.*, **53**, 4379–4386.

136. Raja, E.K., Lill, S.O.N., and Klumpp, D.A. (2012) Friedel–Crafts-type reactions with ureas and thioureas. *Chem. Commun.*, **48**, 8141–8143.

137. Hoi, B. and Janicaud, J. (1945) Friedel–Crafts reaction with ethyl chloroformate. *Bull. Soc. Chim. Fr.*, **12**, 640–642.

138. Coppock, W.H. (1957) New synthesis of aryl esters of aromatic acids. *J. Org. Chem.*, **22**, 325–326.

139. Grunewald, G.L., Sall, D.J., and Monn, J.A. (1988) Conformational and steric aspects of the inhibition of phenylethanolamine N-methyltransferase by benzylamines. *J. Med. Chem.*, **31**, 433–444.

140. Lohaus, G. (1988) 2,4-Dimethoxybenzonitrile. *Org. Synth.*, Coll. Vol. **6**, 465–468.

141. Dowideit, P., Mertens, R., and von Sonntag, C. (1996) Non-hydrolytic decay of formyl chloride into CO and HCl in aqueous solution. *J. Am. Chem. Soc.*, **118**, 11288–11292.

142. Olah, G.A. and Kuhn, S.J. (1960) Formylation with formyl fluoride: a new aldehyde synthesis and formylation method. *J. Am. Chem. Soc.*, **82**, 2380–2382.

143. Coleman, G.H. and Craig, D. (1943) p-Tolualdehyde. *Org. Synth.*, Coll. Vol. **2**, 583–586.

144. Bagno, A., Bukala, J., and Olah, G.A. (1990) Superacid-catalyzed carbonylation of methane, methyl halides, methyl alcohol, and dimethyl ether to methyl acetate and acetic acid. *J. Org. Chem.*, **55**, 4284–4289.

145. Goettmann, F., Fischer, A., Antonietti, M., and Thomas, A. (2006) Metal-free catalysis of sustainable Friedel–Crafts reactions: direct activation of benzene by carbon nitrides to avoid the use of metal chlorides and halogenated compounds. *Chem. Commun.*, 4530–4532.

146. Ziegler, G., Haug, E., Frey, W., and Kantlehner, W. (2001) Orthoamides. Part LVII. Can aromatic aldehydes be prepared from aryl formates via the Fries rearrangement? *Z. Naturforsch., B: Chem. Sci.*, **56**, 1178–1187.

147. Russell, A. and Lockhart, L.B. (1955) 2-Hydroxy-1-naphthaldehyde. *Org. Synth.*, Coll. Vol. **3**, 463–464.

148. Allen, C.F.H. and Leubner, G.W. (1963) Syringic aldehyde. *Org. Synth.*, Coll. Vol. **4**, 866–869.

149. Song, F., Wang, C., Falkowski, J.M., Ma, L., and Lin, W. (2010) Isoreticular chiral metal-organic frameworks for asymmetric alkene epoxidation: tuning catalytic activity by controlling framework catenation and varying open channel sizes. *J. Am. Chem. Soc.*, **132**, 15390–15398.

150. Saito, F., Kuramochi, K., Nakazaki, A., Mizushina, Y., Sugawara, F., and Kobayashi, S. (2006) Synthesis and absolute configuration of (+)-pseudodeflectusin: structural revision of aspergione B. *Eur. J. Org. Chem.*, **2006**, 4796–4799.

151. Sasaki, S., Kusumoto, T., Nomura, I., and Maezaki, H. (2010) Pyrazinooxazepine derivatives. US Patent 2010317651.

152. Boyd, W.J. and Robson, W. (1935) The synthesis of indole-3-aldehyde and its homologues. *Biochem. J.*, **29**, 555–561.

153. Wynberg, H. and Meijer, E.W. (1982) The Reimer–Tiemann reaction. *Org. React.*, **28**, 1–36.
154. Beier, P., Pastyríková, T., and Iakobson, G. (2011) Preparation of SF_5 aromatics by vicarious nucleophilic substitution reactions of nitro(pentafluorosulfanyl)benzenes with carbanions. *J. Org. Chem.*, **76**, 4781–4786.
155. Fuson, R.C. and Miller, T.G. (1952) The von Auwers rearrangement in the naphthalene series. *J. Org. Chem.*, **17**, 316–320.
156. Wynberg, H. and Johnson, W.S. (1959) Synthetic applications of the abnormal Reimer–Tiemann reaction. *J. Org. Chem.*, **24**, 1424–1428.
157. Dodson, R.M. and Webb, W.P. (1951) Polycyclic compounds. I. The Reimer–Tiemann reaction with 1-alkyl-2-naphthols. *J. Am. Chem. Soc.*, **73**, 2767–2769.
158. Kwon, S., Nishimura, Y., Ikeda, M., and Tamura, Y. (1976) An improved procedure for ring expansion of indoles into 3-haloquinolines by use of phase transfer catalysts. *Synthesis*, **1976**, 249.
159. Wenkert, E., Angell, E.C., Ferreira, V.F., Michelotti, E.L., Piettre, S.R., Sheu, J.-H., and Swindell, C.S. (1986) Synthesis of prenylated indoles. *J. Org. Chem.*, **51**, 2343–2351.
160. De Angelis, F., Inesi, A., Feroci, M., and Nicoletti, R. (1995) Reaction of electrogenerated dichlorocarbene with methylindoles. *J. Org. Chem.*, **60**, 445–447.
161. Giordano, C., Minisci, F., Vismara, E., and Levi, S. (1986) A general, selective, and convenient procedure of homolytic formylation of heteroaromatic bases. *J. Org. Chem.*, **51**, 536–537.
162. Edwards, W.R. and Sibille, E.C. (1963) Mixed carboxylic anhydrides in the Friedel–Crafts reaction. *J. Org. Chem.*, **28**, 674–679.
163. Edwards, W.R. and Eckert, R.J. (1966) Friedel–Crafts acylation of thiophene with mixed acetic anhydrides. *J. Org. Chem.*, **31**, 1283–1285.
164. Hidega, K., Csekóa, J., and Hankovszkya, H.O. (1986) Synthesis of nitroxide paramagnetic ketones from nitroxide acid chlorides and anhydrides by Friedel–Crafts acylation. *Synth. Commun.*, **16**, 1839–1847.
165. Soukup, M. (2012) Manufacturing process for sitagliptin from L-aspartic acid. US Patent 2012/123144.
166. Loeffler, H.-P., Thym, S., Koenig, K.-H., and Zeeh, B. (1980) Neue Bis-(N-halogenmethyl)carbamidsäureester und Verfahren zu ihrer Herstellung. DE Patent 2826012.
167. Hoover, F.W., Stevenson, H.B., and Rothrock, H.S. (1963) Chemistry of isocyanic acid. I. Reactions of isocyanic acid with carbonyl compounds. *J. Org. Chem.*, **28**, 1825–1830.
168. Zambron, B., Masnyk, M., Furman, B., and Chmielewski, M. (2009) An entry to 4-aryl-azetidinones via alkylation of nucleophilic arenes using four-membered acyliminium cations. *Tetrahedron*, **65**, 4440–4446.

5 芳烃的亲电卤代反应

5.1 概述

卤代反应是芳烃官能化的一个重要工具。大部分卤代反应是通过卤素对芳基上的氢或其他电正性基团的亲电取代进行的,这些基团包括硅烷基、硼烷基、主族和过渡金属及仲或叔烷基等。卤素原子是邻位/对位定位基团,但仅仅使芳环轻微失活,因此多卤代是常见的副反应。区域选择性也往往是一个问题。

当芳烃容易被氧化时(即具有低的氧化半波电势),占主导地位的往往是另外一种机理,即芳烃和卤代试剂之间的单电子转移。此时,自由基阳离子中间体可以从卤代试剂攫取一个卤负离子得到两个自由基,或者攫取一个卤素自由基得到质子化的芳基卤代物。卤素自由基也可以通过卤代试剂的热或光化学均裂形成,而后加成到芳烃上。另一种可能的机理是经典的自由基链式卤代(图式 5.1

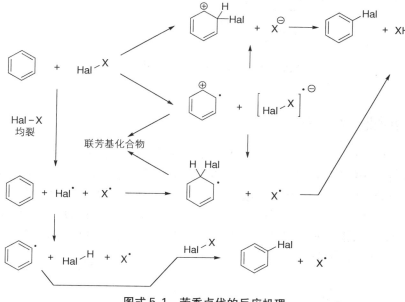

图式 5.1 芳香卤代的反应机理

中最后一个公式)。

大规模的卤代反应通常使用卤素单质或卤化物和氧化剂(氧气、双氧水或电化学氧化),后者更加经济。高温气相卤代或者气相光卤代反应能够在非常大的规模下进行。五氯化磷、氯化亚砜、亚硫酰氯、氯化铜和芳基磺酰氯等试剂也可用于氧化氯代。在小规模的制备反应中,可以使用固体卤代试剂,例如 N-卤代酰胺或酰亚胺,以实现更精确的投料。

当卤间化合物①被用作卤代试剂时,通常是较重的卤素原子与碳原子成键,但也不完全如此(图式 5.2)。反应的选择性取决于其按照离子或自由基机理进行[1]。

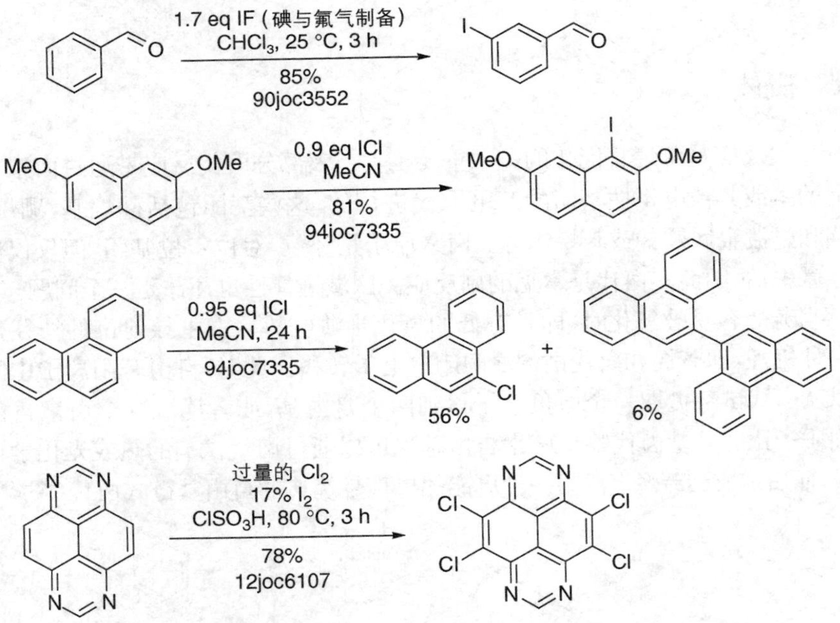

图式 5.2 与混合卤素的卤代反应[1-3]

5.2 典型的副反应

非卤素单质进行的卤化反应结果很难预测,因为反应可以形成多种产物。例如,用 N-卤代琥珀酰亚胺进行的非自由基卤代经常得到 N-芳基琥珀酰亚胺,而非芳基卤化物。同样,磺酰卤不仅可以卤化,也经常作为磺酰化试剂使用(图式 5.3)。N-卤代琥珀酰亚胺不适合进行不活泼底物的卤代,因为 N-卤代琥珀酰亚胺容易异构化为无活性的 3-卤代琥珀酰亚胺[4]。此时,更好的选择是三氯异氰脲酸。

---

① 译者注:是由不同卤素组成的。

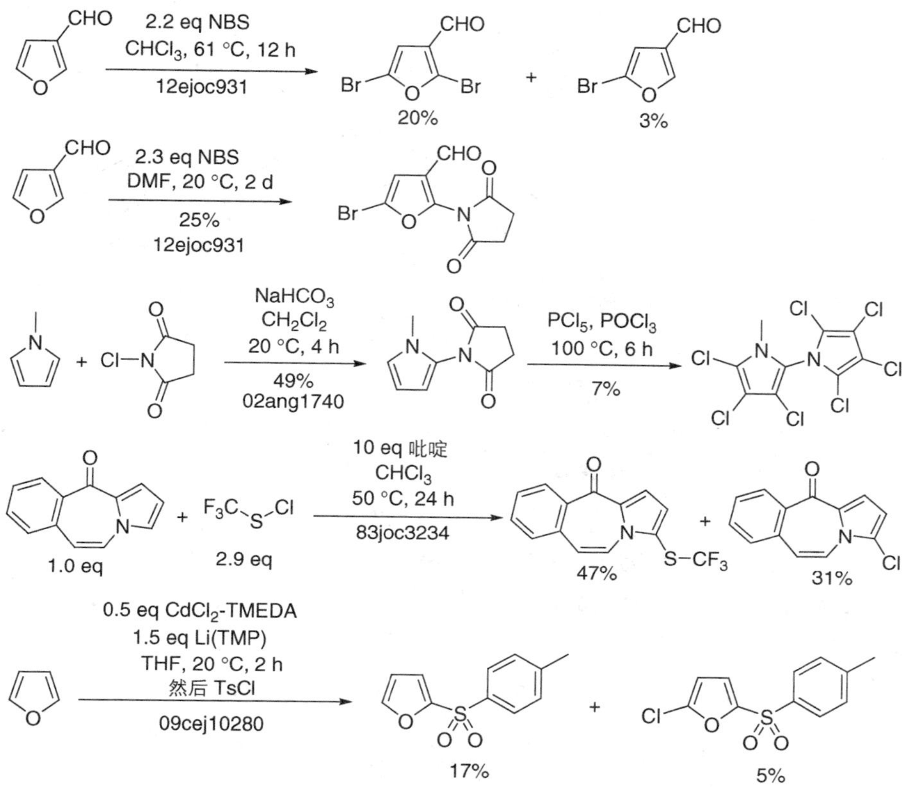

图式 5.3 非卤素单质的卤代反应[5-8]

芳香卤代反应在亲核试剂存在时能导致亲核试剂的芳基化。通常发生在氟化反应中，但也有可能发生在溴化反应或碘化反应中（图式 5.4）。相反，反应体系中的亲核试剂也能够被卤化试剂氧化，使其转化成一个竞争性的亲电试剂。如此的副反应尽管能够通过排除反应体系中其他亲核试剂来避免，然而在氟代反应中二氯甲烷甚至可以作为氯的供体。

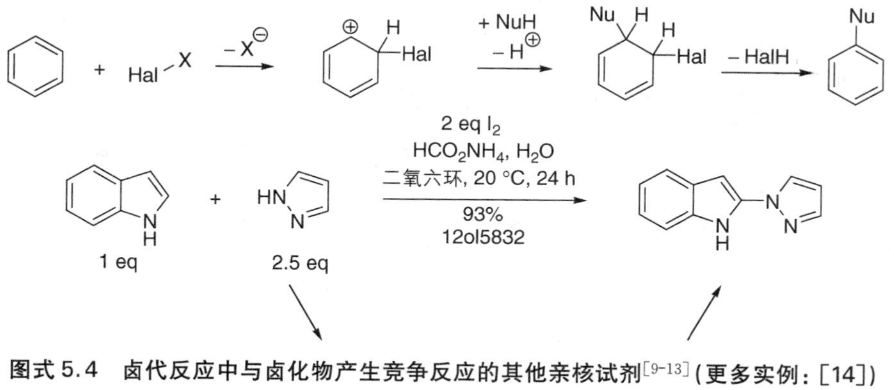

图式 5.4 卤代反应中与卤化物产生竞争反应的其他亲核试剂[9-13]（更多实例：[14]）

109

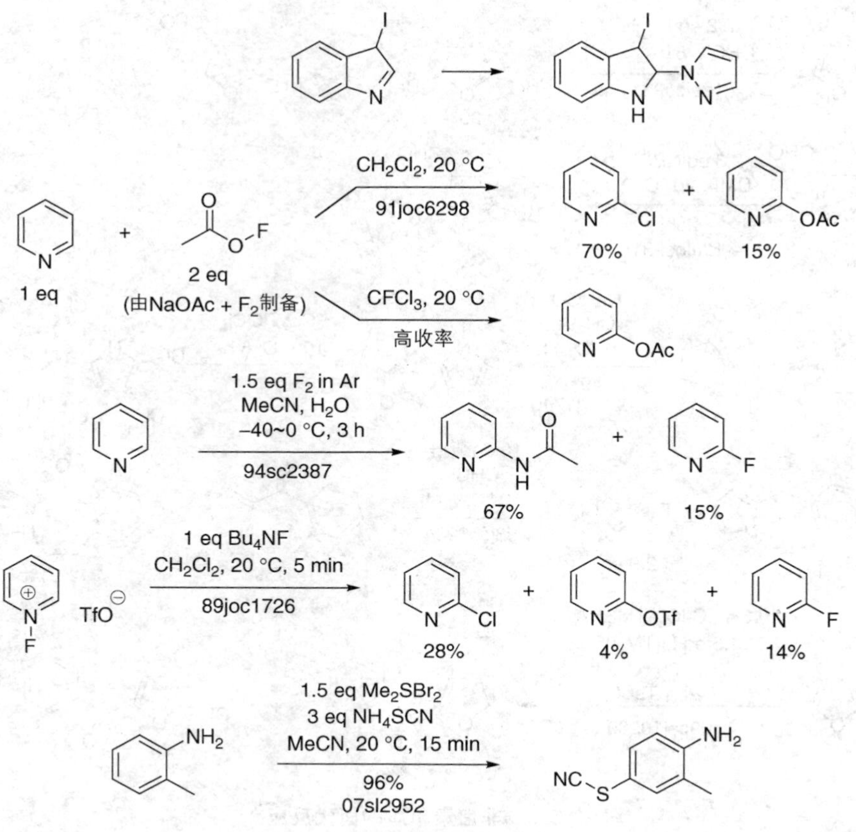

续图式 5.4

除氢原子外，其他取代基也可以在芳族亲电卤代反应中被取代。特别敏感的基团是叔烷基和仲烷基，以及容易形成稳定碳正离子的烷基，如 1-羟烷基或苄基（图式 5.5）。

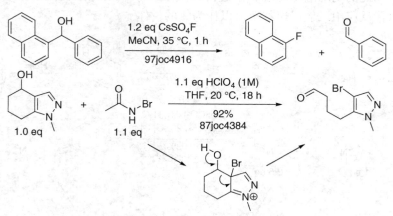

图式 5.5 羟烷基的亲电取代反应[15,16]

芳香族碘代反应可以通过用碘化钾和过硫酸氢钾处理芳烃来实现[17]。但是，过量的过硫酸氢钾可能使芳基碘转换为碘鎓盐，它是一个强氧化剂，可能导致酰胺的霍夫曼(Hofmann)重排[18]、芳族的羟基化、联芳基的形成[19]或者其他氧化反应。

5.3 区域选择性

亲电芳香卤代与其他芳族亲电取代的区域选择性相类似，但经常会得到区域异构体的混合物(图式 5.6)。自由基卤代的区域选择性比离子型卤化低，并且能够引发脂肪族卤代。在反应过程中有少量碘或卤化氢的形成可能改变

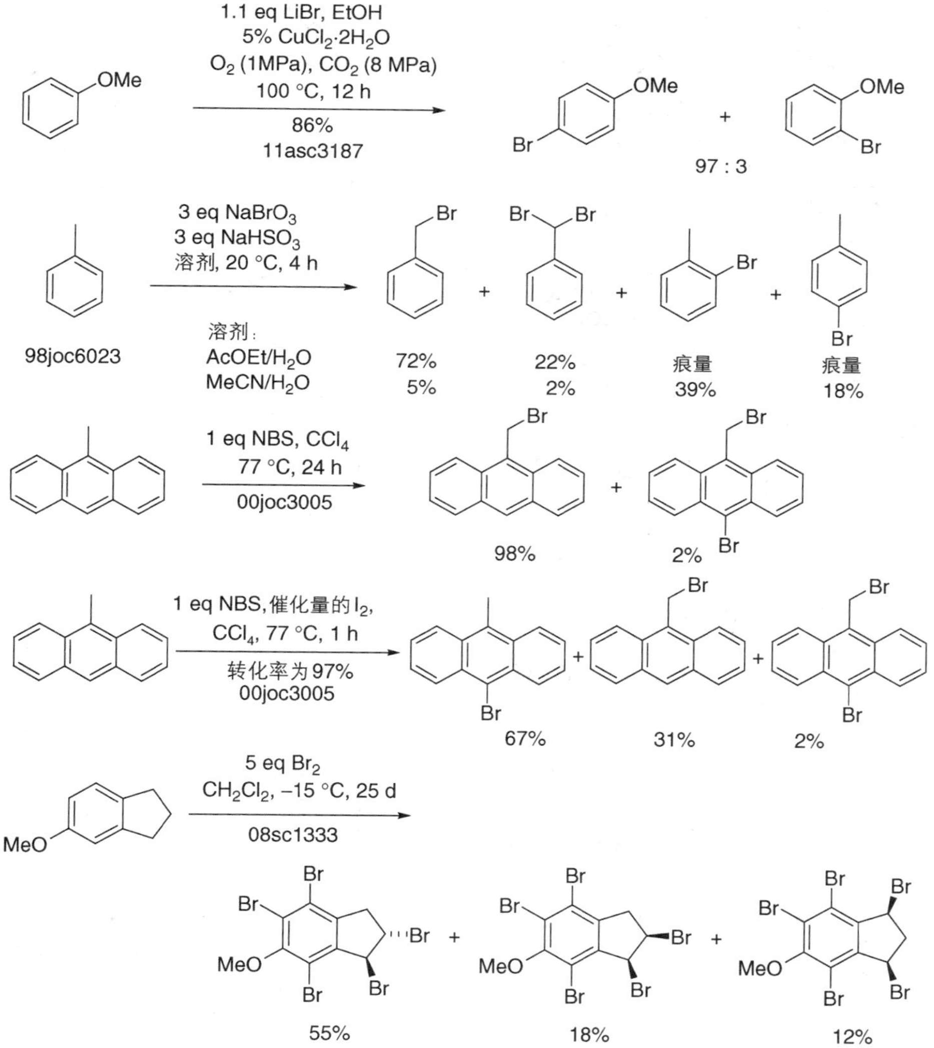

图式 5.6　溴化反应的区域选择性[20-24]

111

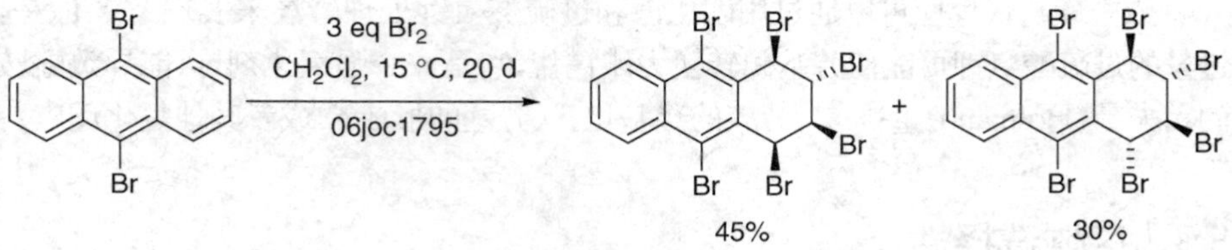

卤代的区域选择性,因此,通过对反应体系进行缓冲(例如,通过使用酸或碱作为溶剂)以保持质子浓度在整个反应过程中大致恒定,就能够使芳香卤代的区域选择性得到优化。

通过形成螯合中间体或分子内卤素转移,有时能够实现高的区域选择性(图式 5.7)。在大多数报道的例子中,卤代被导向到苯基的邻位。

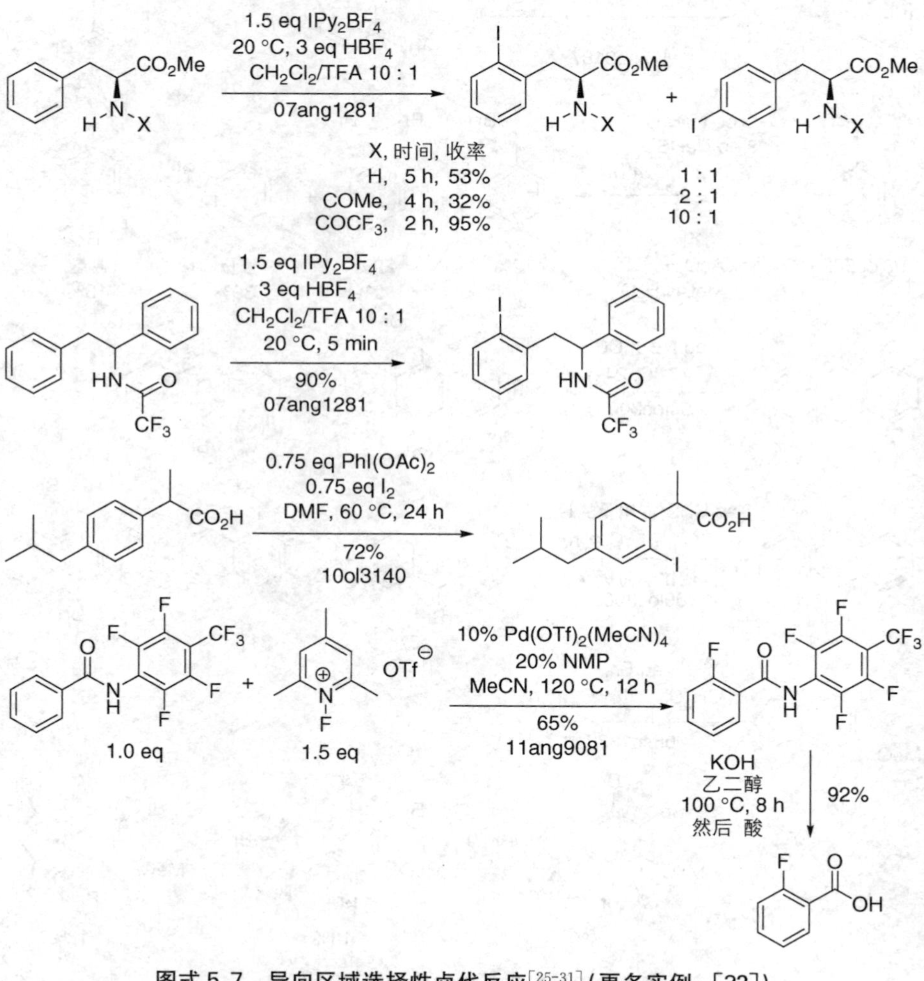

图式 5.7 导向区域选择性卤代反应[25-31](更多实例:[32])

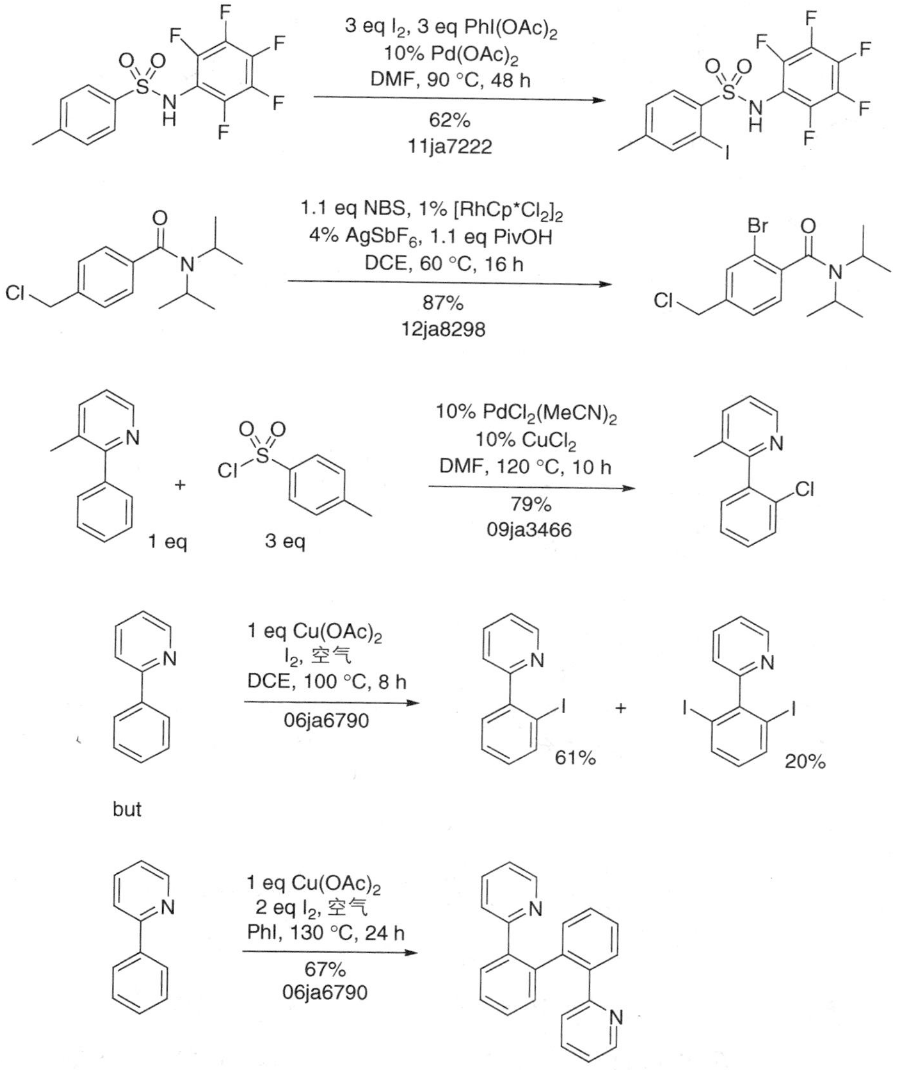

续图式 5.7

5.4 催化

芳香族卤代反应能够通过提高卤化试剂(通常用酸)的亲电性,或者通过增强芳烃的亲核性(通常用碱或通过金属化)进行催化。过渡金属催化的例子已有报道,即在卤代反应之前先进行芳基的金属化(图式 5.8)。一种可能的反应机理是单电子从芳烃转移到金属阳离子,然后卤化物加成到芳烃上,再氧化并脱质子。

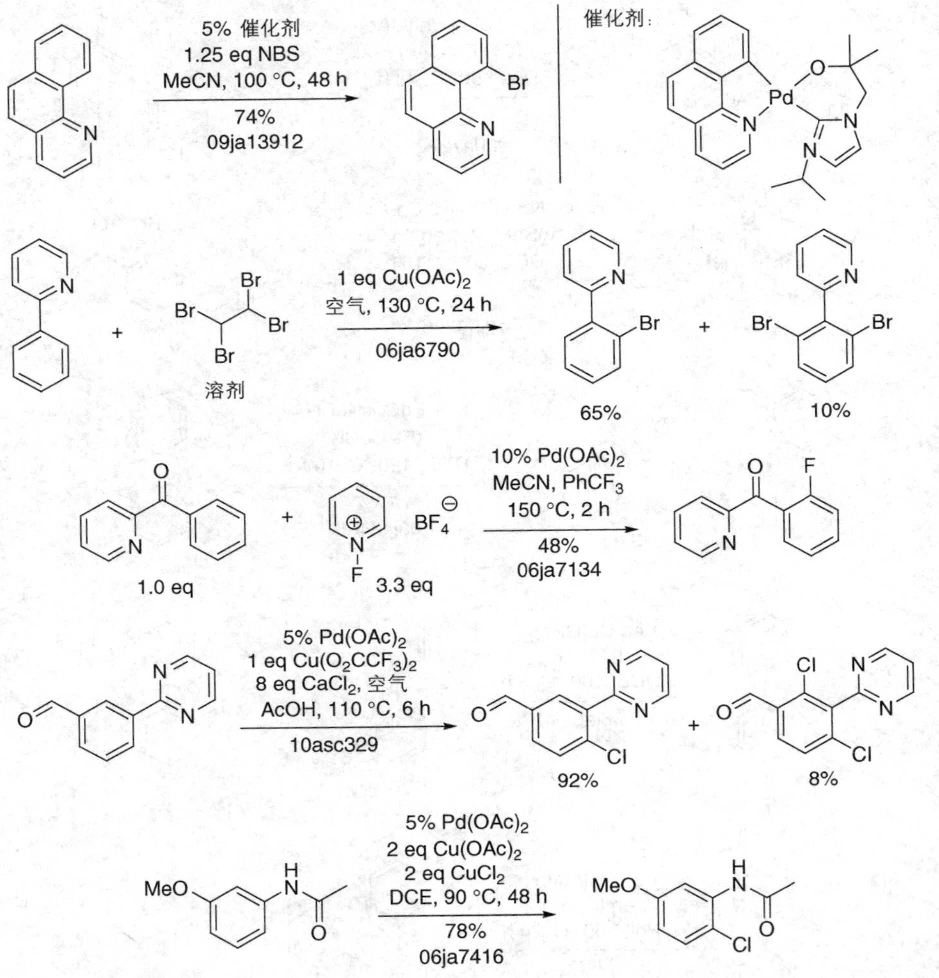

图式 5.8 催化的芳香亲电卤代反应[31,33-36]（更多实例和机理研究：[37-40]）

过渡金属存在的卤代反应可以产生许多预想不到的产物。常见的副反应包括芳烃的氧化二聚[14]、羟基化及各种官能团转化。如果使用二甲基亚砜(DMSO)或 N,N-二甲基甲酰胺(DMF)等作为反应溶剂,强路易斯酸能进一步引起更多的不想要的转化(图式 5.9)。

5.5 氟化反应

氟元素的高反应活性、形成 C—F 键时的放热效应、氟化物的碱性及氟的其他性质导致氟化反应较为复杂。用氟气/氮气可以直接进行亲电氟化,但因反应选择性差而很少使用。一些容易处理的氟化试剂已经被开发,它们可以进行选择性地氟化。

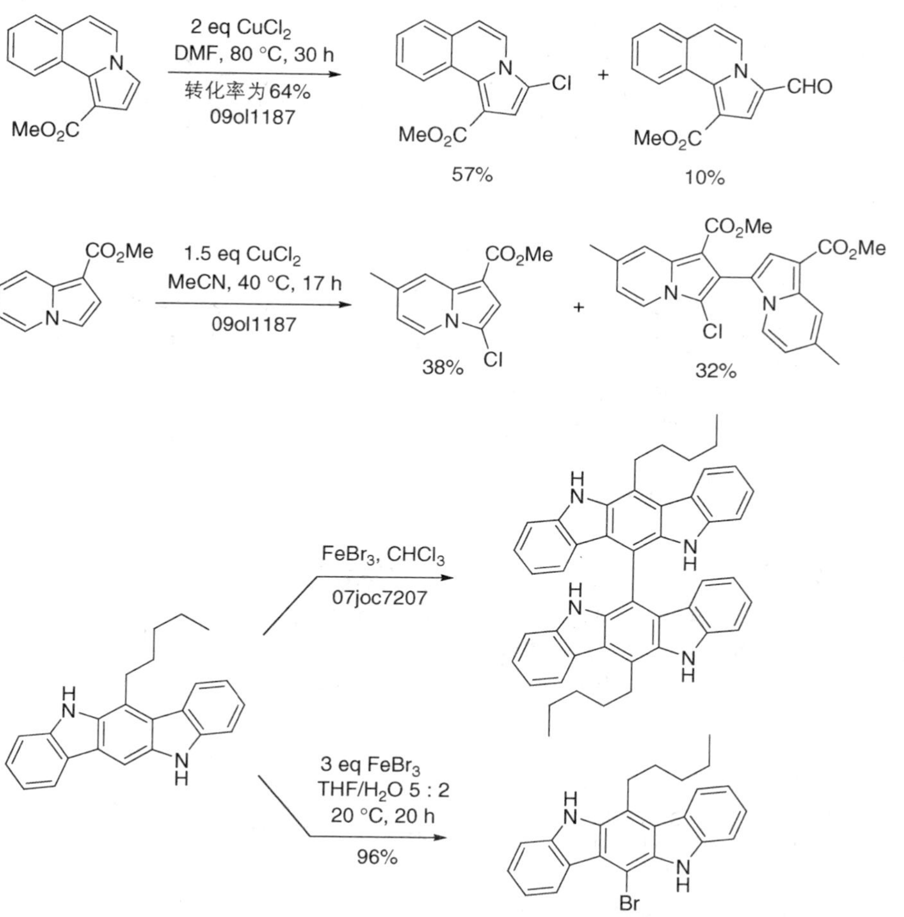

图式5.9 过渡金属存在下卤化反应中甲酰化和芳烃二聚的副反应[41,42]

亲电芳香氟化反应典型的副反应包括多氟化、去芳构化[43]、氟原子对其他卤素或羟基的取代及底物与其他亲核试剂或溶剂的反应(如羟基化或乙酰氧基化,图式5.4)。当其他亲核试剂存在时,用 F^+ 的合成对等体处理芳烃几乎得不到氟代芳烃,而是生成大量的芳基化的亲核试剂(例如参考文献[44]和图式5.10)。亲电氟化反应通常要求无可氧化溶剂和亲核试剂存在。因为氟离子是一个更好的离去基团,多数的 X—F 中间体都不能用于氟化。

苯乙酮与强氟化试剂反应会发生脱乙酰化反应并且得到氟代芳基醚(图式5.11)。氟化硫通常能将苯乙酮转化为1,1-二乙基苯。

氟化硫(SF_4 或 R_2NSF_3)是选择性转化羰基成为偕二氟化合物的试剂。这些试剂一般不太适合芳烃的亲电氟化,但是适用于儿茶酚的氟化(图式5.12)。儿茶酚、间苯二酚(1,3-二羟基苯)或间苯三酚(1,3,5-三羟基苯)等多羟基苯的性质类似于酮类化合物,因此可以与氟化硫进行氧氟交换。

图式 5.10 亲电氟化[43,45-47] (更多实例: [48])

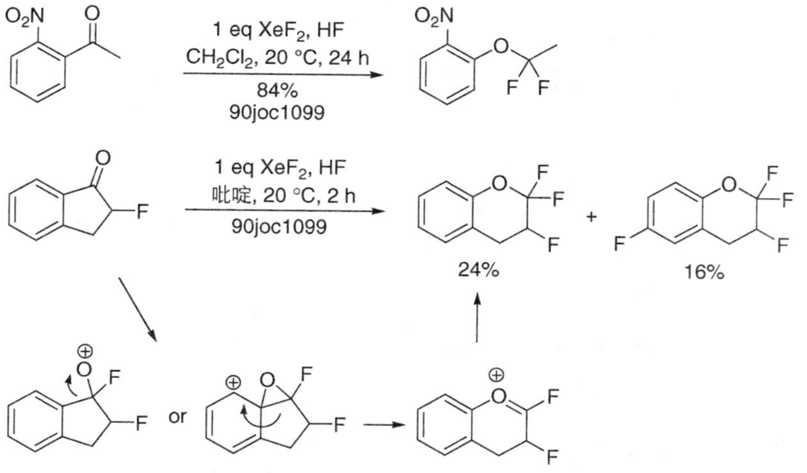

图式 5.11 苯乙酮与 XeF$_2$ 的反应[49]

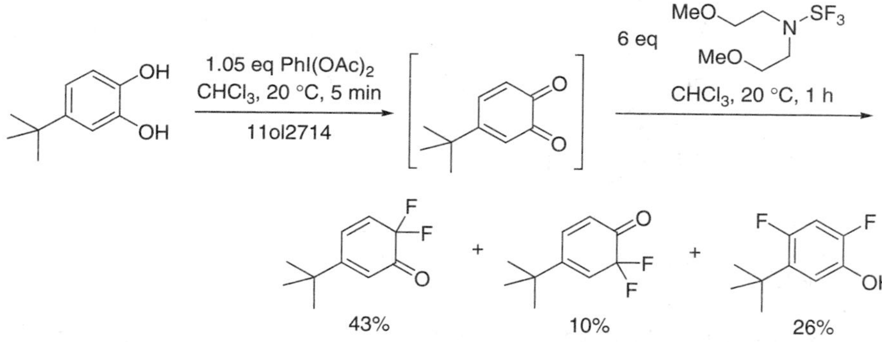

图式 5.12 儿茶酚的氟化反应[50]

5.6 缺电子芳烃

只要不含敏感官能团,缺电子苯可以纯净地卤代(图式 5.13)。强无机酸(硫酸或硝酸)是最适合的溶剂,卤代试剂的活性可以通过卤化铝或其他路易斯酸进一步提高。

多卤代芳烃或其他酸性芳烃可以通过强碱的原位金属化进行卤代(图式 5.14)。对所使用碱的要求是必须不与卤化试剂反应,但能够满足这一要求的碱仅有少数几个。叔醇的盐类似乎非常适合,但必须谨慎操作,因为强碱和亲电试剂混合能够发生剧烈的分解反应。大规模制备时,这样的反应应该通过控制剂量的方式进行,即在反应进行过程中缓慢加入碱或卤代试剂。

图式 5.13　缺电子芳烃的卤代反应[51,52]（更多实例：[53]）

图式 5.14　C—H 酸性芳烃经原位金属化进行卤代[54,55]

5.6.1　吡啶

吡啶缺电子性能与硝基苯接近,故难以卤代。用溴处理吡啶会导致聚合,只有吡啶鎓盐可以（在强烈条件下）以较好收率实现溴代（图式 5.15）。由于吡啶会中和 1 当量酸,故路易斯酸"催化"需要过量的路易斯酸。吡啶的氯代最好在气相中通过热或光化学完成[56]。

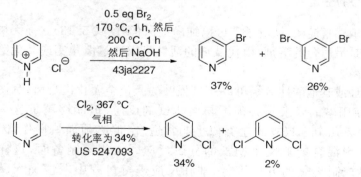

图式 5.15　吡啶的卤代[56-60]（更多实例：[61]）

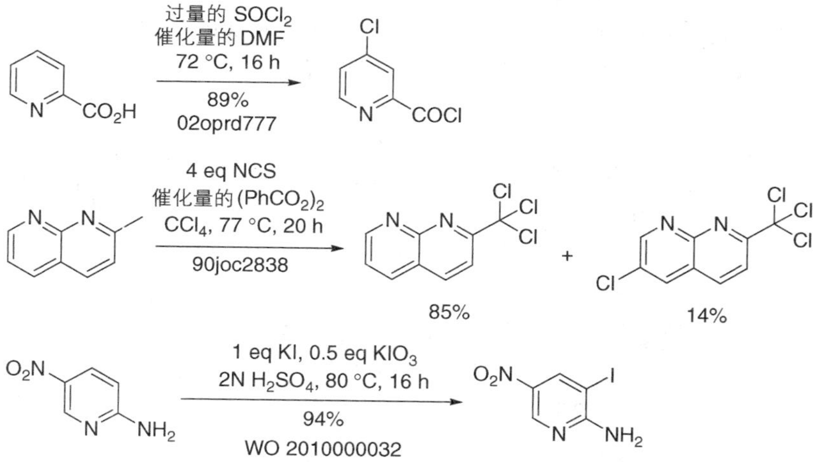

续图式 5.15

通过转化成为其 N-氧化物,可以提高吡啶的反应活性。用三氯氧磷或光气处理吡啶的 N-氧化物可以高收率地得到 2-氯吡啶。同样,N-氨基吡咯通过氯化氢处理可得到 2-氯吡咯(图式 5.16)。

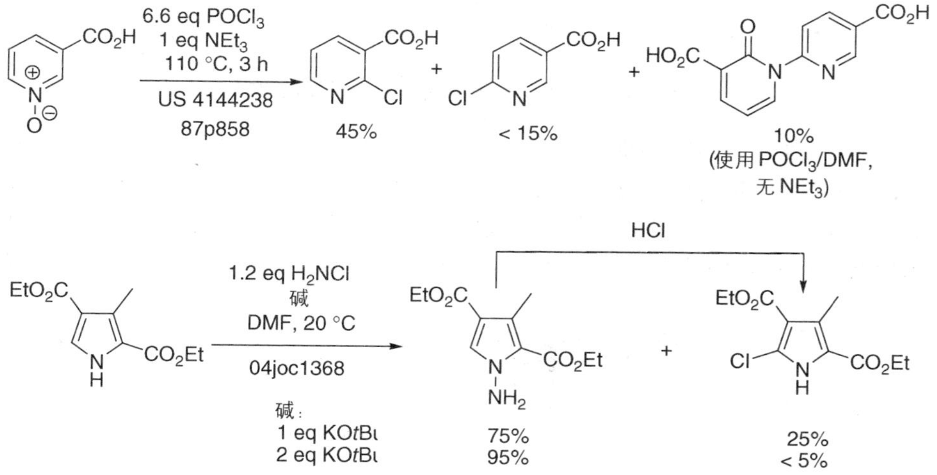

图式 5.16 吡啶的 N-氧化物和吡咯的氯代反应[62-64]

5.6.2 苯甲酸衍生物

苯甲酸衍生物卤代的区域选择性往往较差或不可预测。第一次卤代反应发生在其 3 位(间位)后,通常第二次卤代反应发生在 6 位(邻位)(图式 5.17)。为获得高收率,需要使用大量路易斯酸[65]。羰基通常是间位定位基,但可以通过形成螯合物或分子内卤素正离子转移实现邻位定位。在强酸存在下，形成的苯甲酰阳离子可以显著钝化其反应活性，导致难以进行卤代反应。此外，酯、酰胺或者腈使用强酸处理可发生水解或醇解。

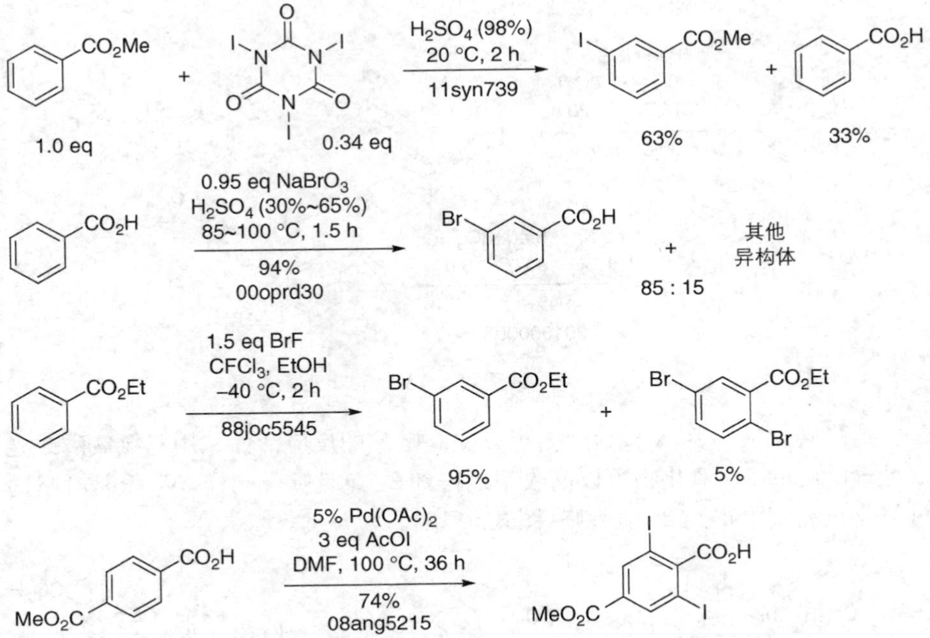

图式 5.17　苯甲酸和苯甲酸酯的卤代反应[66-69]

苯甲酸卤代的另一个问题是其易于脱羧基，特别是在过渡金属存在下[70-73]。用卤素银盐处理苯甲酸通常发生脱羧卤代反应[亨斯狄克（Hunsdiecker）反应]，但该反应也可以在不必形成盐或没有银的情况下发生（图式 5.18）。羟基苯甲酸、萘甲酸、肉桂酸和许多杂芳基羧酸特别容易脱羧基。

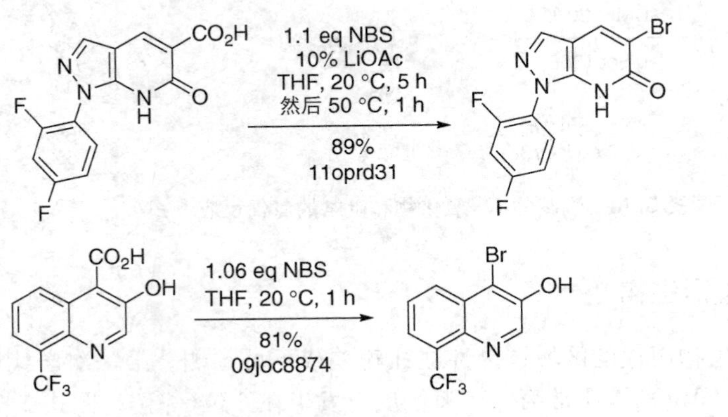

图式 5.18　脱羧卤代反应[74-77]

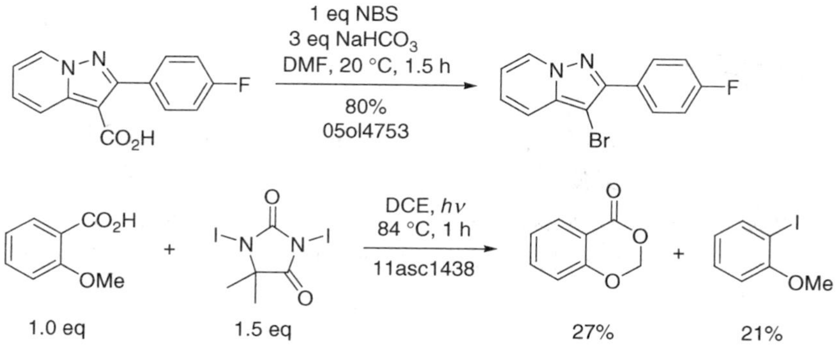

续图式 5.18

苯甲腈的卤代反应不仅困难,而且常常得到的是间位和邻位卤代产物的混合物[65](图式 5.19)。此外,氰基在强酸或强碱条件下容易水合或水解,所得的酰胺会被大多数的卤代试剂进行 N-卤代,用碱处理时它们会发生霍夫曼重排。

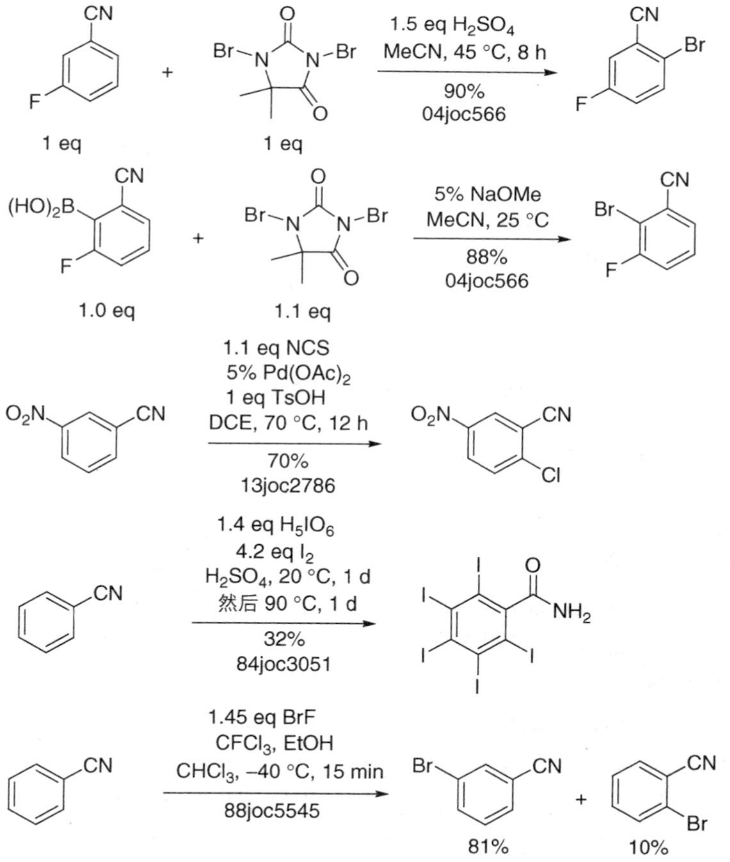

图式 5.19 苯甲腈的卤代反应[68,78-80]（更多实例：[81]）

5.7 富电子芳烃

富电子芳烃和杂芳烃很容易进行卤代，并且难以控制。如果使用了过量的卤代试剂、强氧化剂或使用过于激烈的反应条件，可能发生多卤化、芳烃的羟基化、形成醌或苯醌亚胺(来自芳基醚、酚或苯胺)、去芳构化，甚至 C—C 键的断裂。用氟气进行氟化时常常发生这些副反应，氯代也可引起重排和 C—C 键的断裂(图式 5.20)。

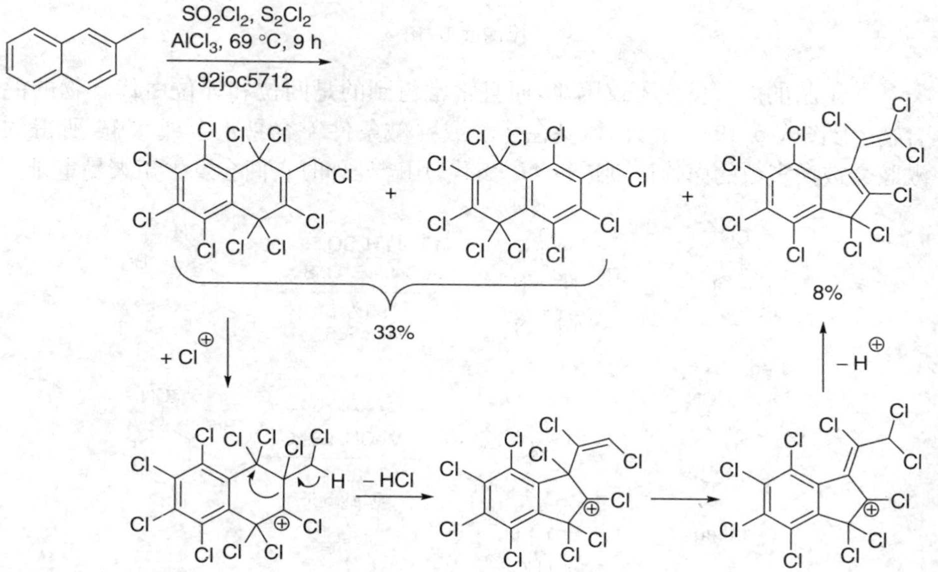

图式 5.20 甲基萘的全氯代[82]

5.7.1 酚类和芳基醚

不受控制的苯酚氯代反应会生成多氯环己烯酮、环己二烯酮和苯醌。在高温下甚至可以形成氯代的二苯并二氧六环。酚和烷氧基苯的纯净的单卤代反应需要在较低反应温度和仅使用稍过量卤代剂的条件下进行。胺或硫醚等捕获剂可用来降低卤代试剂的反应活性以提高其选择性。仔细控制反应条件甚至可对氢醌进行单卤代而不明显形成醌(图式 5.21)。

如果未对反应体系做缓冲处理，烷基酚卤代时很容易被脱烷基(图式 5.22)。类似脱烷基化反应也常见于其他富电子烷基芳烃(图式 5.5)。当没有碱存在时，卤素卤代将产生卤化氢，它能够催化许多不想要的转化。

在卤化过程中，2-芳基(或 2-杂芳基)苯酚可能发生环化反应形成苯并呋喃(图式 5.23)。

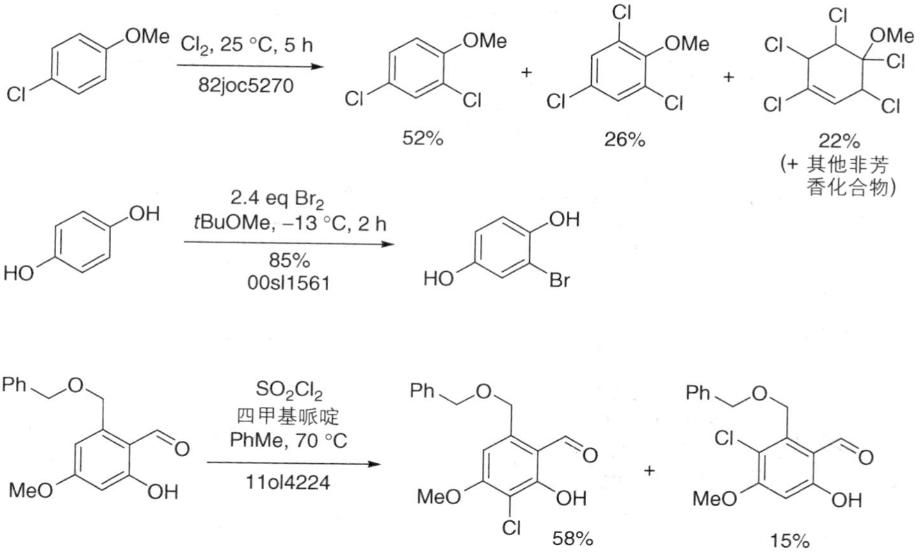

图式 5.21 酚类和芳基醚的卤代[83-85]

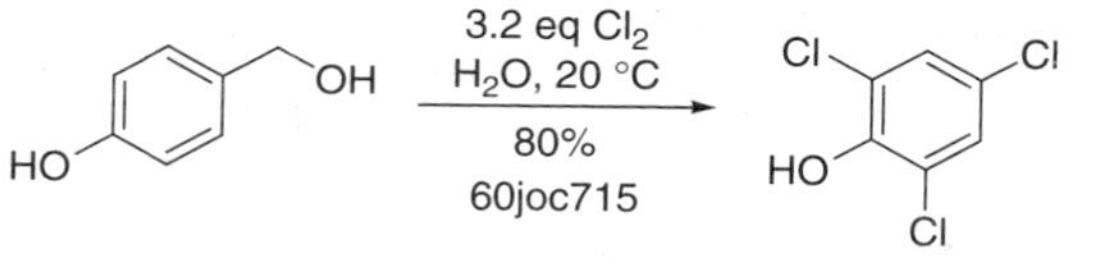

图式 5.22 卤化过程中酚的脱烷基化反应[86]

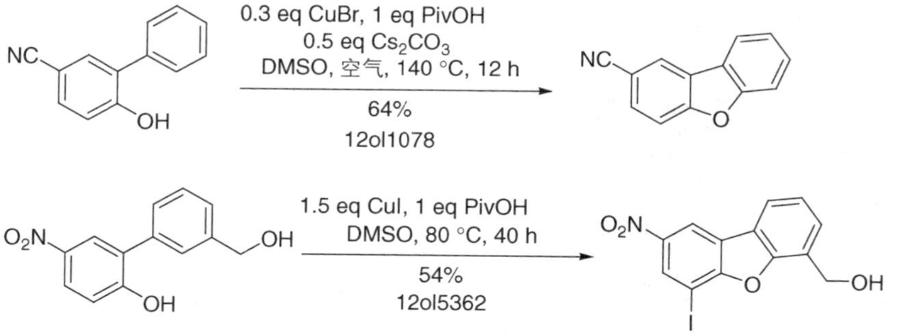

图式 5.23 从 2-芳基苯酚形成苯并呋喃[87,88]

5.7.2 苯胺

在碱性条件下,苯胺很容易卤代,但也能被氧化(成醌)和发生聚合。在酸性条件下常常能获得更好的收率。弱酸(如乙酸)主要导致邻/对位卤代,而在强酸条件下(如以硫酸为溶剂),有时能促进间位卤代,但很少能获得单纯的产物[89,90](图式 5.24)。

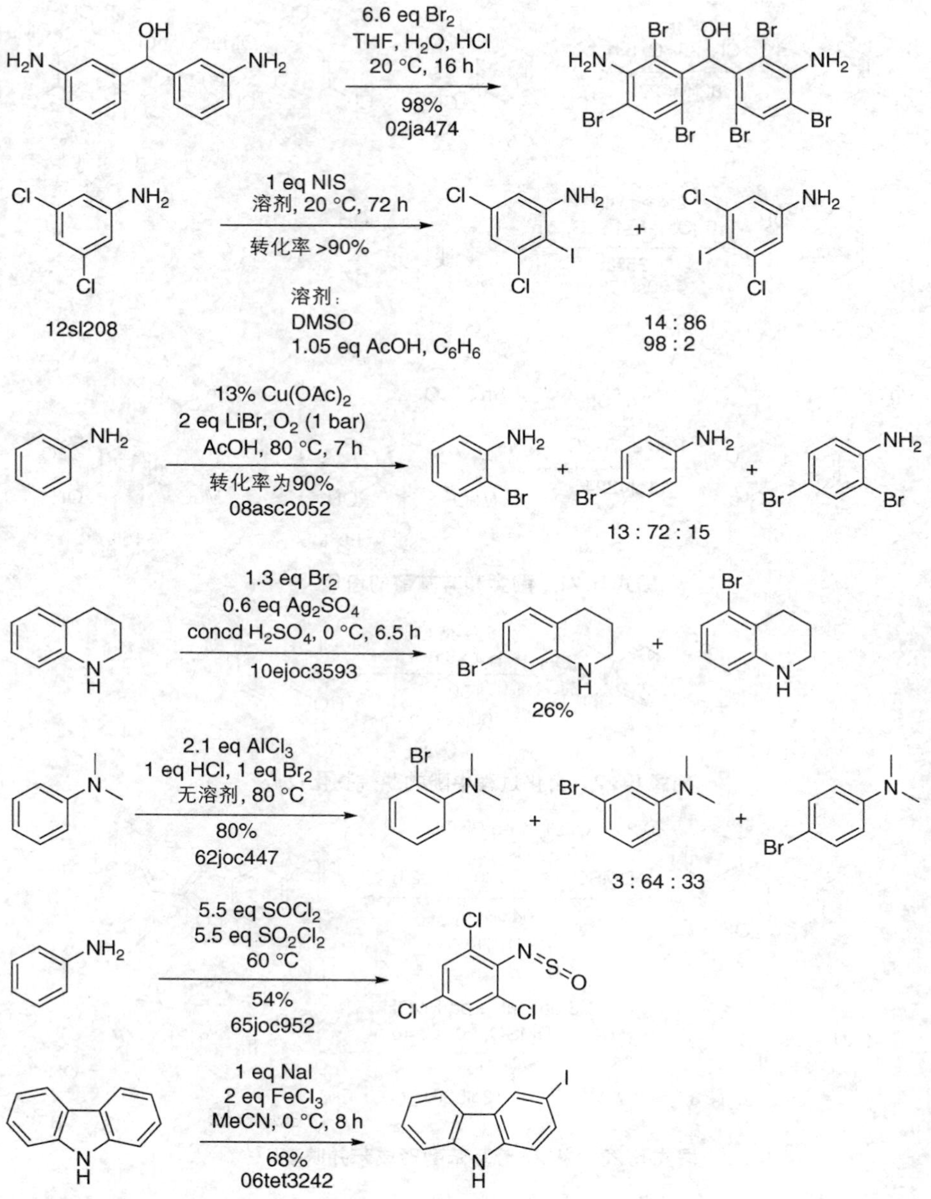

图式5.24 苯胺的卤代反应[91-97]

一些苯胺卤代的典型副反应包括烷基的卤代或脱氢,以及氧化成醌(图式5.25)。事实上,氧化苯胺是生产苯醌的一种旧技术工艺[98]。在乙酸里用双氧水氧化苯胺可得到亚硝基苯和硝基苯[99]。

N-酰化苯胺在大多数反应条件下在邻位/对位卤代。与未酰化的苯胺一样,N-卤代容易产生[103],但该产物本身也是卤化试剂且容易被还原。酰苯胺

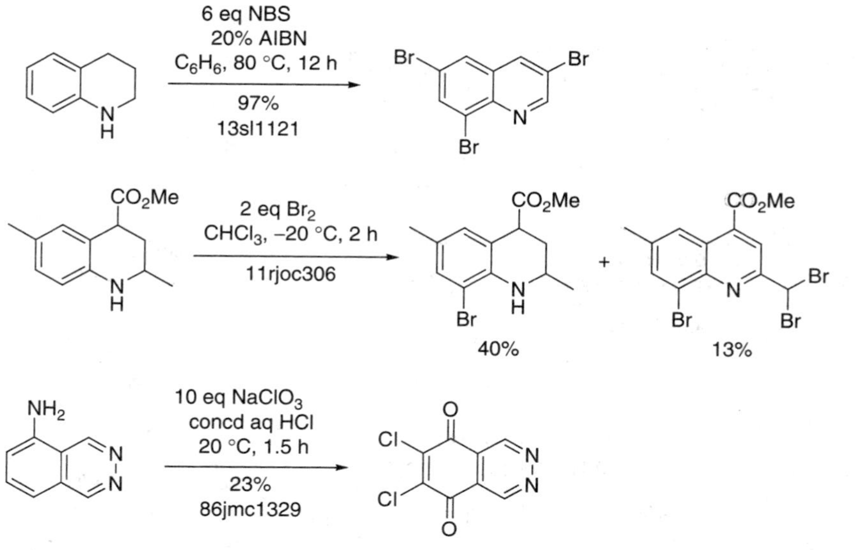

5 芳烃的亲电卤代反应

图式 5.25 苯胺的脱氢和氧化[100-102]

卤代的另外一种竞争过程是环化形成苯并恶唑（图式 5.26）。该环化过程对于硫代酰胺（环化成苯并噻唑）或 *N*-芳基脒（环化成苯并咪唑）是主要反应，因为噻唑和咪唑容易形成非常稳定的芳香杂芳烃（芳香性超过噁唑）。

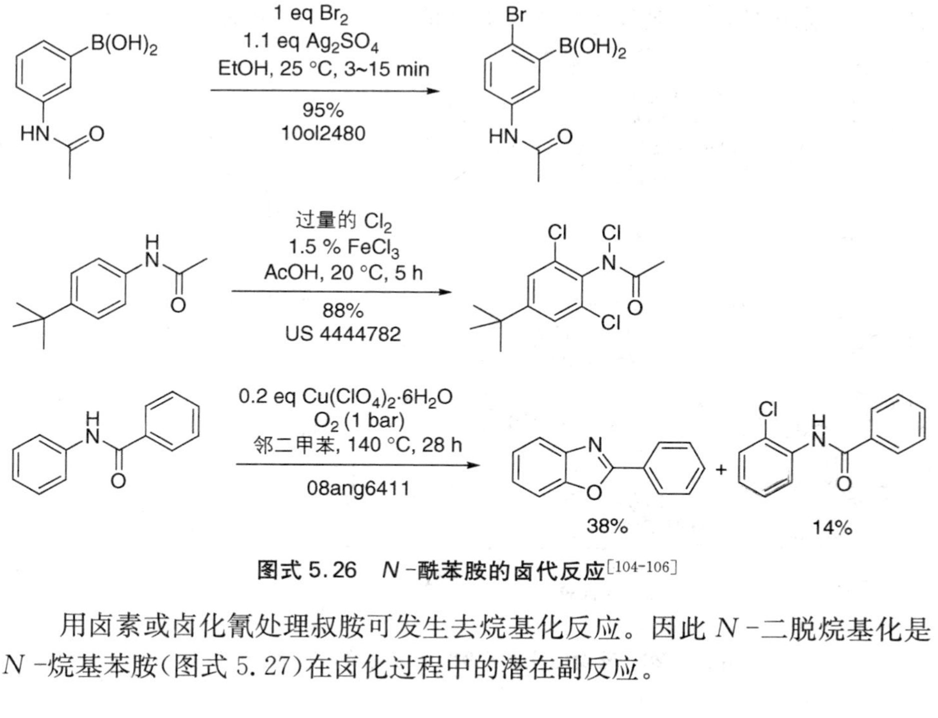

图式 5.26 *N*-酰苯胺的卤代反应[104-106]

用卤素或卤化氰处理叔胺可发生去烷基化反应。因此 *N*-二脱烷基化是 *N*-烷基苯胺（图式 5.27）在卤化过程中的潜在副反应。

125

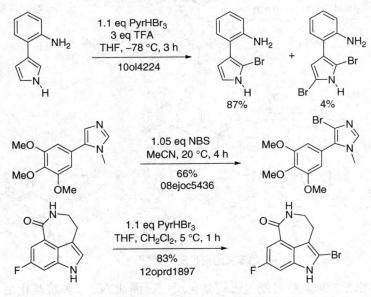

图式 5.27 苯胺在溴代过程中的 N-去烷基化反应[107-109]

5.7.3 唑类

大多数唑类可以被纯净地卤化,但反应条件需要一些优化。吡咯和吲哚具有高度亲核性,卤代速度要比大多数其他富电子芳烃更快(图式 5.28)。

图式 5.28 唑类卤代的实例[110-116](更多实例:[117])

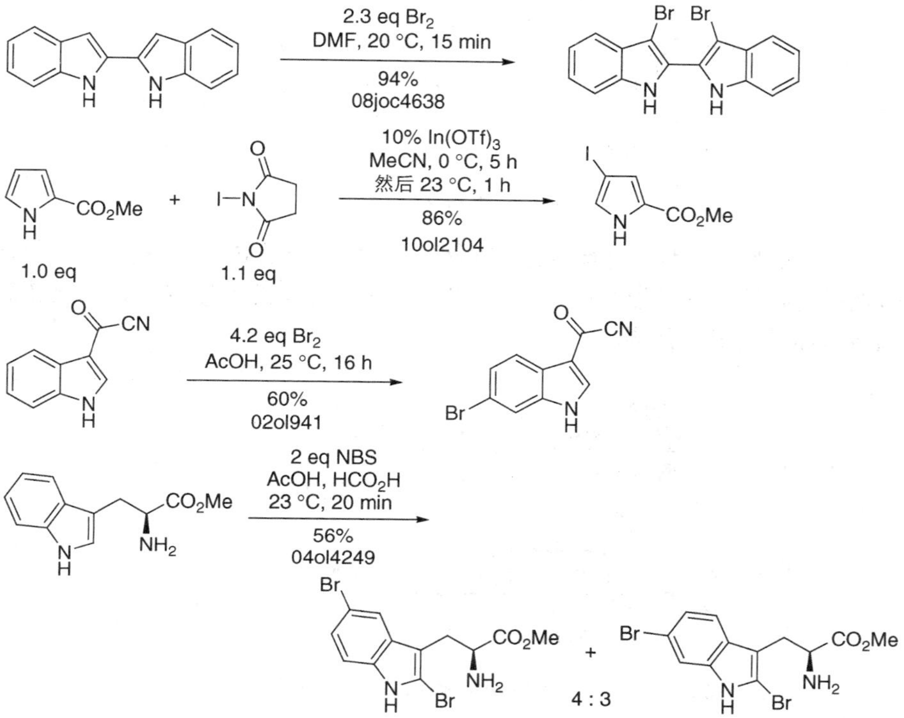

续图式 5.28

如同富电子芳烃一样，由于形成自由基中间体或其他氧化反应产生的副产物而使富电子唑类卤代遭受困扰。吡咯和吲哚的卤代过程中偶尔会观察到羟基化或内酰胺的形成(图式 5.29)。

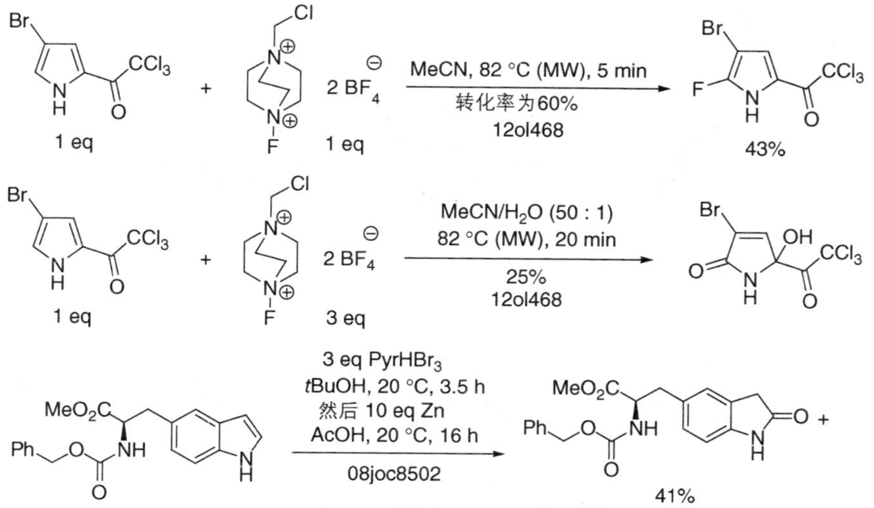

图式 5.29 在唑类卤代过程中形成内酰胺[118,119]

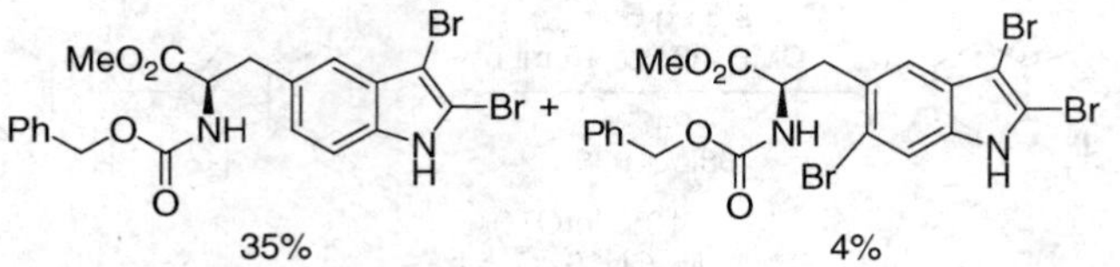

续图式 5.29

唑类区域选择性卤代有时比较困难。一种可能的选择是先进行完全卤代,然后进行选择性还原脱卤(图式 5.30)。这种脱卤的区域选择性高度依赖于底物,而且也很难预测。

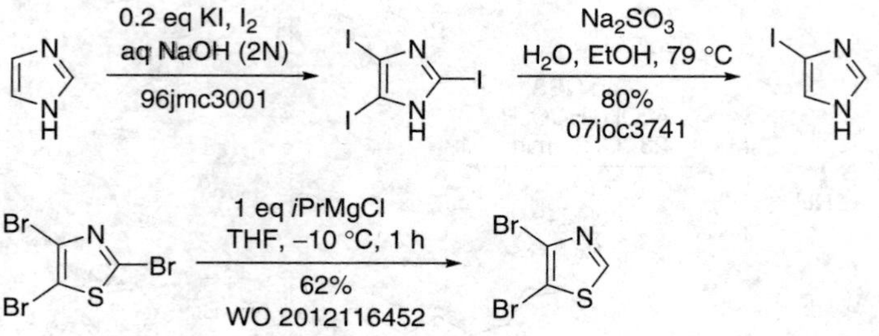

图式 5.30 多卤代咪唑和噻唑进行脱卤的实例[120-122]

带有离去基团的杂原子取代的芳烃(芳基叠氮化物、N-芳基羟胺、N-卤代-N-芳基酰胺及N-芳基肼等)可以重排得到二取代的芳烃[123]。有时唑类和其他杂芳烃也会发生这样的重排(图式 5.31)。

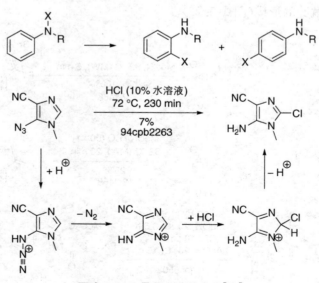

图式 5.31 叠氮咪唑的卤代[124]

侧链卤代是唑类卤代过程中进一步观察到的副反应,它们甚至在没有自由基引发剂的情况下也能发生(图式 5.32)。吲哚在 2 位和 3 位卤化是不稳定的,有时会在形成过程中三聚成苯环(图式 5.33)。

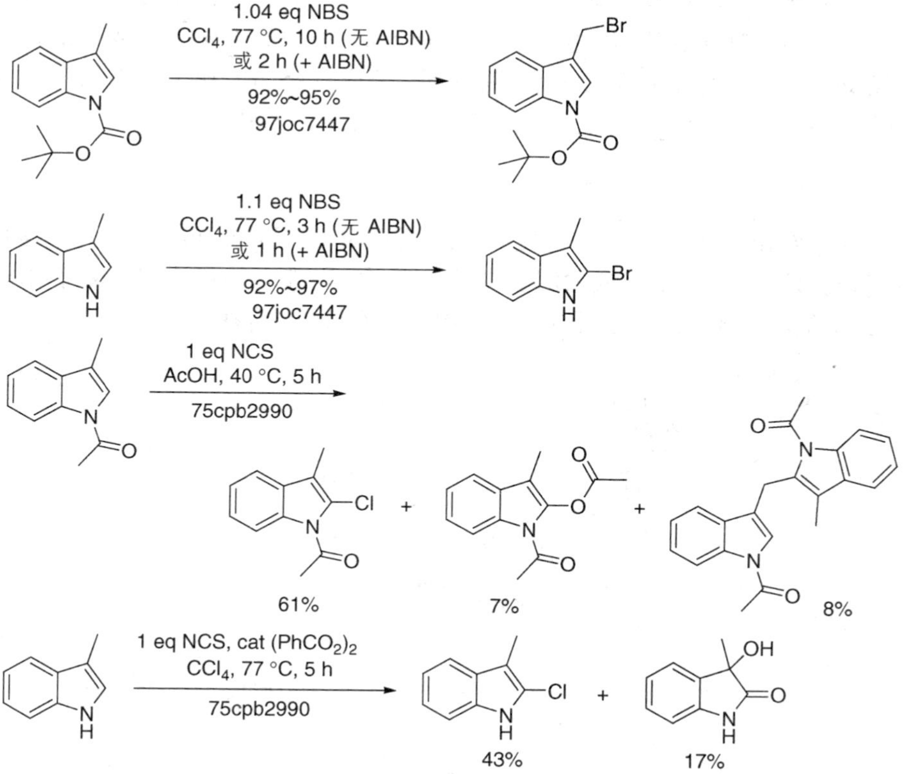

图式 5.32 烷基吲哚在卤代过程中的副反应[125,126]

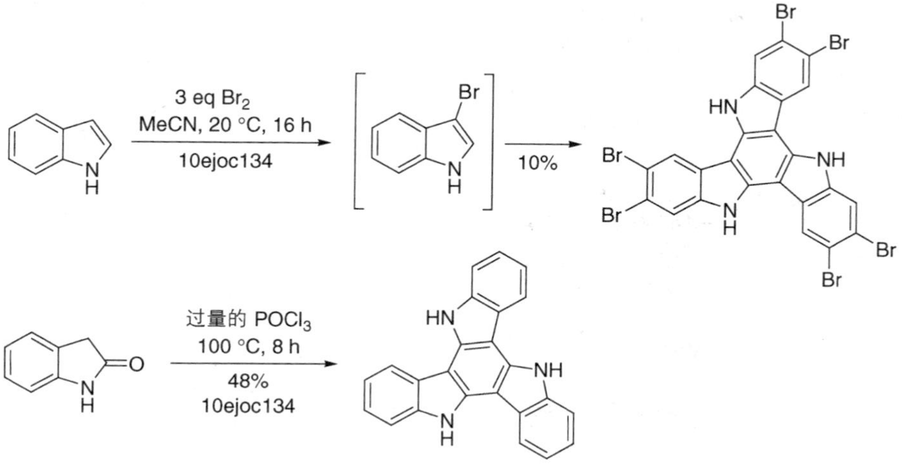

图式 5.33 2-和 3-卤代吲哚的三聚[127]

5.8 敏感的官能团

卤素和大多数卤化剂是强氧化剂，因此卤代反应常常伴随着脱氢和自由基低聚。由于许多官能团都能与卤代试剂反应，故选择性卤代需要仔细控制反应条件。避免过量的卤代试剂、调节 pH 值或者添加捕获试剂以消耗过量的卤代试剂（胺或硫醚）有时足以实现高选择性的芳香族卤代。

5.8.1 烯烃

卤素对烯烃的加成是一个快反应，只有高反应活性的芳烃的卤代反应速度快于烯烃。只有少许这类反应的例子被报道，且收率很低。碘代反应是例外情况：一般的烯烃不与碘发生反应，因此芳香碘化可能在烯烃的存在下进行（图式 5.34）。

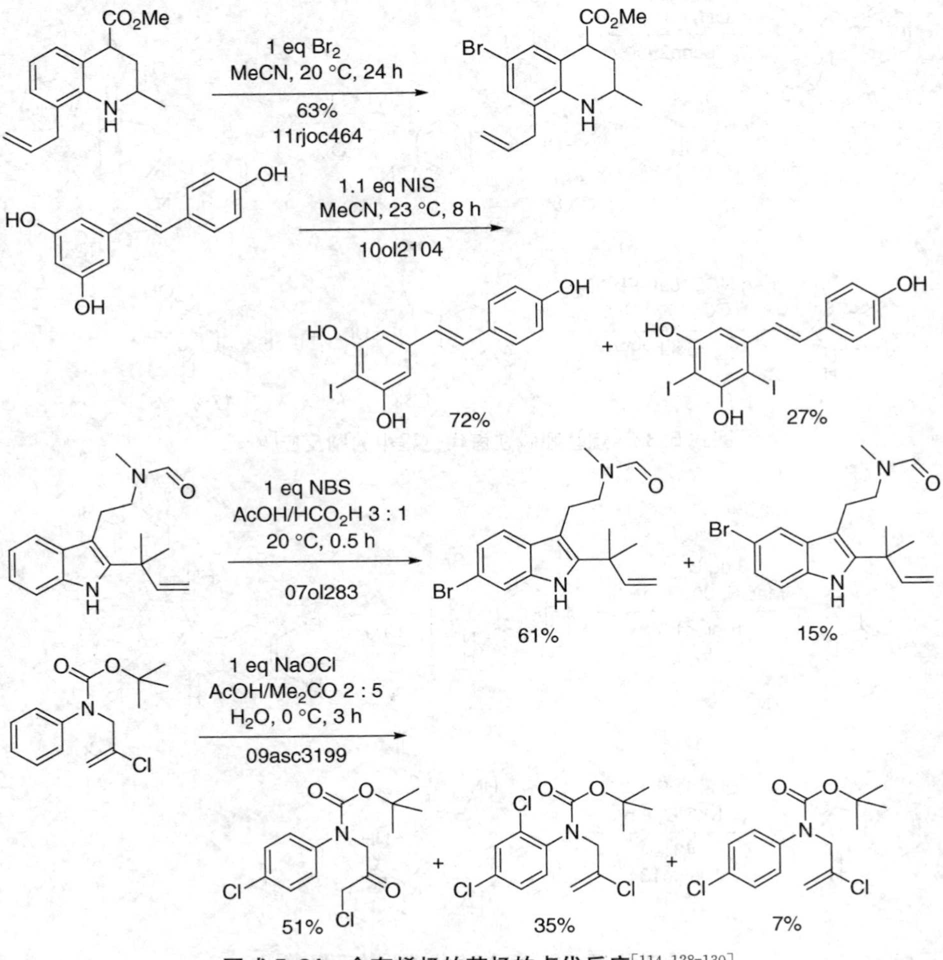

图式 5.34 含有烯烃的芳烃的卤代反应[114,128-130]

在卤代过程中通过 β 消除产生的化合物能够根据使用卤代试剂的不同进一步转化成 1,2-二卤代物或卤代醇(图式 5.35、图式 5.6)。

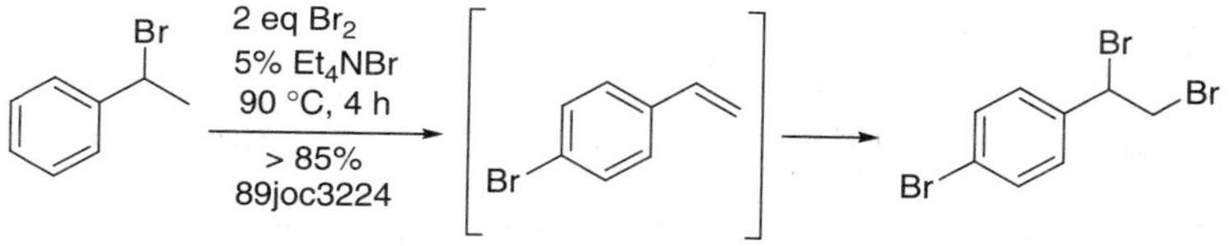

图式 5.35 苄基溴的溴代反应[131]

5.8.2 胺

用卤素处理叔胺通常形成亚胺离子,它们经水解得到醛或酮和去烷基的胺[132,133]。当胺在氧化条件下发生去烷基化时,被解离的烷基能够被转化成活泼的烷基化或酰化试剂。这些新形成的活性中间体可能进一步对富电子芳烃烷基化或酰化(图式 5.36)。

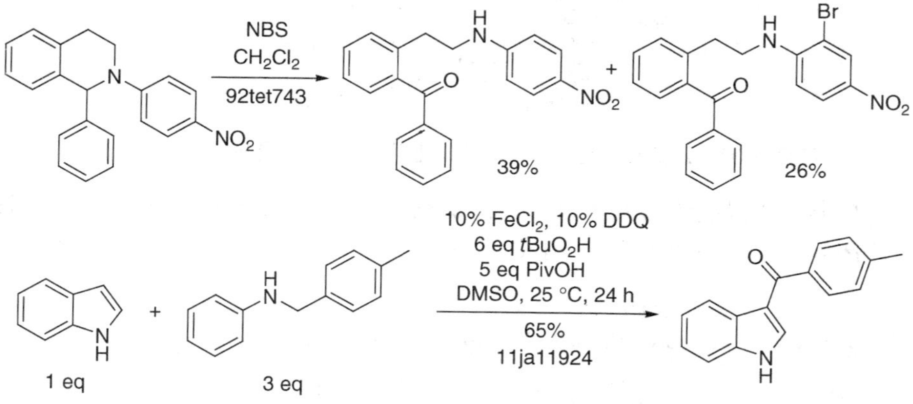

图式 5.36 胺的氧化裂解[134,135]

5.8.3 醚

富电子苄醚对氧化剂是敏感的,其在卤代反应过程中能够转化成醛或酮。这些反应通过自由基或者攫氢进行(图式 5.37)。

5.8.4 硫醇和硫醚

大多数硫醇是强还原剂并且能被卤素氧化成对称的二硫醚。过量的卤素可导致硫醇的硫原子卤代。硫基卤化物和二硫醚都可以与富电子芳烃反应生成芳基硫醚,在被卤代试剂或路易斯酸活化时尤其如此。

在水的存在下用卤代试剂处理硫醚时,它们通常会转化成为亚砜而后生成砜。通过使用乙酸作为溶剂或者避免质子溶剂,有时能够避免这一反应。

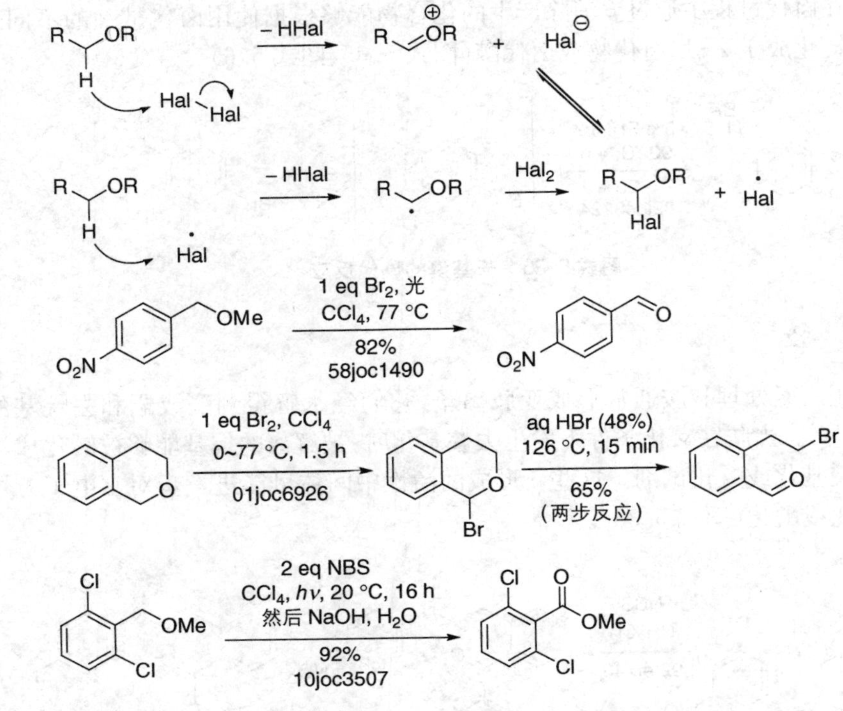

图式 5.37 用卤代试剂处理醚时的氧化反应[136-138]（更多实例：[139]）

硫醚和亚砜能够与大多数卤化试剂反应，产生的亲电中间体可以与各种官能团发生不可预知的反应（图式 5.38）。

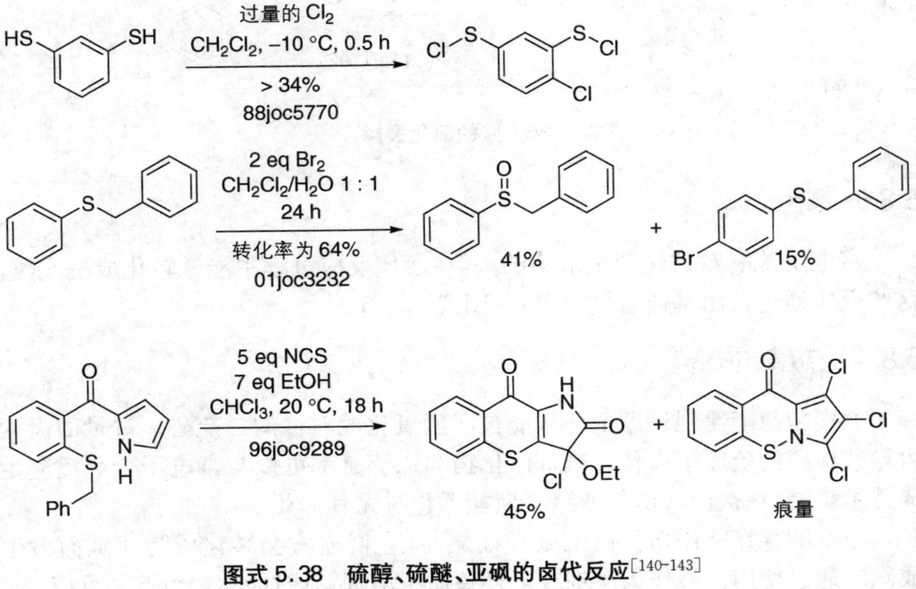

图式 5.38 硫醇、硫醚、亚砜的卤代反应[140-143]

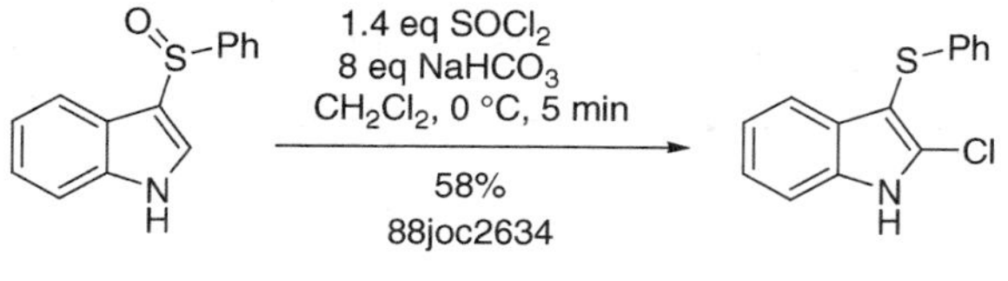

5.8.5 醛、酮和其他的 C—H 酸性化合物

醛衍生的缩醛、缩醛胺、肟、腙和水合物是强的氢给体并且能够被卤代试剂氧化。因此，醛的氧化有时可以通过从反应体系中除去水、醇和胺进行阻止。尽管如此，醛的氧化仍是卤代反应中一个常见的副反应（图式 5.39）。

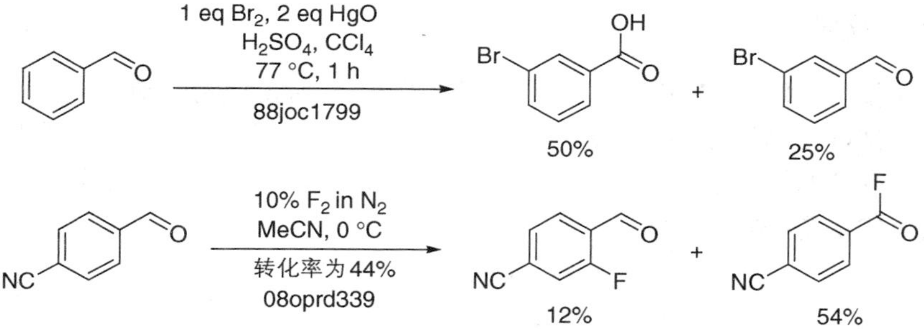

图式 5.39 苯甲醛的卤代反应[144,145]

烷基酮类和其他含有酸性 C—H 的化合物很容易在碳上卤代，在强酸性 C—H 的化合物中这些反应是可逆的。富电子的芳烃，例如酚或酰化苯胺，卤代速度通常比酮更快（图式 5.40）。

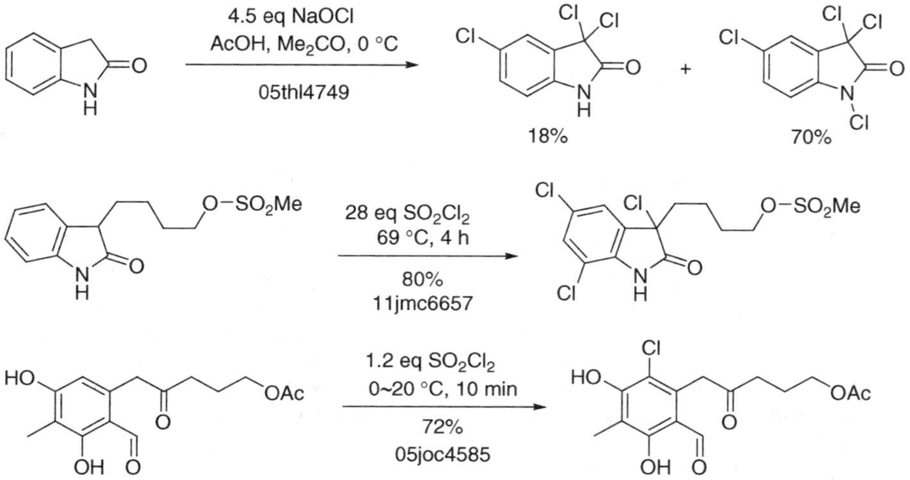

图式 5.40 酸性 C—H 的酮和酰胺的卤代反应[146-150]

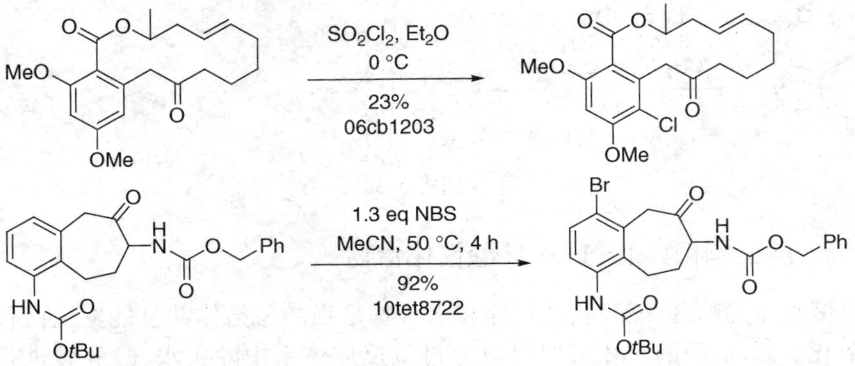

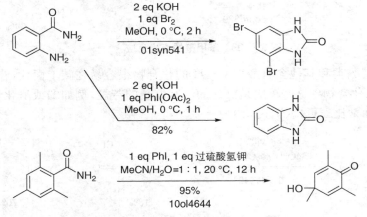

续图式 5.40

5.8.6 酰胺

酰胺容易发生 N-卤代。对于氨衍生的酰胺,所得产物经碱处理脱卤化氢发生重排,得到异氰酸酯(霍夫曼重排)。使用某些氧化剂,霍夫曼重排甚至可以在没有碱的情况下发生。在水存在下,过量的氧化剂可以将胺中间体进一步氧化成苯醌(图式 5.41)。

图式 5.41 苯甲酰胺的卤代和氧化反应[18,151]

参考文献

1. Turner, D.E., O'Malley, R.F., Sardella, D.J., Barinelli, L.S., and Kaul, P. (1994) The reaction of iodine monochloride with polycyclic aromatic compounds: polar and electron transfer pathways. *J. Org. Chem.*, **59**, 7335–7340.

2. Rozen, S. and Zamir, D. (1990) A novel aromatic iodination method using F_2. *J. Org. Chem.*, **55**, 3552–3555.

3. Geib, S., Martens, S.C., Zschieschang, U., Lombeck, F., Wadepohl, H., Klauk, H., and Gade, L.H. (2012) 1,3,6,8-Tetraazapyrenes: synthesis, solid-state structures, and properties as redox-active materials. *J. Org. Chem.*, **77**, 6107–6116.

4. Yamazaki, T., Matoba, K., Imoto, S., and Terashima, M. (1976) Studies on O-alkylated imides. II. Some reactions

of O-ethyl succinimide and O-ethyl 4,4-dimethylglutarimide. *Chem. Pharm. Bull.*, **24**, 3011–3018.

5. Singh, A.P., Lee, K.M., Kim, K., Jun, T., and Churchill, D.G. (2012) Metal-free intermolecular C_{fur}–N_{succ} bond coupling of highly substituted 3-furancarbaldehydes and their use in *meso*-substituted BODIPY synthesis. *Eur. J. Org. Chem.*, **2012**, 931–939.

6. Wu, J., Vetter, W., Gribble, G.W., Schneekloth, J.S., Blank, D.H., and Görls, H. (2002) Structure and synthesis of the natural heptachloro-1′-methyl-1,2′-bipyrrole (Q1). *Angew. Chem. Int. Ed.*, **41**, 1740–1743.

7. Bélanger, P.C., Atkinson, J.G., Rooney, C.S., Britcher, S.F., and Remy, D.C. (1983) Synthesis of 2-, 3-, and 9-substituted 11-oxo-11H-pyrrolo [2,1-b][3]benzazepines. *J. Org. Chem.*, **48**, 3234–3241.

8. Snégaroff, K., L'Helgoual'ch, J.-M., Bentabed-Ababsa, G., Nguyen, T.T., Chevallier, F., Yonehara, M., Uchiyama, M., Derdour, A., and Mongin, F. (2009) Deprotonative metalation of functionalized aromatics using mixed lithium–cadmium, lithium–indium, and lithium–zinc species. *Chem. Eur. J.*, **15**, 10280–10290.

9. Wu, W.-B. and Huang, J.-M. (2012) Highly regioselective C–N bond formation through C–H azolation of indoles promoted by iodine in aqueous media. *Org. Lett.*, **14**, 5832–5835.

10. Hebel, D. and Rozen, S. (1991) Utilizing acetyl hypofluorite for chlorination, bromination, and etherification of the pyridine system. *J. Org. Chem.*, **56**, 6298–6301.

11. Kiselyov, A.S. and Strekowski, L. (1994) Carboxamidation of pyridines by the system of elemental fluorine-carbonitrile-water: a useful alternative to the Chichibabin amination. *Synth. Commun.*, **24**, 2387–2392.

12. Umemoto, T. and Tomizawa, G. (1989) Preparation of 2-fluoropyridines via base-induced decomposition of N-fluoropyridinium salts. *J. Org. Chem.*, **54**, 1726–1731.

13. Bhalerao, D.S. and Akamanchi, K.G. (2007) Efficient and novel method for thiocyanation of aromatic and heteroaromatic compounds using bromodimethylsulfonium bromide and ammonium thiocyanate. *Synlett*, 2952–2956.

14. Li, Y.-X., Ji, K.-G., Wang, H.-X., Ali, S., and Liang, Y.-M. (2011) Iodine-induced regioselective C–C and C–N bonds formation of N-protected indoles. *J. Org. Chem.*, **76**, 744–747.

15. Stavber, S., Košir, I., and Zupan, M. (1997) Reactions of alcohols with cesium fluoroxysulfate. *J. Org. Chem.*, **62**, 4916–4920.

16. LeTourneau, M.E. and Peet, N.P. (1987) Functionalized pyrazoles from indazol-4-ols. *J. Org. Chem.*, **52**, 4384–4387.

17. Narender, N., Srinivasu, P., Kulkarni, S.J., and Raghavan, K.V. (2002) Regioselective oxyiodination of aromatic compounds using potassium iodide and oxone®. *Synth. Commun.*, **32**, 2319–2324.

18. Zagulyaeva, A.A., Banek, C.T., Yusubov, M.S., and Zhdankin, V.V. (2010) Hofmann rearrangement of carboxamides mediated by hypervalent iodine species generated in situ from iodobenzene and oxone: reaction scope and limitations. *Org. Lett.*, **12**, 4644–4647.

19. Pradal, A., Toullec, P.Y., and Michelet, V. (2011) Gold-catalyzed oxidative acyloxylation of arenes. *Org. Lett.*, **13**, 6086–6089.

20. Liu, A.-H., He, L.-N., Hua, F., Yang, Z.-Z., Huang, C.-B., Yu, B., and Lia, B. (2011) *In situ* acidic carbon dioxide/ethanol system for selective oxybromination of aromatic ethers catalyzed by copper chloride. *Adv. Synth. Catal.*, **353**, 3187–3195.

21. Kikuchi, D., Sakaguchi, S., and Ishii, Y. (1998) An alternative method for the selective bromination of alkylbenzenes using $NaBrO_3$/$NaHSO_3$ reagent. *J. Org. Chem.*, **63**, 6023–6026.

22. Duan, S., Turk, J., Speigle, J., Corbin, J., Masnovi, J., and Baker, R.J. (2000) Halogenations of anthracenes and dibenz[a,c]anthracene with

N-bromosuccinimide and N-chlorosuccinimide. *J. Org. Chem.*, **65**, 3005–3009.

23. Tutar, A., Berkil, K., Hark, R.R., and Balci, M. (2008) Bromination of 5-methoxyindane: synthesis of new benzoindenone derivatives and ready access to 7H-benzo[c]fluoren-7-one skeleton. *Synth. Commun.*, **38**, 1333–1345.

24. Cakmak, O., Erenler, R., Tutar, A., and Celik, N. (2006) Synthesis of new anthracene derivatives. *J. Org. Chem.*, **71**, 1795–1801.

25. Barluenga, J., Alvarez-Gutiérrez, J.M., Ballesteros, A., and González, J.M. (2007) Direct *ortho* iodination of β- and γ-aryl alkylamine derivatives. *Angew. Chem. Int. Ed.*, **46**, 1281–1283.

26. Mei, T.-S., Wang, D.-H., and Yu, J.-Q. (2010) Expedient drug synthesis and diversification via *ortho*-C–H iodination using recyclable PdI_2 as the precatalyst. *Org. Lett.*, **12**, 3140–3143.

27. Chan, K.S.L., Wasa, M., Wang, X., and Yu, J.-Q. (2011) Palladium(II)-catalyzed selective monofluorination of benzoic acids using a practical auxiliary: a weak-coordination approach. *Angew. Chem. Int. Ed.*, **50**, 9081–9084.

28. Dai, H.-X., Stepan, A.F., Plummer, M.S., Zhang, Y.-H., and Yu, J.-Q. (2011) Divergent C–H functionalizations directed by sulfonamide pharmacophores: late-stage diversification as a tool for drug discovery. *J. Am. Chem. Soc.*, **133**, 7222–7228.

29. Schröder, N., Wencel-Delord, J., and Glorius, F. (2012) High-yielding, versatile, and practical [Rh(III)Cp*]-catalyzed *ortho* bromination and iodination of arenes. *J. Am. Chem. Soc.*, **134**, 8298–8301.

30. Zhao, X., Dimitrijević, E., and Dong, V.M. (2009) Palladium-catalyzed C–H bond functionalization with arylsulfonyl chlorides. *J. Am. Chem. Soc.*, **131**, 3466–3467.

31. Chen, X., Hao, X.-S., Goodhue, C.E., and Yu, J.Q. (2006) Cu(II)-catalyzed functionalizations of aryl C–H bonds using O_2 as an oxidant. *J. Am. Chem. Soc.*, **128**, 6790–6791.

32. Kalyani, D., Dick, A.R., Anani, W.Q., and Sanford, M.S. (2006) Scope and selectivity in palladium-catalyzed directed C–H bond halogenation reactions. *Tetrahedron*, **62**, 11483–11498.

33. Arnold, P.L., Sanford, M.S., and Pearson, S.M. (2009) Chelating N-heterocyclic carbene alkoxide as a supporting ligand for $Pd^{II/IV}$ C–H bond functionalization catalysis. *J. Am. Chem. Soc.*, **131**, 13912–13913.

34. Hull, K.L., Anani, W.Q., and Sanford, M.S. (2006) Palladium-catalyzed fluorination of carbon–hydrogen bonds. *J. Am. Chem. Soc.*, **128**, 7134–7135.

35. Song, B., Zheng, X., Mo, J., and Xua, B. (2010) Palladium-catalyzed monoselective halogenation of C–H bonds: efficient access to halogenated arylpyrimidines using calcium halides. *Adv. Synth. Catal.*, **352**, 329–335.

36. Wan, X., Ma, Z., Li, B., Zhang, K., Cao, S., Zhang, S., and Shi, Z. (2006) Highly selective C–H functionalization/halogenation of acetanilide. *J. Am. Chem. Soc.*, **128**, 7416–7417.

37. Powers, D.C., Benitez, D., Tkatchouk, E., Goddard, W.A., and Ritter, T. (2010) Bimetallic reductive elimination from dinuclear Pd(III) complexes. *J. Am. Chem. Soc.*, **132**, 14092–14103.

38. Powers, D.C., Xiao, D.Y., Geibel, M.A.L., and Ritter, T. (2010) On the mechanism of palladium-catalyzed aromatic C–H oxidation. *J. Am. Chem. Soc.*, **132**, 14530–14536.

39. Stowers, K.J. and Sanford, M.S. (2009) Mechanistic comparison between Pd-catalyzed ligand-directed C–H chlorination and C–H acetoxylation. *Org. Lett.*, **11**, 4584–4587.

40. Li, J.-J., Mei, T.-S., and Yu, J.-Q. (2008) Synthesis of indolines and tetrahydroisoquinolines from arylethylamines by Pd^{II}-catalyzed C–H activation reactions. *Angew. Chem. Int. Ed.*, **47**, 6452–6455.

41. Xia, J.-B. and You, S.-L. (2009) Synthesis of 3-haloindolizines by copper(II) halide mediated direct functionalization of indolizines. *Org. Lett.*, **11**, 1187–1190.

42. Gu, R., Hameurlaine, A., and Dehaen, W. (2007) Facile one-pot synthesis of 6-monosubstituted and 6,12-disubstituted 5,11-dihydroindolo[3,2-b]carbazoles and preparation of various functionalized derivatives. *J. Org. Chem.*, **72**, 7207–7213.
43. Zweig, A., Fischer, R.G., and Lancaster, J.E. (1980) New method for selective monofluorination of aromatics using silver difluoride. *J. Org. Chem.*, **45**, 3597–3603.
44. Syvret, R.G., Butt, K.M., Nguyen, T.P., Bulleck, V.L., and Rieth, R.D. (2002) Novel process for generating useful electrophiles from common anions using Selectfluor® fluorination agent. *J. Org. Chem.*, **67**, 4487–4493.
45. Visser, G.W.M., Bakker, C.N.M., Halteren, B.W.V., Herscheid, J.D.M., Brinkman, G.A., and Hoekstra, A. (1986) Fluorination and fluorodemercuration of aromatic compounds with acetyl hypofluorite. *J. Org. Chem.*, **51**, 1886–1889.
46. Lin, R., Ding, S., Shi, Z., and Jiao, N. (2011) An efficient difluorohydroxylation of indoles using selectfluor as a fluorinating reagent. *Org. Lett.*, **13**, 4498–4501.
47. Tian, T., Zhong, W.-H., Meng, S., Meng, X.-B., and Li, Z.-J. (2013) Hypervalent iodine mediated *para*-selective fluorination of anilides. *J. Org. Chem.*, **78**, 728–732.
48. Itoh, N., Sakamoto, T., Miyazawa, E., and Kikugawa, Y. (2002) Introduction of a hydroxy group at the para position and N-iodophenylation of N-arylamides using phenyliodine(III) bis(trifluoroacetate). *J. Org. Chem.*, **67**, 7424–7428.
49. Zajc, B. and Zupan, M. (1990) Fluorination with xenon difluoride. 37. Room-temperature rearrangement of aryl-substituted ketones to difluoro-substituted ethers. *J. Org. Chem.*, **55**, 1099–1102.
50. Nemoto, H., Nishiyama, T., and Akai, S. (2011) Nucleophilic deoxyfluorination of catechols. *Org. Lett.*, **13**, 2714–2717.
51. Jian, H. and Tour, J.M. (2005) Preparative fluorous mixture synthesis of diazonium-functionalized oligo(phenylene vinylene)s. *J. Org. Chem.*, **70**, 3396–3424.
52. Dewkar, G.K., Narina, S.V., and Sudalai, A. (2003) $NaIO_4$-mediated selective oxidative halogenation of alkenes and aromatics using alkali metal halides. *Org. Lett.*, **5**, 4501–4504.
53. Rajesh, K., Somasundaram, M., Saiganesh, R., and Balasubramanian, K.K. (2007) Bromination of deactivated aromatics: a simple and efficient method. *J. Org. Chem.*, **72**, 5867–5869.
54. Guilarte, V., Castroviejo, M.P., García-García, P., Fernández-Rodríguez, M.A., and Sanz, R. (2011) Approaches to the synthesis of 2,3-dihaloanilines. Useful precursors of 4-functionalized-1H-indoles. *J. Org. Chem.*, **76**, 3416–3437.
55. Popov, I., Do, H.-Q., and Daugulis, O. (2009) In situ generation and trapping of aryllithium and arylpotassium species by halogen, sulfur, and carbon electrophiles. *J. Org. Chem.*, **74**, 8309–8313.
56. Toomey, J.E. (1993) Chlorination process. US Patent 5247093.
57. McElvain, S.M. and Goese, M.A. (1943) The halogenation of pyridine. *J. Am. Chem. Soc.*, **65**, 2227–2233.
58. Bankston, D., Dumas, J., Natero, R., Riedl, B., Monahan, M.-K., and Sibley, R. (2002) A scaleable synthesis of BAY 43-9006: a potent Raf kinase inhibitor for the treatment of cancer. *Org. Process Res. Dev.*, **6**, 777–781.
59. Newkome, G.R., Theriot, K.J., Majestic, V.K., Spruell, P.A., and Baker, G.R. (1990) Functionalization of 2-methyl- and 2,7-dimethyl1-1,8-naphthyridine. *J. Org. Chem.*, **55**, 2838–2842.
60. Deadman, J.J., Jones, E.D., Le, G.T., Rhodes, D.I., Thien-Thong, N., and van de Graff, N.A. (2010) Compounds having antiviral properties. WO Patent 2010000032.
61. Giblin, G.M.P., Billinton, A., Briggs, M., Brown, A.J., Chessell, I.P., Clayton, N.M., Eatherton, A.J., Goldsmith, P., Haslam, C., Johnson, M.R., Mitchell, W.L., Naylor, A., Perboni, A., Slingsby,

B.P., and Wilson, A.W. (2009) Discovery of 1-[4-(3-chlorophenylamino)-1-methyl-1H-pyrrolo[3,2-c]pyridin-7-yl]-1-morpholin-4-ylmethanone (GSK554418A), a brain penetrant 5-azaindole CB_2 agonist for the treatment of chronic pain. *J. Med. Chem.*, **52**, 5785–5788.

62. Said, A. (1979) Process for the production of pure white 2-chloronicotinic acid. US Patent 4144238.
63. Kretzschmar, E. and Dietz, G. (1987) Untersuchung eines Nebenproduktes bei der Herstellung von 2-Chlornicotinsäure. *Pharmazie*, **42**, 858.
64. Hynes, J., Doubleday, W.W., Dyckman, A.J., Godfrey, J.D., Grosso, J.A., Kiau, S., and Leftheris, K. (2004) N-Amination of pyrrole and indole heterocycles with monochloramine (NH_2Cl). *J. Org. Chem.*, **69**, 1368–1371.
65. Pearson, D.E., Stamper, W.E., and Suthers, B.R. (1963) The swamping catalyst effect. V. The halogenation of aromatic acid derivatives. *J. Org. Chem.*, **28**, 3147–3149.
66. da Ribeiro, R.S., Esteves, P.M., and de Mattos, M.C.S. (2011) Superelectrophilic iodination of deactivated arenes with triiodoisocyanuric acid. *Synthesis*, 739–744.
67. Groweiss, A. (2000) Use of sodium bromate for aromatic bromination: research and development. *Org. Process Res. Dev.*, **4**, 30–33.
68. Rozen, S., Brand, M., and Lidor, R. (1988) Aromatic bromination using BrF with no Friedel–Crafts catalyst. *J. Org. Chem.*, **53**, 5545–5547.
69. Mei, T.-S., Giri, R., Maugel, N., and Yu, J.-Q. (2008) Pd^{II}-catalyzed monoselective *ortho* halogenation of C–H bonds assisted by counter cations: a complementary method to directed *ortho* lithiation. *Angew. Chem. Int. Ed.*, **47**, 5215–5219.
70. Cahiez, G., Moyeux, A., Gager, O., and Poizat, M. (2013) Copper-catalyzed decarboxylation of aromatic carboxylic acids: en route to milder reaction conditions. *Adv. Synth. Catal.*, **355**, 790–796.
71. Dickstein, J.S., Curto, J.M., Gutierrez, O., Mulrooney, C.A., and Kozlowski, M.C. (2013) Mild aromatic palladium-catalyzed protodecarboxylation: kinetic assessment of the decarboxylative palladation and the protodepalladation steps. *J. Org. Chem.*, **78**, 4744–4761.
72. Arroniz, C., Ironmonger, A., Rassias, G., and Larrosa, I. (2013) Direct *ortho*-arylation of *ortho*-substituted benzoic acids: overriding Pd-catalyzed protodecarboxylation. *Org. Lett.*, **15**, 910–913.
73. Goossen, L.J., Rodríguez, N., Melzer, B., Linder, C., Deng, G., and Levy, L.M. (2007) Biaryl synthesis via Pd-catalyzed decarboxylative coupling of aromatic carboxylates with aryl halides. *J. Am. Chem. Soc.*, **129**, 4824–4833.
74. Milburn, R.R., Thiel, O.R., Achmatowicz, M., Wang, X., Zigterman, J., Bernard, C., Colyer, J.T., DiVirgilio, E., Crockett, R., Correll, T.L., Nagapudi, K., Ranganathan, K., Hedley, S.J., Allgeier, A., and Larsen, R.D. (2011) Development of a practical synthesis of a pyrazolopyridinone-based p38 MAP kinase inhibitor. *Org. Process Res. Dev.*, **15**, 31–43.
75. Janz, K. and Kaila, N. (2009) Bromodecarboxylation of quinoline salicylic acids: increasing the diversity of accessible substituted quinolines. *J. Org. Chem.*, **74**, 8874–8877.
76. Stevens, K.L., Jung, D.K., Alberti, M.J., Badiang, J.G., Peckham, G.E., Veal, J.M., Cheung, M., Harris, P.A., Chamberlain, S.D., and Peel, M.R. (2005) Pyrazolo[1,5-a]pyridines as p38 kinase inhibitors. *Org. Lett.*, **7**, 4753–4756.
77. Kulbitski, K., Nisnevich, G., and Gandelman, M. (2011) Metal-free efficient, general and facile iododecarboxylation method with biodegradable co-products. *Adv. Synth. Catal.*, **353**, 1438–1442.
78. Szumigala, R.H., Devine, P.N., Gauthier, D.R., and Volante, R.P. (2004) Facile synthesis of 2-bromo-3-fluorobenzonitrile: an application and study of the halodeboronation of aryl boronic acids. *J. Org. Chem.*, **69**, 566–569.

79. Du, B., Jiang, X., and Sun, P. (2013) Palladium-catalyzed highly selective *ortho*-halogenation (I, Br, Cl) of arylnitriles via sp^2 C–H bond activation using cyano as directing group. *J. Org. Chem.*, **78**, 2786–2791.
80. Mattern, D.L. (1984) Direct aromatic periodination. *J. Org. Chem.*, **49**, 3051–3053.
81. Mattern, D.L. and Chen, X. (1991) Direct polyiodination of benzenesulfonic acid. *J. Org. Chem.*, **56**, 5903–5907.
82. García, R., Riera, J., Carilla, J., Juliá, L., Molins, E., and Miravitlles, C. (1992) Extensive chlorination of methylnaphthalenes, Friedel–Crafts alkylation of pentachlorobenzene by heptachloro(chloromethyl)naphthalenes, and related results. *J. Org. Chem.*, **57**, 5712–5719.
83. Watson, W.D. (1982) Formation of nonaromatic products in the chlorination of simple substituted aromatic ethers. *J. Org. Chem.*, **47**, 5270–5276.
84. Barrero, A.F., Alvarez-Manzaneda, E.J., Chahboun, R., and González Díaz, C. (2000) New routes toward drimanes and *nor*-drimanes from (−)-sclareol. *Synlett*, 1561–1564.
85. Yu, M. and Snider, B.B. (2011) Syntheses of chloroisosulochrin and isosulochrin and biomimetic elaboration to maldoxin, maldoxone, dihydromaldoxin, and dechlorodihydromaldoxin. *Org. Lett.*, **13**, 4224–4227.
86. Sarkanen, K.V. and Dence, C.W. (1960) Reactions of *p*-hydroxybenzyl alcohol derivatives and their methyl ethers with molecular chlorine. *J. Org. Chem.*, **25**, 715–720.
87. Zhao, J., Wang, Y., He, Y., Liu, L., and Zhu, Q. (2012) Cu-catalyzed oxidative C(sp^2)–H cycloetherification of *o*-arylphenols for the preparation of dibenzofurans. *Org. Lett.*, **14**, 1078–1081.
88. Zhao, J., Zhang, Q., Liu, L., He, Y., Li, J., Li, J., and Zhu, Q. (2012) CuI-mediated sequential iodination/cycloetherification of *o*-arylphenols: synthesis of 2- or 4-iododibenzofurans and mechanistic studies. *Org. Lett.*, **14**, 5362–5365.
89. Berrier, C., Jacquesy, J.C., and Renoux, A. (1990) Bromination of anilines in superacids. *Bull. Soc. Chim. Fr.*, 93–97.
90. Bieler, N., Ellinger, S., Furrer, M., Ladnak, V., Müller, C., Schmid, L., and Zur Täschler, C. (2012) Process for the selective *meta*-chlorination of alkylanilines. WO Patent 2012022460.
91. Tomioka, H., Watanabe, T., Hattori, M., Nomura, N., and Hirai, K. (2002) Generation, reactions, and kinetics of sterically congested triplet diphenylcarbenes. Effects of bromine groups. *J. Am. Chem. Soc.*, **124**, 474–482.
92. Shen, H. and Vollhardt, K.P.C. (2012) Remarkable switch in the regiochemistry of the iodination of anilines by *N*-iodosuccinimide: synthesis of 1,2-dichloro-3,4-diiodobenzene. *Synlett*, 208–214.
93. Menini, L., da Cruz Santos, J.C., and Gusevskayaa, E.V. (2008) Copper-catalyzed oxybromination and oxychlorination of primary aromatic amines using LiBr or LiCl and molecular oxygen. *Adv. Synth. Catal.*, **350**, 2052–2058.
94. Kolmakov, K., Belov, V.N., Wurm, C.A., Harke, B., Leutenegger, M., Eggeling, C., and Hell, S.W. (2010) A versatile route to red-emitting carbopyronine dyes for optical microscopy and nanoscopy. *Eur. J. Org. Chem.*, **2010**, 3593–3610.
95. Suthers, B.R., Riggins, P.H., and Pearson, D.E. (1962) The swamping catalyst effect. IV. The halogenation of anilines. *J. Org. Chem.*, **27**, 447–451.
96. Ottmann, G. and Hooks, H. (1965) Chlorination of aromatic *N*-sulfinylamines. *J. Org. Chem.*, **30**, 952–954.
97. Mohanakrishnan, A.K., Prakash, C., and Ramesh, N. (2006) A simple iodination protocol via in situ generated ICl using NaI/FeCl$_3$. *Tetrahedron*, **62**, 3242–3247.
98. Vliet, E.B. (1941) Quinone. *Org. Synth.*, Coll. Vol. **1**, 482–484, reference 2 cited therein.
99. Holmes, R.R. and Bayer, R.P. (1960) A simple method for the direct oxidation of aromatic amines to nitroso

compounds. *J. Am. Chem. Soc.*, **82**, 3454–3456.

100. Akrawi, O.A., Mohammed, H.H., and Langer, P. (2013) Synthesis and Suzuki–Miyaura reactions of 3,6,8-tribromoquinoline: a structural revision. *Synlett*, 1121–1124.

101. Zhuravleva, Y.A., Zimichev, A.V., Zemtsova, M.N., and Klimochkin, Y.N. (2011) Bromination of methyl 2-methyl-6-*R*-1,2,3,4-tetrahydroquinoline-4-carboxylates. *Russ. J. Org. Chem.*, **47**, 306–307.

102. Shaikh, I.A., Johnson, F., and Grollman, A.P. (1986) Streptonigrin. 1. Structure–activity relationships among simple bicyclic analogues. Rate dependence of DNA degradation on quinone reduction potential. *J. Med. Chem.*, **29**, 1329–1340.

103. Gassman, P.G. and Campbell, G.A. (1971) The mechanism of the chlorination of anilines and related aromatic amines. The involvement of nitrenium ions. *J. Am. Chem. Soc.*, **93**, 2567–2569.

104. Al-Zoubi, R.M. and Hall, D.G. (2010) Mild silver(I)-mediated regioselective iodination and bromination of arylboronic acids. *Org. Lett.*, **12**, 2480–2483.

105. DeMarinis, R.M. and Hieble, P. (1984) 2-(4-*Tert*-butyl-2,6-dichlorophenylimino)imidazolidine and use as an antihypertension agent. US Patent 4444782.

106. Ueda, S. and Nagasawa, H. (2008) Synthesis of 2-arylbenzoxazoles by copper-catalyzed intramolecular oxidative C–O coupling of benzanilides. *Angew. Chem. Int. Ed.*, **47**, 6411–6413.

107. Doyle, M.P., Van Lente, M.A., Mowat, R., and Fobare, W.F. (1980) Alkyl nitrite–metal halide deamination reactions. 7. Synthetic coupling of electrophilic bromination with substitutive deamination for selective synthesis of multiply brominated aromatic compounds from arylamines. *J. Org. Chem.*, **45**, 2570–2575.

108. Mekh, M.A., Ozeryanskii, V.A., and Pozharskii, A.F. (2006) 5,6-Bis(dimethylamino)acenaphthylene as an activated alkene and "proton sponge" in halogenation reactions. *Tetrahedron*, **62**, 12288–12296.

109. Potturi, H.K., Gurung, R.K., and Hou, Y. (2012) Nitromethane with IBX/TBAF as a nitrosating agent: synthesis of nitrosamines from secondary or tertiary amines under mild conditions. *J. Org. Chem.*, **77**, 626–631.

110. Gu, Z. and Zakarian, A. (2010) Total synthesis of rhazinilam: axial-to-point chirality transfer in an enantiospecific Pd-catalyzed transannular cyclization. *Org. Lett.*, **12**, 4224–4227.

111. Bellina, F., Cauteruccio, S., Di Fiore, A., and Rossi, R. (2008) Regioselective synthesis of 4,5-diaryl-1-methyl-1*H*-imidazoles including highly cytotoxic derivatives by Pd-catalyzed direct C-5 arylation of 1-methyl-1*H*-imidazole with aryl bromides. *Eur. J. Org. Chem.*, **2008**, 5436–5445.

112. Gillmore, A.T., Badland, M., Crook, C.L., Castro, N.M., Critcher, D.J., Fussell, S.J., Jones, K.J., Jones, M.C., Kougoulos, E., Mathew, J.S., McMillan, L., Pearce, J.E., Rawlinson, F.L., Sherlock, A.E., and Walton, R. (2012) Multikilogram scale-up of a reductive alkylation route to a novel PARP inhibitor. *Org. Process Res. Dev.*, **16**, 1897–1904.

113. Qi, T., Qiu, W., Liu, Y., Zhang, H., Gao, X., Liu, Y., Lu, K., Du, C., Yu, G., and Zhu, D. (2008) Synthesis, structures, and properties of disubstituted heteroacenes on one side containing both pyrrole and thiophene rings. *J. Org. Chem.*, **73**, 4638–4643.

114. Zhou, C.-Y., Li, J., Peddibhotla, S., and Romo, D. (2010) Mild arming and derivatization of natural products via an In(OTf)$_3$-catalyzed arene iodination. *Org. Lett.*, **12**, 2104–2107.

115. Miyake, F.Y., Yakushijin, K., and Horne, D.A. (2002) Synthesis of marine sponge bisindole alkaloids dihydrohamacanthins. *Org. Lett.*, **4**, 941–943.

116. Miyake, F.Y., Yakushijin, K., and Horne, D.A. (2004) A concise synthesis of spirotryprostatin A. *Org. Lett.*, **6**, 4249–4251.

117. Miyake, F.Y., Yakushijin, K., and Horne, D.A. (2004) Preparation

and synthetic applications of 2-halotryptamines: synthesis of elacomine and isoelacomine. *Org. Lett.*, **6**, 711–713.
118. Troegel, B. and Lindel, T. (2012) Microwave-assisted fluorination of 2-acylpyrroles: synthesis of fluorohymenidin. *Org. Lett.*, **14**, 468–471.
119. Han, X., Civiello, R.L., Fang, H., Wu, D., Gao, Q., Chaturvedula, P.V., Macor, J.E., and Dubowchik, G.M. (2008) Catalytic asymmetric syntheses of tyrosine surrogates. *J. Org. Chem.*, **73**, 8502–8510.
120. Zhang, X., Yao, X.-T., Dalton, J.T., Shams, G., Lei, L., Patil, P.N., Feller, D.R., Hsu, F.-L., George, C., and Miller, D.D. (1996) Medetomidine analogs as α_2-adrenergic ligands. 2. Design, synthesis, and biological activity of conformationally restricted naphthalene derivatives of medetomidine. *J. Med. Chem.*, **39**, 3001–3013.
121. Lovely, C.J., Du, H., Sivappa, R., Bhandari, M.R., He, Y., and Dias, H.V.R. (2007) Preparation and Diels–Alder chemistry of 4-vinylimidazoles. *J. Org. Chem.*, **72**, 3741–3749.
122. Wu, F., Lu, E., and Barden, C. (2012) Antimicrobial/adjuvant compounds and methods. WO Patent 2012116452.
123. Smith, P.A.S. and Brown, B.B. (1951) The reaction of aryl azides with hydrogen halides. *J. Am. Chem. Soc.*, **73**, 2438–2441.
124. Saito, T., Asahi, Y., Nakajima, S., and Fujii, T. (1994) Purines. LXIV. Synthesis of 9-methyl-2-azaadenine 1-oxide, its O-methyl derivative, and 1-substituted 5-azidoimidazole-4-carboxamides. *Chem. Pharm. Bull.*, **42**, 2263–2268.
125. Liu, R., Zhang, P., Gan, T., and Cook, J.M. (1997) Regiospecific bromination of 3-methylindoles with NBS and its application to the concise synthesis of optically active unusual tryptophans present in marine cyclic peptides. *J. Org. Chem.*, **62**, 7447–7456.
126. Hino, T., Nakamura, T., and Nakagawa, M. (1975) Halogenation of 1-substituted skatoles. Preparation of 3-bromomethylindoles. *Chem. Pharm. Bull.*, **23**, 2990–2997.
127. Franceschin, M., Ginnari-Satriani, L., Alvino, A., Ortaggi, G., and Bianco, A. (2010) Study of a convenient method for the preparation of hydrosoluble fluorescent triazatruxene derivatives. *Eur. J. Org. Chem.*, **2010**, 134–141.
128. Zhuravleva, Y.A., Zimichev, A.V., Zemtsova, M.N., and Klimochkin, Y.N. (2011) Unusual bromination of methyl 8-allyl-2-methyl-1,2,3,4-tetrahydroquinoline-4-carboxylate. *Russ. J. Org. Chem.*, **47**, 464–465.
129. Lindel, T., Bräuchle, L., Golz, G., and Böhrer, P. (2007) Total synthesis of flustramine C via dimethylallyl rearrangement. *Org. Lett.*, **9**, 283–286.
130. Pace, V., Martínez, F., Fernández, M., Sinisterra, J.V., and Alcántara, A.R. (2009) Highly efficient synthesis of new α-arylamino-α'-chloropropan-2-ones via oxidative hydrolysis of vinyl chlorides promoted by calcium hypochlorite. *Adv. Synth. Catal.*, **351**, 3199–3206.
131. Dakka, J. and Sasson, Y. (1989) Bromination of α-substituted alkylbenzenes: synthesis of (p-bromophenyl)acetylene. *J. Org. Chem.*, **54**, 3224–3226.
132. Chen, C.-K., Hortmann, A.G., and Marzabadi, M.R. (1988) ClO_2 oxidation of amines: synthetic utility and a biomimetic synthesis of elaeocarpidine. *J. Am. Chem. Soc.*, **110**, 4829–4831.
133. Deno, N.C. and Fruit, R.E. (1968) The oxidative cleavage of amines by aqueous bromine at 25 °C. *J. Am. Chem. Soc.*, **90**, 3502–3506.
134. Hedley, K.A. and Stanforth, S.P. (1992) Ring-opening reactions of N-aryl-1,2,3,4-tetrahydroisoquinoline derivatives. *Tetrahedron*, **48**, 743–750.
135. Wu, W. and Su, W. (2011) Mild and selective Ru-catalyzed formylation and Fe-catalyzed acylation of free (N–H) indoles using anilines as the carbonyl source. *J. Am. Chem. Soc.*, **133**, 11924–11927.
136. Markees, D.G. (1958) Reaction of benzyl methyl ethers with bromine and N-bromosuccinimide. *J. Org. Chem.*, **23**, 1490–1492.
137. Page, P.C.B., Rassias, G.A., Barros, D., Ardakani, A., Buckley, B., Bethell, D., Smith, T.A.D., and Slawin, A.M.Z. (2001) Functionalized iminium salt

138. Mayhoub, A.S., Talukdar, A., and Cushman, M. (2010) An oxidation of benzyl methyl ethers with NBS that selectively affords either aromatic aldehydes or aromatic methyl esters. *J. Org. Chem.*, **75**, 3507–3510.
139. Adimurthy, S. and Patoliya, P.U. (2007) N-Bromosuccinimide: a facile reagent for the oxidation of benzylic alcohols to aldehydes. *Synth. Commun.*, **37**, 1571–1577.
140. Miura, Y. and Ohana, T. (1988) Thioaminyl diradicals: an electron spin resonance spectroscopic study. *J. Org. Chem.*, **53**, 5770–5772.
141. Bravo, A., Dordi, B., Fontana, F., and Minisci, F. (2001) Oxidation of organic sulfides by Br_2 and H_2O_2. Electrophilic and free-radical processes. *J. Org. Chem.*, **66**, 3232–3234.
142. Xia, M., Chen, S., and Bates, D.K. (1996) Reactions of benzyl aryl sulfides with excess active halogen reagents. *J. Org. Chem.*, **61**, 9289–9292.
143. Garcia, J., Ortiz, C., and Greenhouse, R. (1988) The regiospecific Pummerer-like introduction of chlorine atoms into pyrrol-3-yl and indol-3-yl sulfoxides. *J. Org. Chem.*, **53**, 2634–2637.
144. Khan, S.A., Munawar, M.A., and Siddiq, M. (1988) Monobromination of deactivated active rings using bromine, mercuric oxide, and strong acid. *J. Org. Chem.*, **53**, 1799–1800.
145. Chambers, R.D., Sandford, G., Trmcic, J., and Okazoe, T. (2008) Elemental fluorine. Part 21. Direct fluorination of benzaldehyde derivatives. *Org. Process Res. Dev.*, **12**, 339–344.
146. Meketa, M.L., Mahajan, Y.R., and Weinreb, S.M. (2005) An efficacious method for the halogenation of β-dicarbonyl compounds under mildly acidic conditions. *Tetrahedron Lett.*, **46**, 4749–4751.
147. Volk, B., Gacsályi, I., Pallagi, K., Poszávácz, L., Gyönös, I., Szabó, E., Bakó, T., Spedding, M., Simig, G., and Szénási, G. (2011) Optimization of (arylpiperazinylbutyl)oxindoles exhibiting selective 5-HT_7 receptor antagonist activity. *J. Med. Chem.*, **54**, 6657–6669.
148. Wei, W.-G. and Yao, Z.-J. (2005) Synthesis studies toward chloroazaphilone and vinylogous γ-pyridones: two common natural product core structures. *J. Org. Chem.*, **70**, 4585–4590.
149. Proisy, N., Sharp, S.Y., Boxall, K., Connelly, S., Roe, S.M., Prodromou, C., Slawin, A.M.Z., Pearl, L.H., Workman, P., and Moody, C.J. (2006) Inhibition of Hsp90 with synthetic macrolactones: synthesis and structural and biological evaluation of ring and conformational analogs of radicicol. *Chem. Biol.*, **13**, 1203–1215.
150. Roux, L., Charrier, C., Defoin, A., Bisseret, P., and Tarnus, C. (2010) Iodine(III)-mediated ring expansion: an efficient and green pathway in the synthesis of a key precursor for the design of aminopeptidase (APN or CD13) inhibitors. *Tetrahedron*, **66**, 8722–8728.
151. Prakash, O., Batra, H., Kaur, H., Sharma, P.K., Sharma, V., Singh, S.P., and Moriarty, R.M. (2001) Hypervalent iodine oxidative rearrangement of anthranilamides, salicylamides and some β-substituted amides: a new and convenient synthesis of 2-benzimidazolones, 2-benzoxazolones and related compounds. *Synthesis*, 541–543.

6 通过亲电反应形成芳烃 C—N 键

6.1 芳烃的硝化反应

6.1.1 机理

芳烃的亲电硝化反应通常是由硝酸和硫酸混合后产生 NO_2^+ 而实现的，它是大多数古老炸药制备的关键步骤。乙酸酐、三氟乙酸酐或其他非均相酸性催化剂可被用于替代硫酸作为脱水剂[1]。分离出来的硝鎓盐，例如四氟硼酸硝，以及 O_3/NO_2[2] 也是有用的硝化试剂。在离子液体[3]和共晶熔体[4]中进行硝化的实例也有报道。使用 NO^+ 进行亚硝化反应常常也会分离得到硝基芳烃副产物[5]（图式 6.1）。

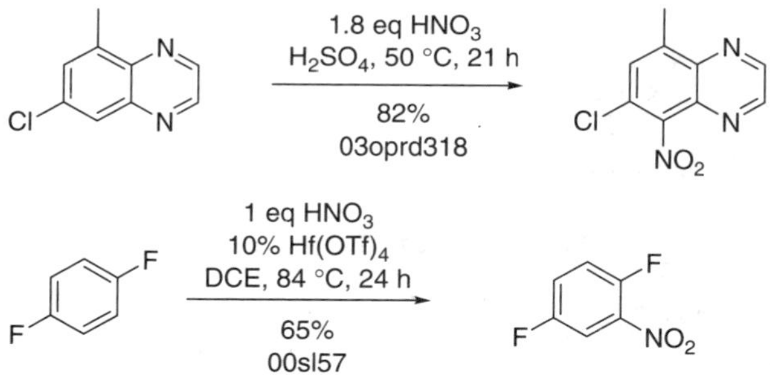

图式 6.1 芳香单硝化的实例[6,7]

硝化反应与其他芳香亲电取代反应的机理是相同的。由于硝基的强吸电子效应，通常认为不发生多次硝化，但也会得到少量多次硝化的副产物（图式 6.2）。多次硝化需要苛刻的反应条件，而且常常因为产品的氧化降解而失败。例如，五硝基苯或六硝基苯不能通过直接硝化制备，而只能通过苯胺的硝化，接着氧化或取代活化的氨基来制备[8]。

亲电硝化试剂不仅可以进攻芳环上未取代的位置，而且还可以取代其他

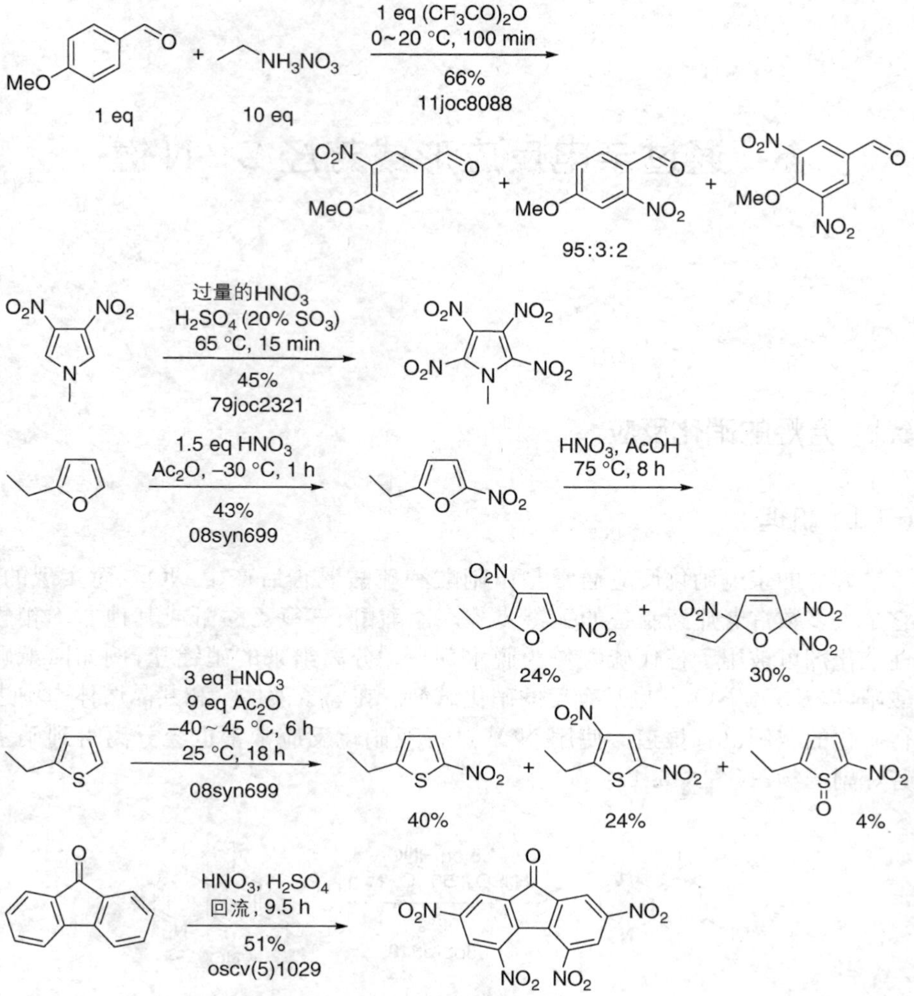

图式 6.2 多次芳香硝化的实例[9-12]

非氢的基团[13]（图式 6.3），它们包括三烷基硅基、二烷基硼、磺酸基、烷基巯基、羧酸、卤化物[14,15]、烷基[16]、甲酰基[17,18]及其他酰基。芳烃硝化很少用于结构复杂的中间体的原因之一是其缺乏选择性。

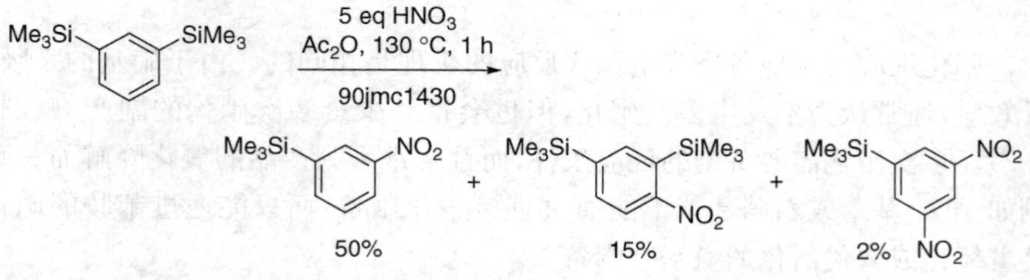

图式 6.3 硝基对各种官能团的本位取代[8,19-25]

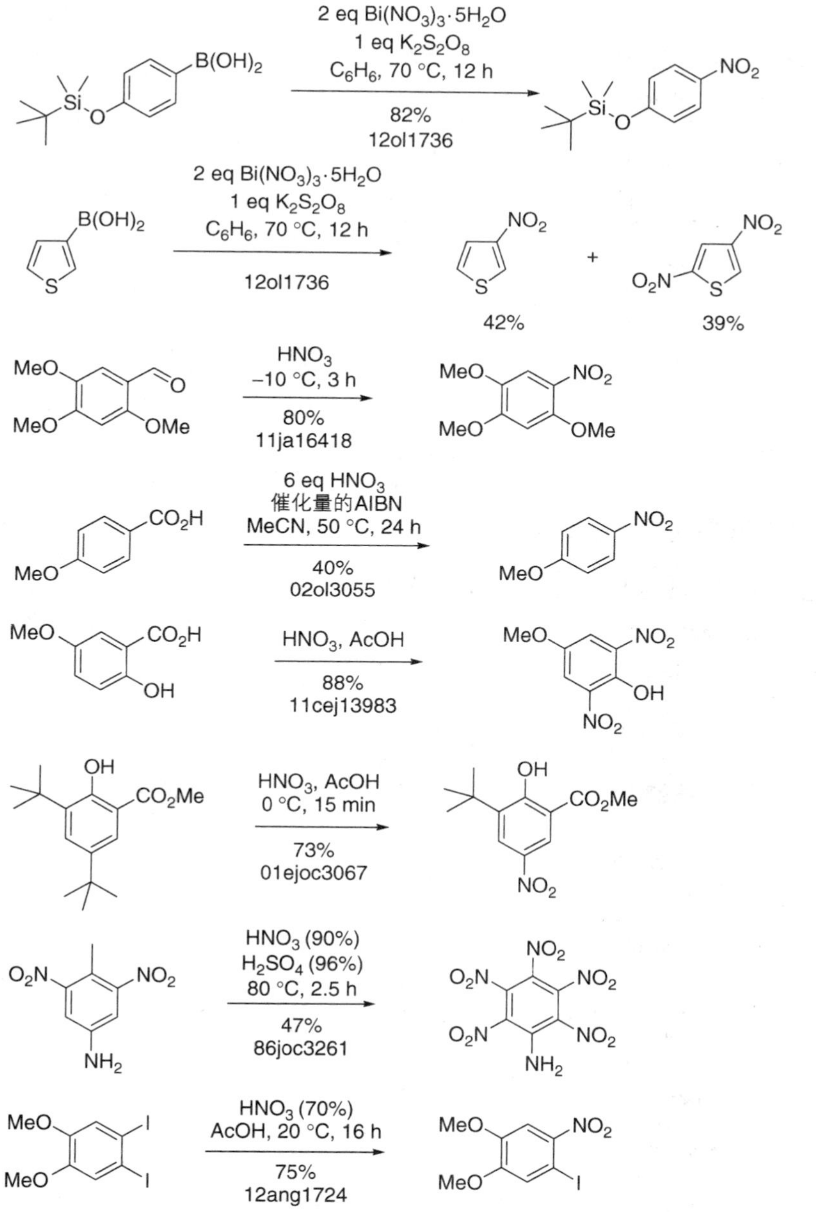

续图式 6.3

在亲电硝化反应条件下，许多官能团可以被氧化。NO_2^+是一种强氧化剂，它能够将富电子的芳烃转化成芳基自由基阳离子。

芳烃也可以用二氧化氮进行硝化[26]。这个反应也通过芳基自由基阳离子进行（图式 6.4），并且能够导致许多副产物，例如侧链硝化产物、联芳基及二芳基甲烷或乙烷。当用硝酸乙酸酐进行硝化时，偶尔也会观察到苄基硝化[27]。

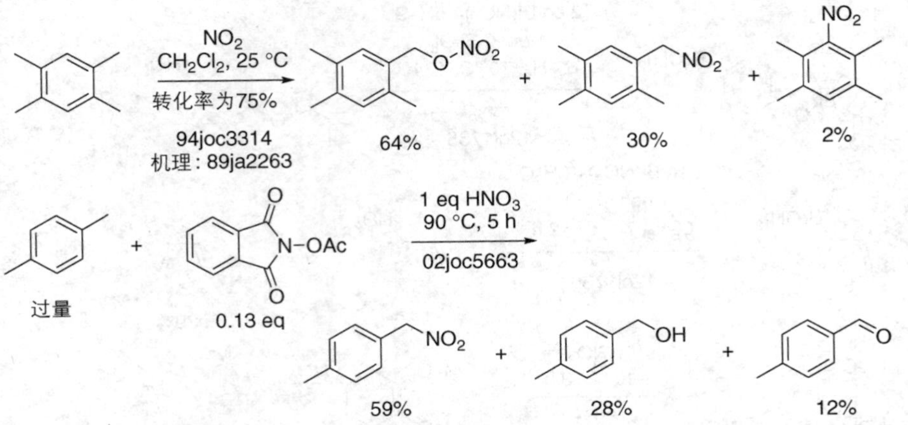

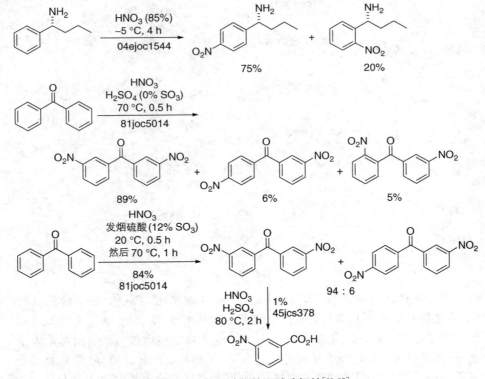

图式 6.4 苄基硝化[26,28,29]

6.1.2 区域选择性

硝化反应的区域选择性通常比傅-克酰化低,而且反应异构体并不总是能够通过重结晶容易除去。有时通过使用发烟硫酸做溶剂可以增强亲电芳香硝化的区域选择性(图式 6.5)。

图式 6.5 芳香硝化的区域选择性[30-32]

因为亚硝化的区域选择性常常比硝化高,并且亚硝基芳烃可以高收率地氧化成硝基芳烃(例如用过酸或过氧化氢),亚硝化/氧化过程可能是对直接硝化的一种有用替代方法[33](图式 6.6)。亚硝化常常导致硝基芳烃作为副产物,甚至主产物(如参考文献[5]中所述),要实现高收率的亚硝化需要仔细优化条件。亚硝酸也可以脱除叔胺的烷基[34]。

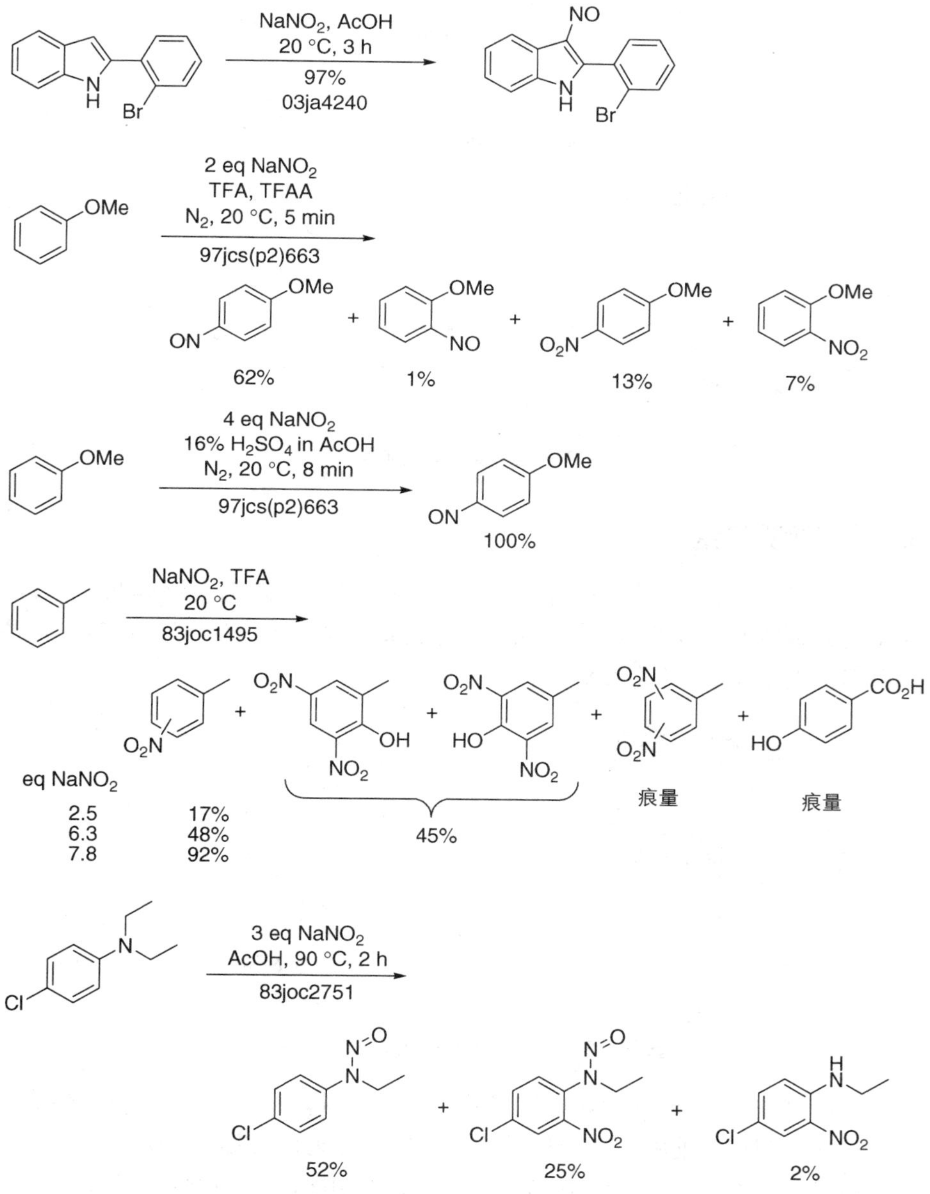

图式 6.6　芳烃与亚硝酸的反应[35-38]

6.1.3 催化

亲电芳香硝化能被提高 NO_2^+ 浓度的酸催化。一些过渡金属催化的反应实例已经被报道,反应可能是通过芳香金属化进行的(图式 6.7)。

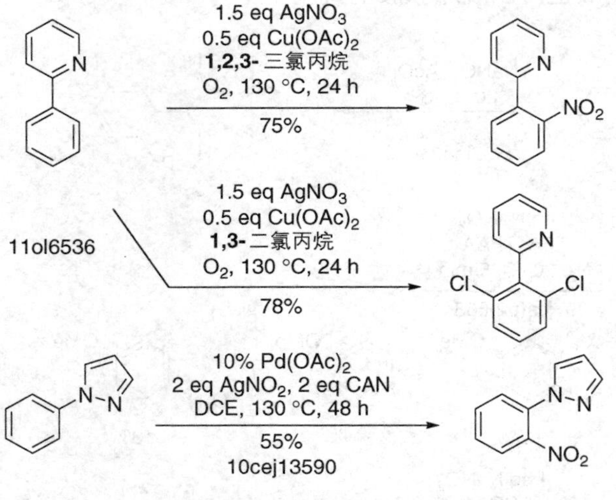

图式 6.7 芳烃的催化硝化[39,40]

6.1.4 缺电子芳烃

缺电子芳烃通常也能被硝化。由于亲电硝化是高度放热的反应,其发生热失控和爆炸十分常见,故小心控制反应温度非常关键。在高温下,芳烃可能发生完全氧化降解(图式 6.8)。

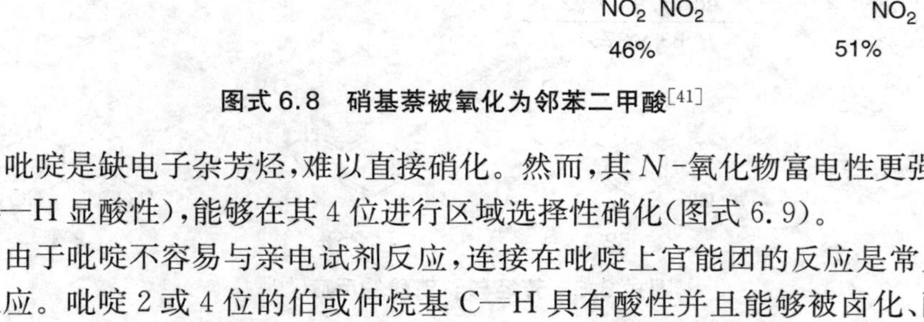

图式 6.8 硝基萘被氧化为邻苯二甲酸[41]

吡啶是缺电子杂芳烃,难以直接硝化。然而,其 N-氧化物富电性更强(而且 C—H 显酸性),能够在其 4 位进行区域选择性硝化(图式 6.9)。

由于吡啶不容易与亲电试剂反应,连接在吡啶上官能团的反应是常见的副反应。吡啶 2 或 4 位的伯或仲烷基 C—H 具有酸性并且能够被卤化、酰化

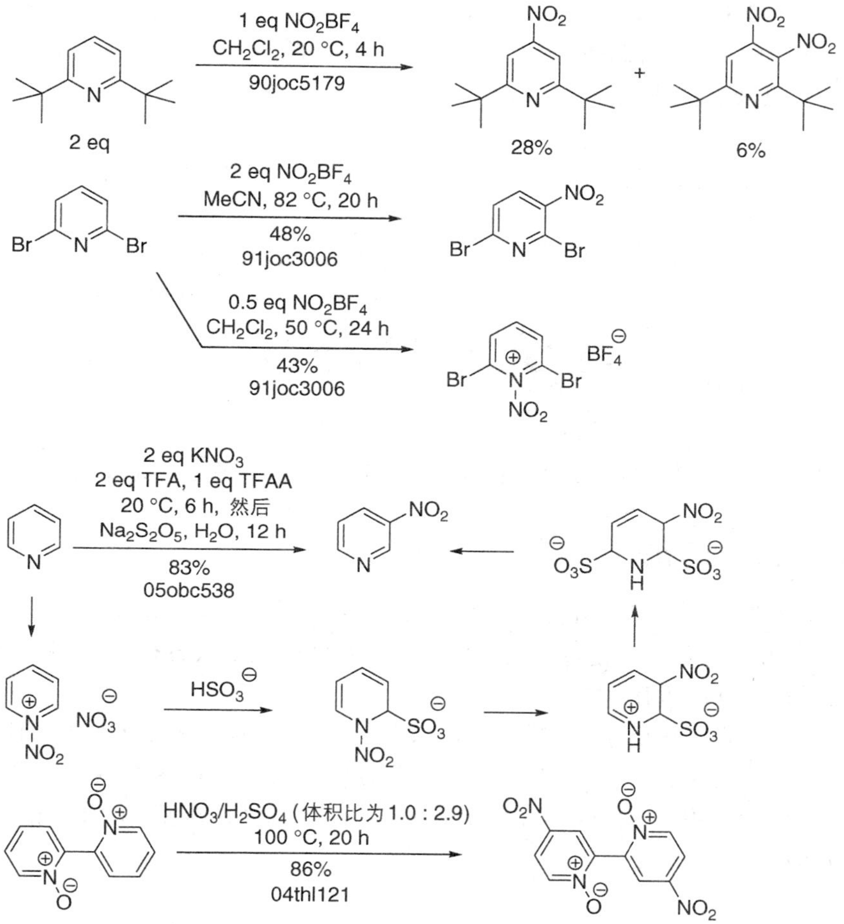

图式 6.9 吡啶和吡啶 N-氧化物的硝化反应[42-45]

或硝化。生成的多卤代烷基或多硝基烷基是很好的离去基团,它们能够被亲核试剂取代(例如参考文献[46]所述)。强反应条件可以将烷基吡啶氧化成吡啶羧酸。如果加热太剧烈,一些杂芳环羧酸能够发生部分或完全脱羧[47](图式 6.10)。

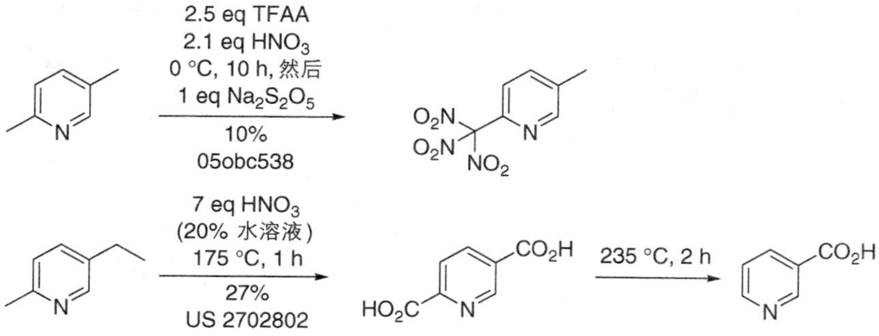

图式 6.10 烷基吡啶与硝酸的反应[44,48]

酰胺 NH 基团氮原子可以被硝化[49]，但该反应很少观察到。酰苯胺通常可在碳上干净只进行硝化（图式6.11）。N-硝基酰胺是强氧化剂，在后处理中容易被弱还原剂还原。

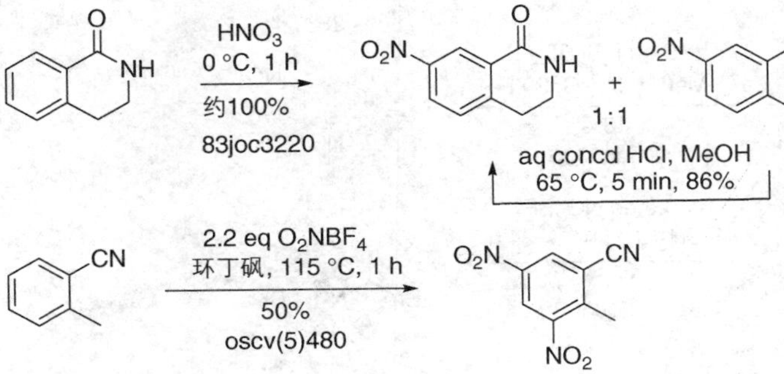

图式 6.11 苯甲酰胺和苯甲腈的硝化[50,51]

6.1.5 富电子芳烃

为了避免许多潜在副反应的发生，活泼芳烃的亲电硝化反应通常在低温下进行，并且仅仅使用稍过量的硝化试剂。副反应包括去芳构化、多硝化和芳烃及其取代基的氧化降解。

亲电芳香硝化反应中，烷基可以被脱氢或氧化为烷基亚硝酸酯、酮或羧酸。事实上，在硝酸中加热烷基芳烃或杂芳烃是将其转化成为羧酸的最廉价和最环保的方法之一。能形成稳定碳正离子的烷基（仲烷基、烯丙基和苄基）特别敏感（图式6.12）。

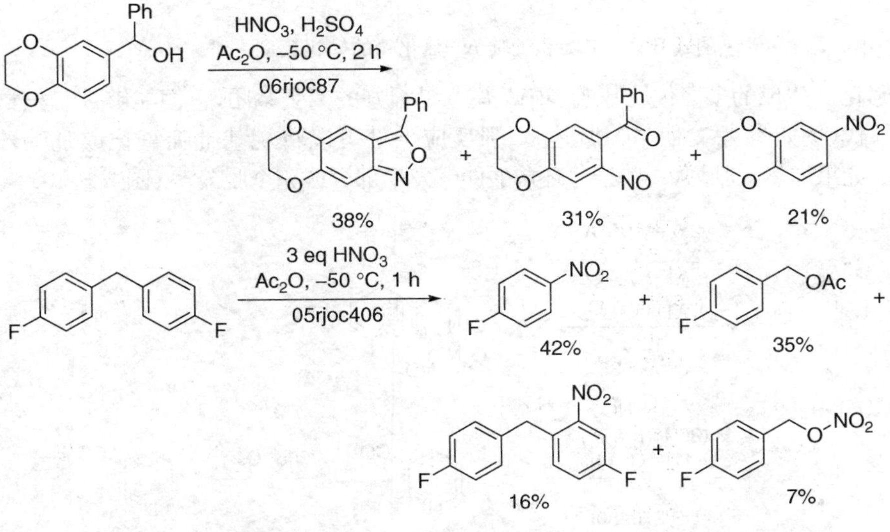

图式 6.12 用硝酸使烷基芳烃去烷基和氧化[52-55]

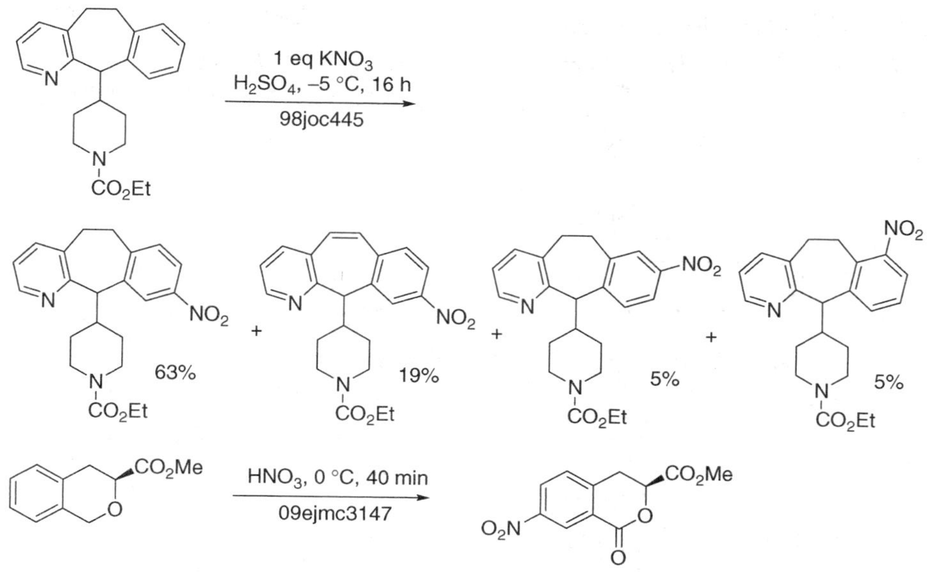

续图式 6.12

在硝酸/乙酸酐中硝化烷基苯时可以得到乙酰氧基苯或酚[56]。机理研究[57]表明这些产物是由乙酰硝酸酐对芳烃的 1,4-加成后消去一分子亚硝酸而产生（图式 6.13）。通过使用 NO_2BF_4 或硝酸/硫酸体系作为硝化剂可以防止酚的形成。硝酸和水的混合物能引起大量的富电子芳烃羟化产物[58]。

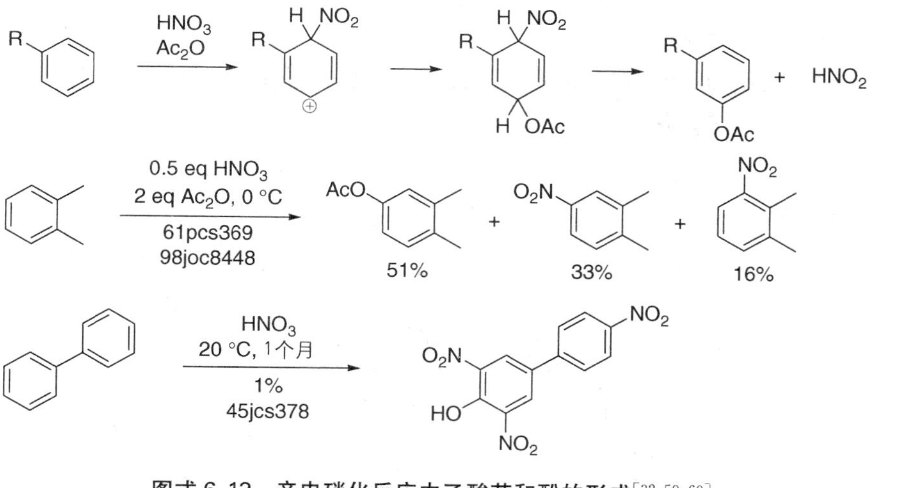

图式 6.13 亲电硝化反应中乙酸苯和酚的形成[32,59,60]

6.1.5.1 苯胺

在强酸存在下苯胺硝化大多数情况发生在间位，假如苯胺没有被完全质子化则会产生邻/对位硝基苯胺（图式 6.14）。苯胺也可以在氮上被硝化（产生有爆炸性的硝胺),但是要求氨基没有被质子化。大多数适用于硝化的强酸都可以避免苯胺的 N-硝化,只有低碱性的苯胺(如多硝基苯胺)会进行 N-硝化。在还原剂(如二氧化氮)的存在下,芳族硝胺可以被还原和脱水形成重氮盐(图式 6.14 中最后一个反应)。

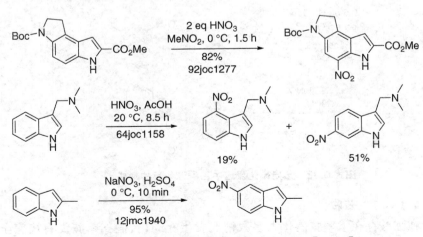

图式 6.14 苯胺的硝化反应[61,62]

6.1.5.2 吲哚

吲哚硝化的结果难以预测。按照精准的取代方式和反应条件,吲哚的所有位置都可能被硝化(图式 6.15)。

图式 6.15 吲哚的硝化反应[63-68](更多实例:[47,69])

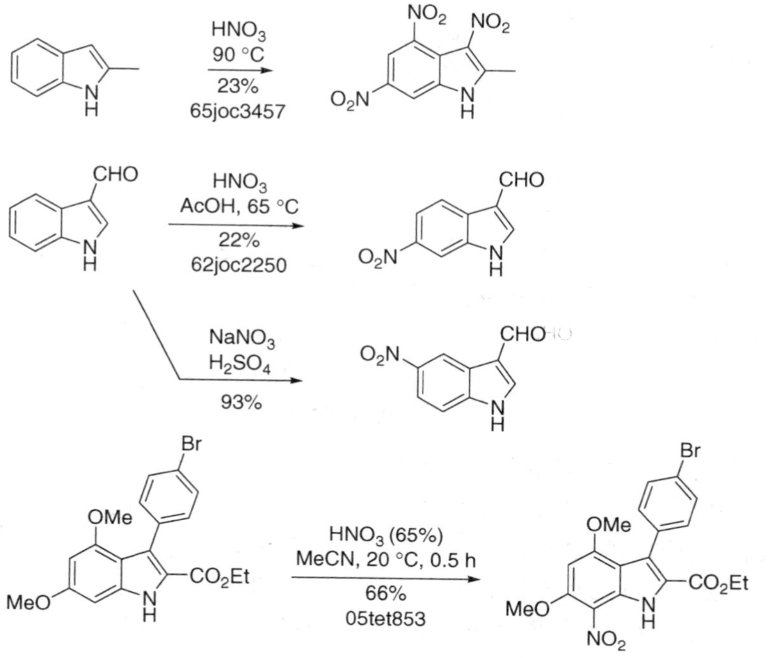

续图式 6.15

因为吲哚容易经单电子转移而氧化,所以在用亲电试剂处理吲哚时,氧化二聚是一个典型的副反应(图式 6.16)。

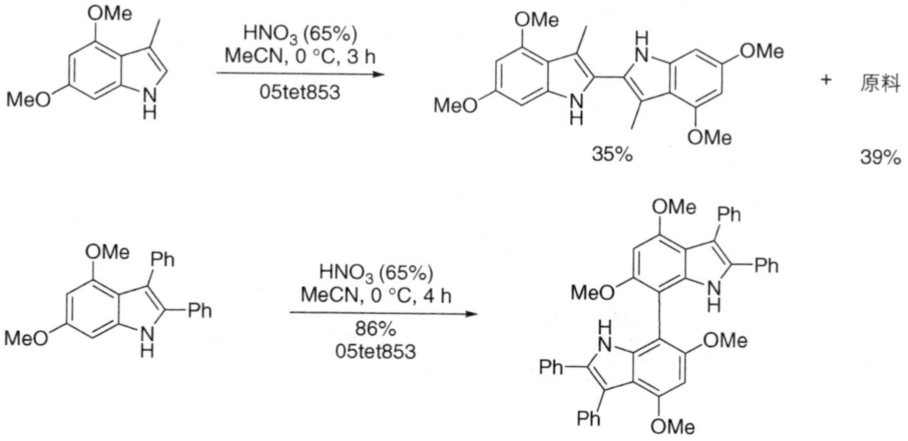

图式 6.16 硝酸引发的吲哚氧化二聚[68]

6.1.5.3 酚

酚与硝酸的硝化反应通常是邻/对位选择性。重要的副反应包括多重硝化和氧化成醌(图式 6.17)。

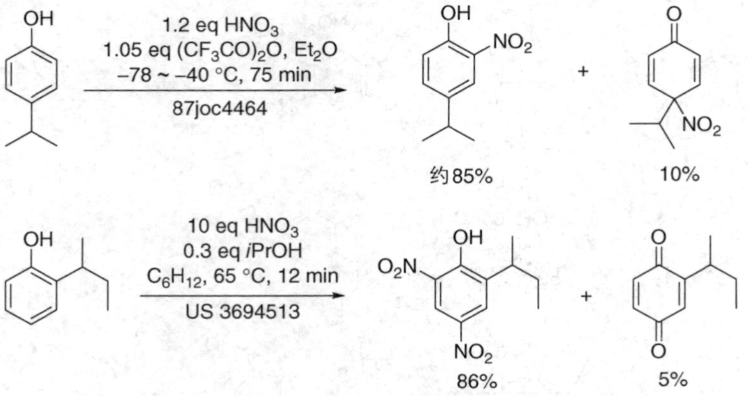

图式 6.17 酚的硝化反应[70,71]

用亚硝酸处理酚可以得到预期的苯醌肟（亚硝基苯酚的互变异构体），但也有许多其他产物。如果酚在氧化时容易形成醌（二羟基苯和氨基苯酚），则亚硝酸盐可以通过对醌的共轭加成形成间硝基苯酚。过量亚硝酸还可以将亚硝基苯酚进一步氧化成硝基苯酚（图式 6.18）。此外，亚硝基苯酚也是亲电胺化试剂，它们能够胺化其他富电子芳烃。这种反应有时可通过缩短反应时间和降低温度来防止。

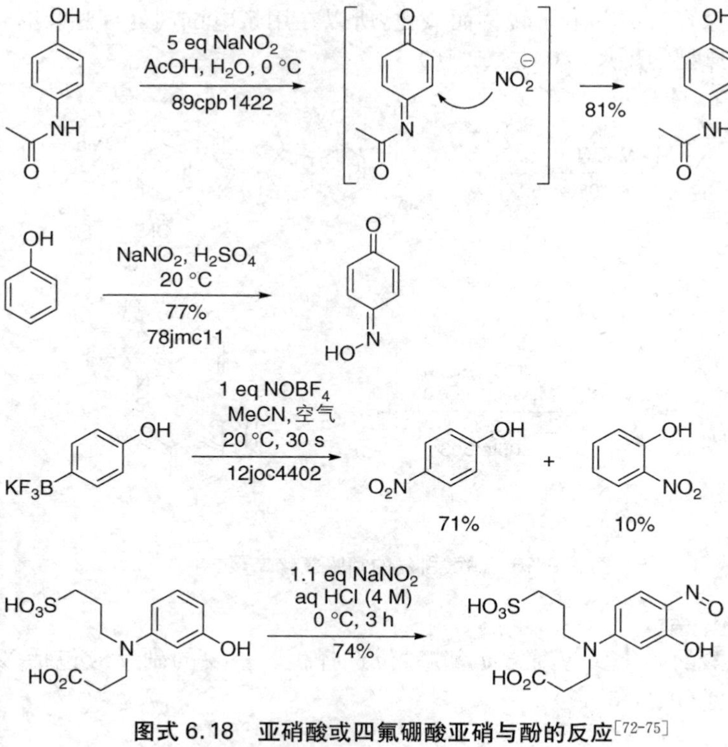

图式 6.18 亚硝酸或四氟硼酸亚硝与酚的反应[72-75]

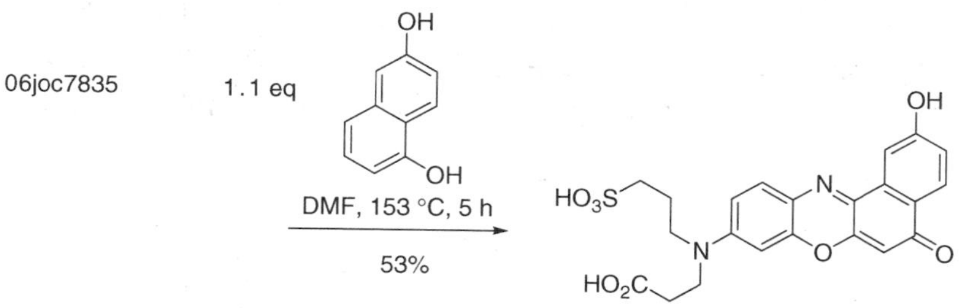

6.2 芳烃的亲电胺化反应

亲电胺化自身存在问题的原因是芳烃的反应活性因胺化而增加。通过使用大大过量的芳烃有时可以增加单胺化产物的收率,但这并不总是实用的。近年来,已经开发出许多用于亲电芳香胺化的有用方法(图式 6.19)。

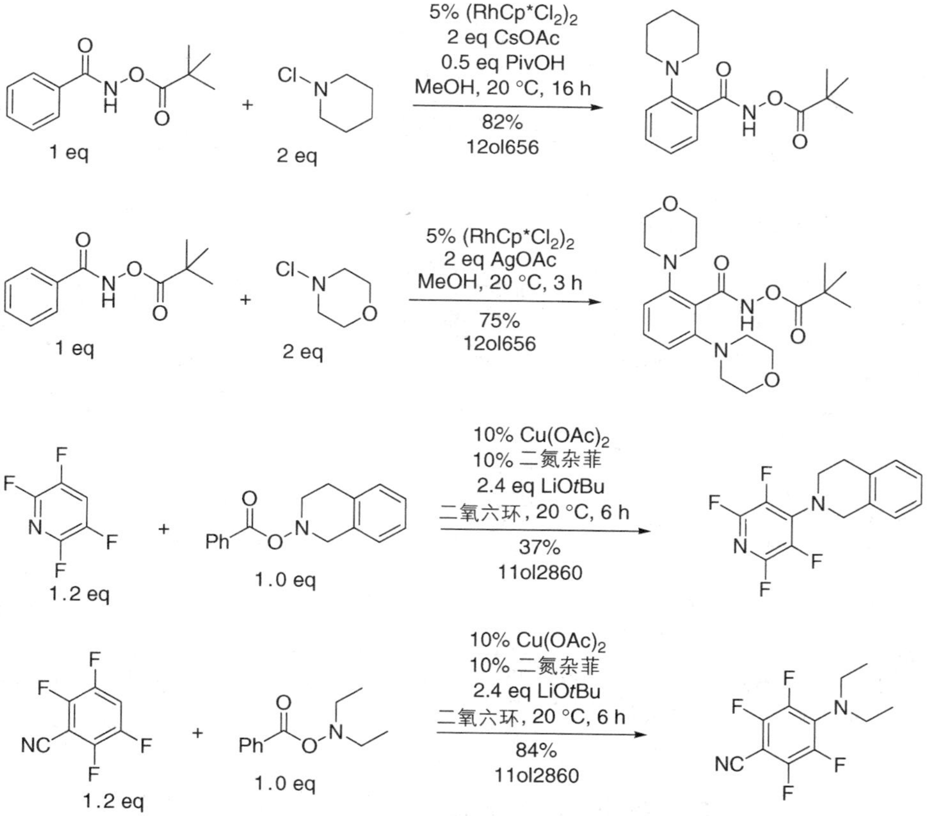

图式 6.19 分子间的亲电芳香胺化反应[76,77]（更多实例：[78,79]）

芳香族未取代的位置进行胺化也可以通过用卤素和胺处理富电子芳烃(图式 6.20 和图式 6.21),或者在氧化剂存在下通过强缺电子芳烃与胺的反应来实现。过渡金属催化的各种反应可能通过芳烃金属化进行(图式 6.20)。

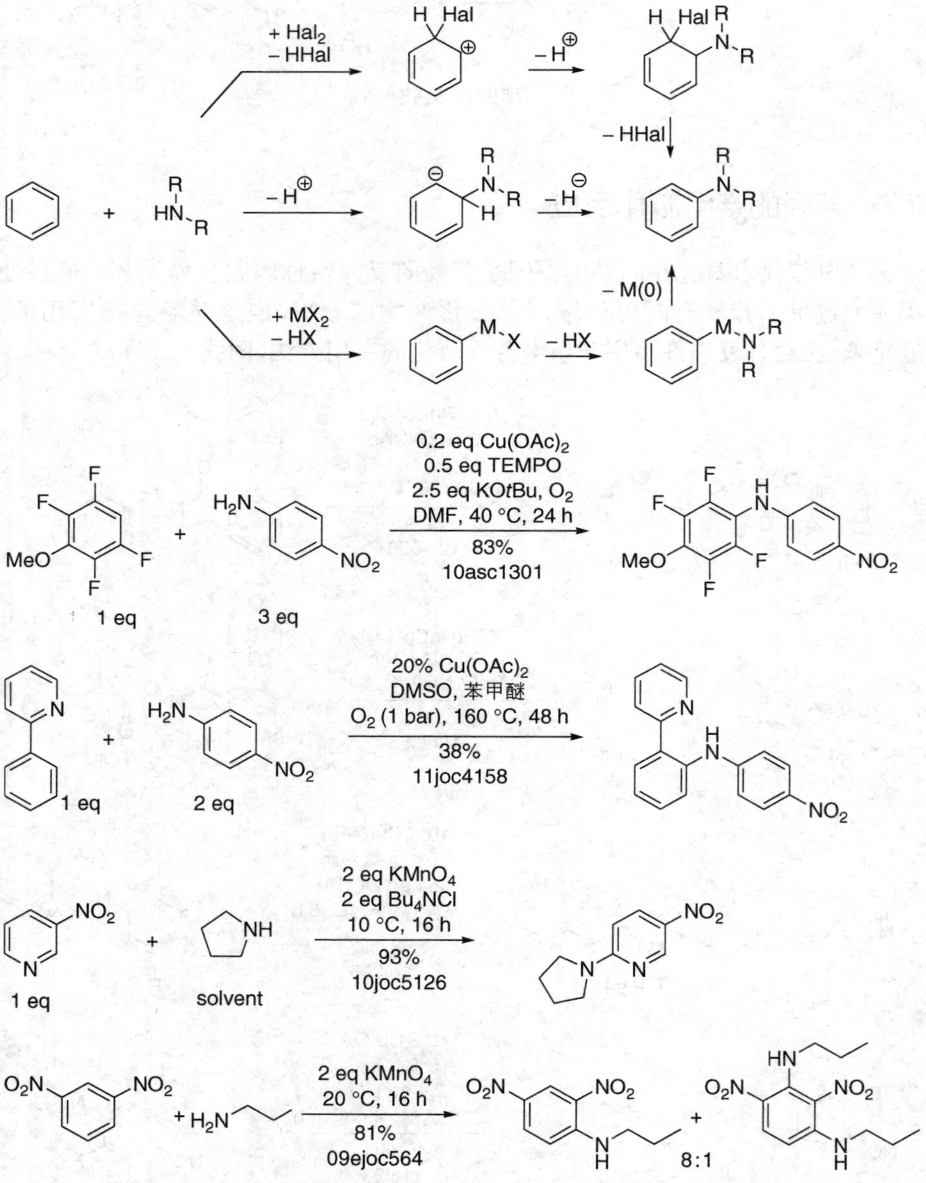

图式 6.20 与胺的氧化芳香胺化[80-83] (Hal:卤素,X:离去基团,R:烷基、芳基;更多实例:[84,85])

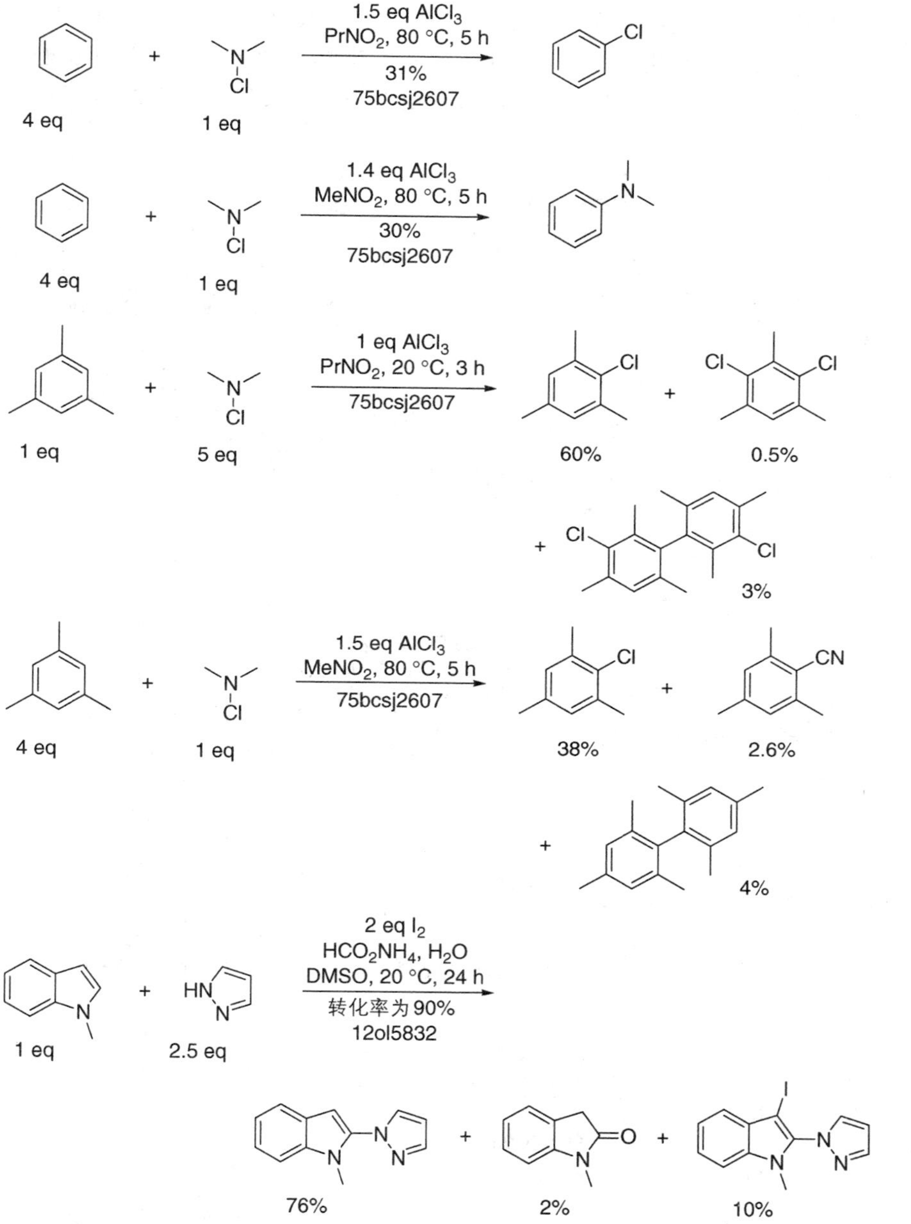

图式 6.21　芳烃亲电胺化的副反应[86,87]（更多实例：[88]）

6.2.1 典型的副反应

　　N-卤代胺能被用作亲电胺化试剂或卤化试剂。同样，取代的羟胺可以引发胺化和羟基化反应。另外，N-卤代胺或羟胺还是氧化剂，能进行脱氢、氧化二聚或形成自由基（图式 6.21）。

苯胺容易进行氧化,在氧化剂的存在下可以低聚。当用氧化剂处理二芳基胺时,它们可以环化成咔唑,可以使苯胺衍生物环化成吲哚、苯并咪唑或其他杂环(图式 6.22)。这可能是大部分亲电芳香胺化反应收率低的另一原因。

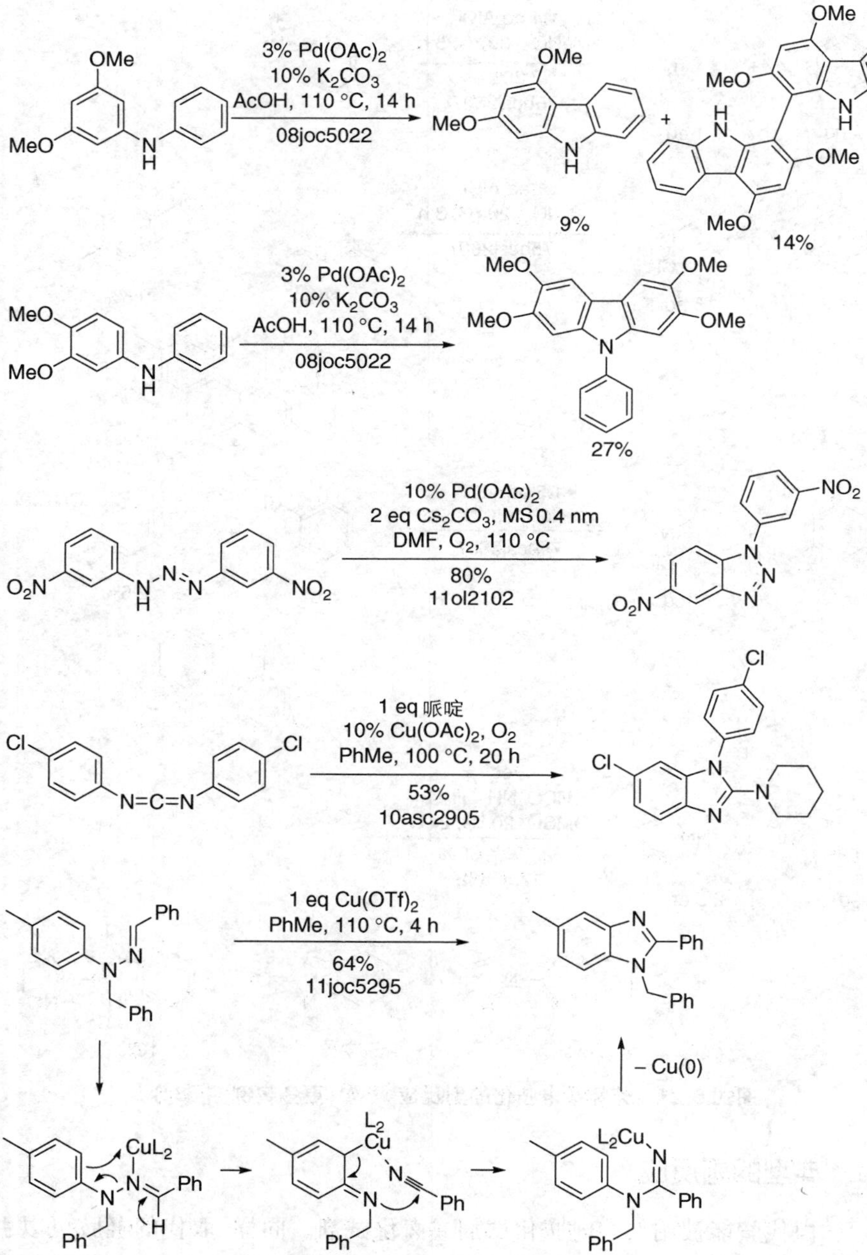

图式 6.22 苯胺衍生物的氧化转化[89-92]

6.3 芳烃的亲电酰胺化反应

由于酰苯胺是不那么容易被氧化的苯胺,亲电形成 RCON—Ar 键的反应往往比亲电芳香胺化纯净。许多成功的芳香酰胺化反应实例已被报道(图式 6.23)。如同胺化,这些反应可以通过氮上带有离去基团的酰胺或者酰胺与额外的氧化剂完成。当使用过渡金属催化剂时,该反应可以通过芳烃的碳金属化或酰胺的氮金属化进行。

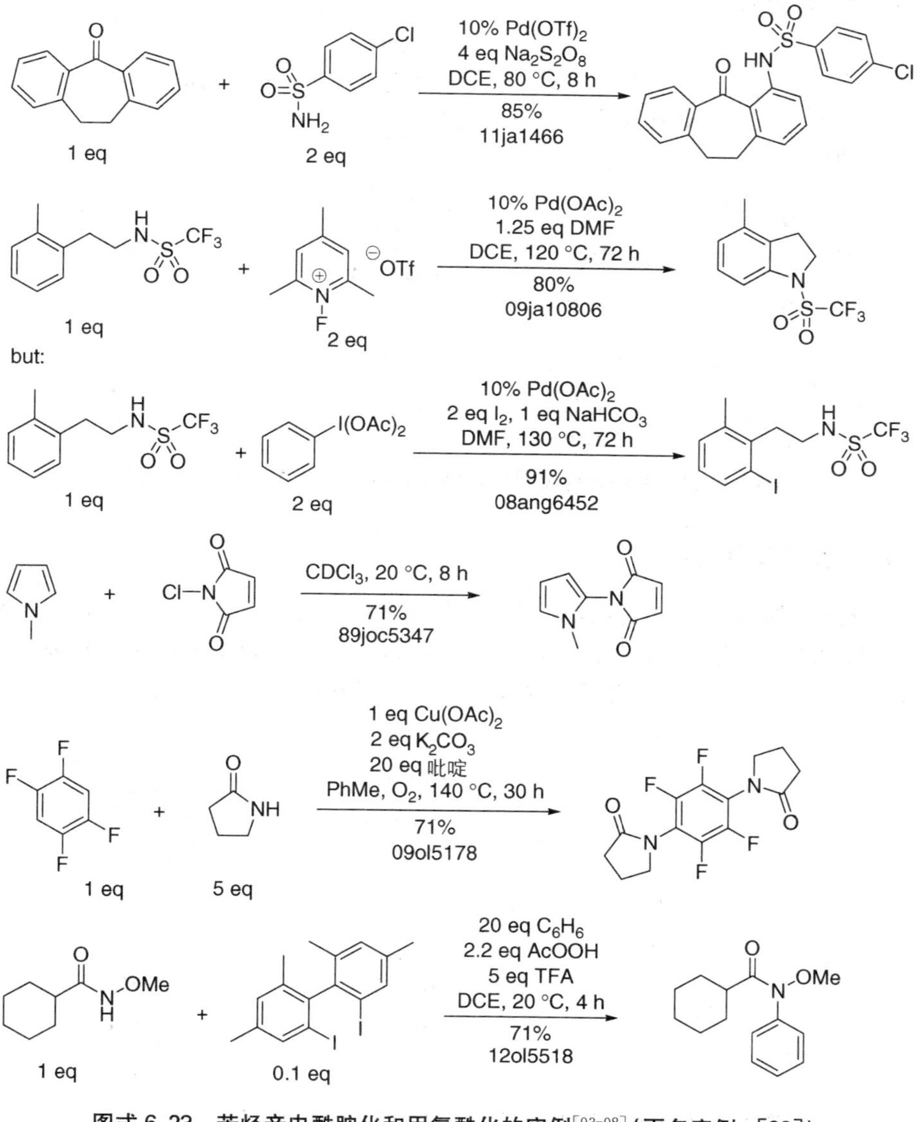

图式 6.23 芳烃亲电酰胺化和甲氨酰化的实例[93-98] (更多实例: [99])

续图式 6.23

分子内芳香酰胺化通常比分子间反应更加容易进行。与酰胺化密切相关的反应有氮烯的芳基化、通过 N-芳基咪肼环化制备苯并咪唑，以及硝基化合物的分子内胺化(图式 6.24)。

图式 6.24　与叠氮化物、肟和硝基烯烃的芳烃亲电胺化[100-102]

6.3.1　典型的副反应

亲电酰胺化类似于卤代，类似的副反应也能被预测。由于酰胺不是优良的亲核试剂，其他亲核试剂甚至卤化物都可以与酰胺进行竞争反应。此外，在所需的反应条件下，产物常常容易发生卤代或羟基化反应(图式 6.25)。

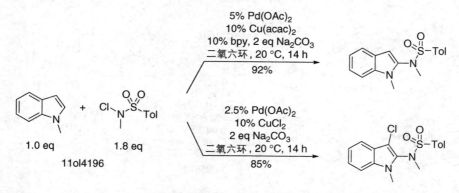

图式 6.25　芳烃酰胺化反应中卤代和乙酰氧基化副反应[103-105]

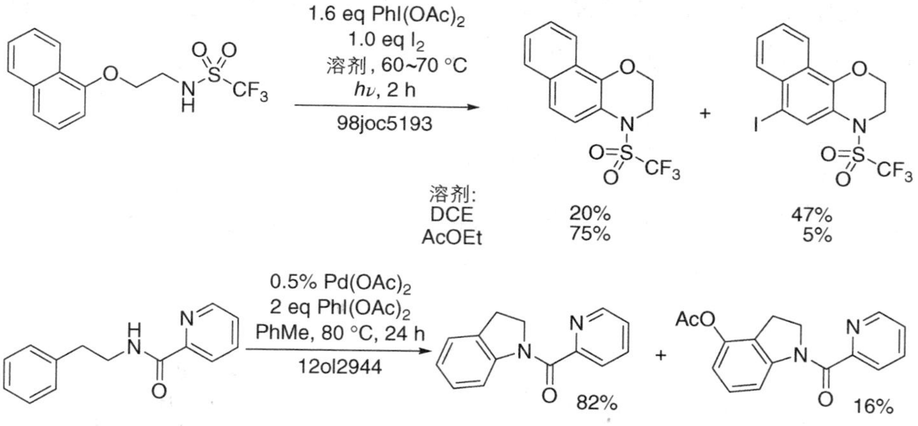

6 通过亲电反应形成芳烃 C—N 键

[图式：略]

续图式 6.25

二乙酰碘苯等高价碘化合物常用作酰胺化反应的额外氧化剂，它们可能首先与芳香烃反应，随后进行碘苯的芳族亲核取代（图式 6.26）。只有当在苯基上的芳族亲核取代反应速度明显慢于起始芳烃氧化时，这种反应才能产生所需产物。此外，高价碘化合物偶尔也可用作碘代试剂。

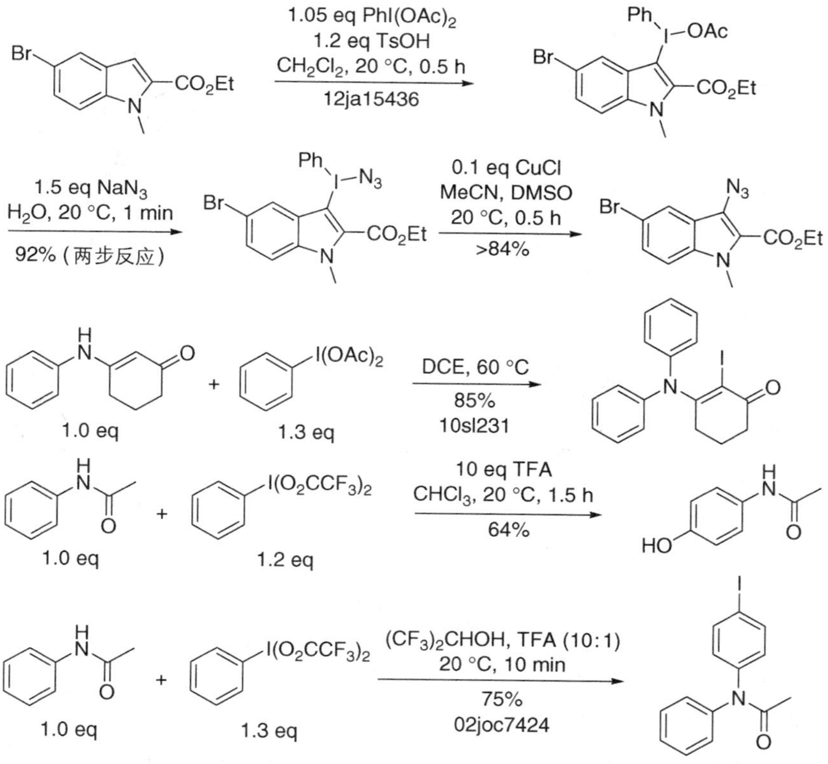

图式 6.26 与高价碘试剂形成 C—N 键[106-108]

如同卤代和硝化,在芳烃亲电酰胺化过程中,可预测的反应是苄基酰胺化和其他非氢基团的取代(图式 6.27)。

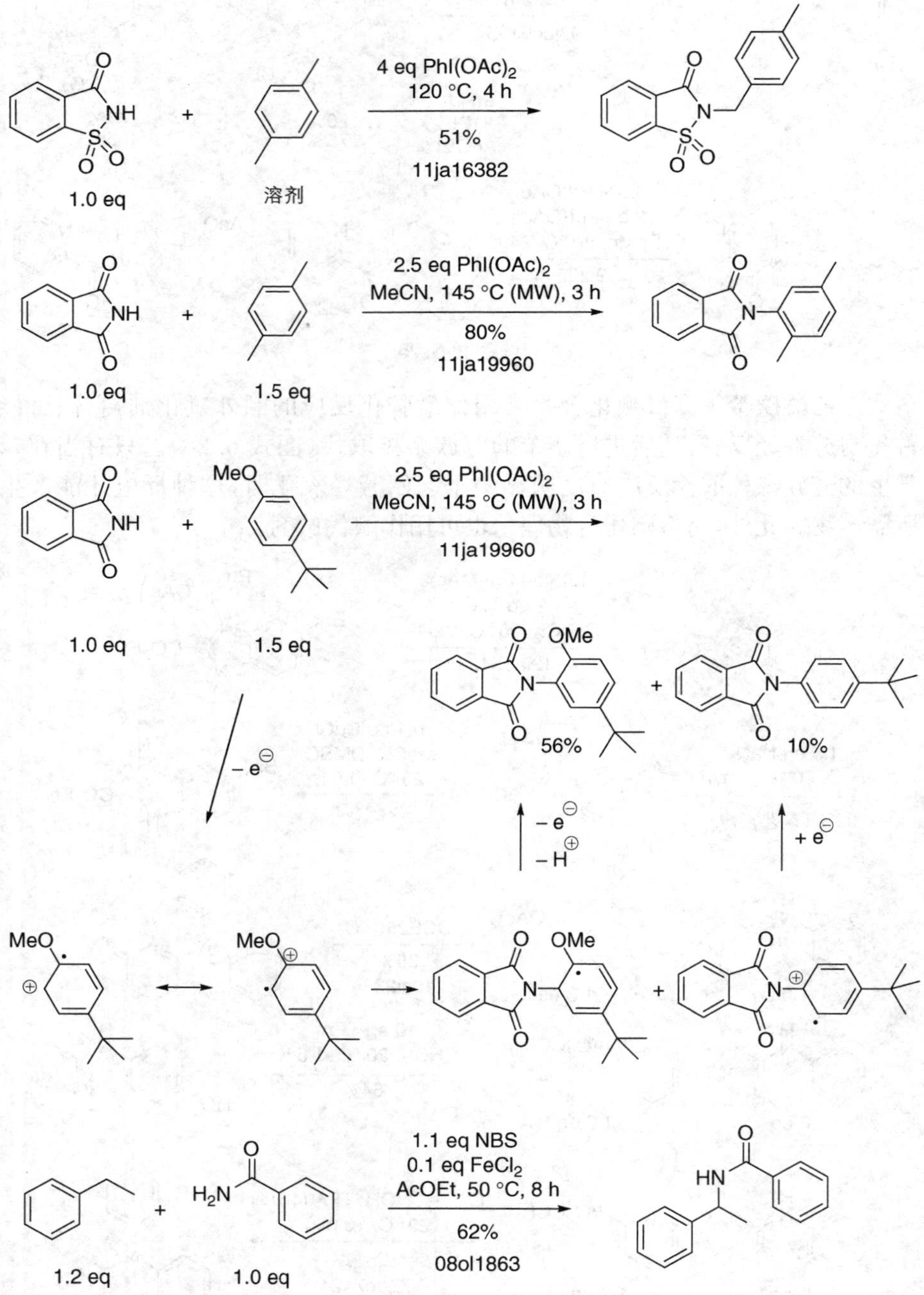

图式 6.27 芳烃酰胺化的副反应:苄酰胺化和甲氧基的取代[109-111]

在经 N-金属化酰胺实现的分子内酰胺化中有时候可以观察到脂肪族非苄位的酰胺化(图式 6.28)。

图式 6.28 分子内的脂肪族酰胺化[112](更多实例:[113,114])

参考文献

1. Choudary, B.M., Kantam, M.L., Sateesh, M., Rao, K.K., and Raghavan, K.V. (2002) Process for the production of nitroaromatic compounds from aromatic hydrocarbons using modified clay catalysts. US Patent 6376726.
2. Nose, M., Suzuki, H., and Suzuki, H. (2001) Nonacid nitration of benzenedicarboxylic and naphthalenecarboxylic acid esters. *J. Org. Chem.*, **66**, 4356–4360.
3. Laali, K.K. and Gettwert, V.J. (2001) Electrophilic nitration of aromatics in ionic liquid solvents. *J. Org. Chem.*, **66**, 35–40.
4. Mascal, M., Yin, L., Edwards, R., and Jarosh, M. (2008) Aromatic nitration in liquid $Ag_{0.51}K_{0.42}Na_{0.07}NO_3$. *J. Org. Chem.*, **73**, 6148–6151.
5. Koley, D., Colón, O.C., and Savinov, S.N. (2009) Chemoselective nitration of phenols with *tert*-butyl nitrite in solution and on solid support. *Org. Lett.*, **11**, 4172–4175.
6. Marterer, W., Prikoszovich, W., Wiss, J., and Prashad, M. (2003) The nitration of 8-methylquinoxalines in mixed acid. *Org. Process Res. Dev.*, **7**, 318–323.
7. Barrett, A.G.M., Braddock, D.C., Ducray, R., McKinnell, R.M., and Waller, F.J. (2000) Lanthanide triflate and triflide catalyzed atom economic nitration of fluoroarenes. *Synlett*, 57–58.
8. Atkins, R.L., Hollins, R.A., and Wilson, W.S. (1986) Synthesis of polynitro compounds. Hexasubstituted benzenes. *J. Org. Chem.*, **51**, 3261–3266.
9. Aridoss, G. and Laali, K.K. (2011) Ethylammonium nitrate (EAN)/Tf_2O and EAN/TFAA: ionic liquid based systems for aromatic nitration. *J. Org. Chem.*, **76**, 8088–8094.
10. Doddi, G., Mencarelli, P., Razzini, A., and Stegel, F. (1979) Nitration of 1-R-pyrroles: formation of polynitro-1-R-pyrroles and orienting effects in the reactions of 3-nitro-1-R-pyrroles. *J. Org. Chem.*, **44**, 2321–2323.
11. Katritzky, A.R., Vakulenko, A.V., Sivapackiam, J., Draghici, B., and Damavarapub, R. (2008) Synthesis of dinitro-substituted furans, thiophenes, and azoles. *Synthesis*, 699–706.

12. Newman, M.S. and Boden, H. (1973) 2,4,5,7-tetranitrofluorenone. *Org. Synth.*, Coll. Vol. **5**, 1029–1030.
13. Moodie, R.B. and Schofield, K. (1976) Ipso attack in aromatic nitration. *Acc. Chem. Res.*, **9**, 287–292.
14. Takahashi, K., Ohta, M., Shoji, Y., Kasai, M., Kunishiro, K., Miike, T., Kanda, M., and Shirahase, H. (2010) Novel Acyl-CoA: cholesterol acyltransferase inhibitor: indoline-based sulfamide derivatives with low lipophilicity and protein binding ratio. *Chem. Pharm. Bull.*, **58**, 1057–1065.
15. Duggan, S.A., Fallon, G., Langford, S.J., Lau, V.-L., Satchell, J.F., and Paddon-Row, M.N. (2001) Crown-linked porphyrin systems. *J. Org. Chem.*, **66**, 4419–4426.
16. Yamato, T., Kamimura, H., and Furukawa, T. (1997) Medium-sized cyclophanes. 43. First evidence for *anti*−*syn*-ring inversion under the nitration of 5,13-di-*tert*-butyl-8,16-dimethoxy[2.2]metacyclophane. *J. Org. Chem.*, **62**, 7560–7564.
17. Balogh, M., Pennetreau, P., Hermecz, I., and Gerstmans, A. (1990) Clay-supported iron(III) nitrate: a multifunctional reagent. Oxidation and nitration of nitrogen bridgehead compounds. *J. Org. Chem.*, **55**, 6198–6202.
18. Carpenter, M.S., Easter, W.M., and Wood, T.F. (1951) Nitro musks. I. Isomers, homologs, and analogs of musk ambrette. *J. Org. Chem.*, **16**, 586–617.
19. Yamakawa, T., Kagechika, H., Kawachi, E., Hashimoto, Y., and Shudo, K. (1990) Retinobenzoic acids. 5. Retinoidal activities of compounds having a trimethylsilyl or trimethylgermyl group(s) in human promyelocytic leukemia cells HL-60. *J. Med. Chem.*, **33**, 1430–1437.
20. Manna, S., Maity, S., Rana, S., Agasti, S., and Maiti, D. (2012) *ipso*-Nitration of arylboronic acids with bismuth nitrate and perdisulfate. *Org. Lett.*, **14**, 1736–1739.
21. Donohoe, T.J., Jonesa, C.R., and Barbosa, L.C.A. (2011) Total synthesis of (±)-streptonigrin: *de novo* construction of a pentasubstituted pyridine using ring-closing metathesis. *J. Am. Chem. Soc.*, **133**, 16418–16421.
22. Das, J.P., Sinha, P., and Roy, S. (2002) A nitro-Hunsdiecker reaction: from unsaturated carboxylic acids to nitrostyrenes and nitroarenes. *Org. Lett.*, **4**, 3055–3058.
23. Dantas de Araujo, A., Christensen, C., Buchardt, J., Kent, S.B.H., and Alewood, P.F. (2011) Synthesis of tripeptide mimetics based on dihydroquinolinone and benzoxazinone scaffolds. *Chem. Eur. J.*, **17**, 13983–13986.
24. Scheurer, A., Mosset, P., Bauer, W., and Saalfrank, R.W. (2001) A practical route to regiospecifically substituted (*R*)- and (*S*)-oxazolylphenols. *Eur. J. Org. Chem.*, **16**, 3067–3074.
25. Stupi, B.P., Li, H., Wang, J., Wu, W., Morris, S.E., Litosh, V.A., Muniz, J., Hersh, M.N., and Metzker, M.L. (2012) Stereochemistry of benzylic carbon substitution coupled with ring modification of 2-nitrobenzyl groups as key determinants for fast-cleaving reversible terminators. *Angew. Chem. Int. Ed.*, **51**, 1724–1727.
26. Bosch, E. and Kochi, J.K. (1994) Thermal and photochemical nitration of aromatic hydrocarbons with nitrogen dioxide. *J. Org. Chem.*, **59**, 3314–3325.
27. Occhipinti, G., Liguori, L., Tsoukala, A., and Bjørsvik, H.-R. (2010) A switchable oxidation process leading to various versatile pharmaceutical intermediates. *Org. Process Res. Dev.*, **14**, 1379–1384.
28. Nishiwaki, Y., Sakaguchi, S., and Ishii, Y. (2002) An efficient nitration of light alkanes and the alkyl side-chain of aromatic compounds with nitrogen dioxide and nitric acid catalyzed by *N*-hydroxyphthalimide. *J. Org. Chem.*, **67**, 5663–5668.
29. Masnovi, J.M., Sankararaman, S., and Kochi, J.K. (1989) Direct observation of the kinetic acidities of transient aromatic cation radicals. The mechanism of electrophilic side-chain nitration of the methylbenzenes. *J. Am. Chem. Soc.*, **111**, 2216–2263.
30. Dalmolen, J., van der Sluis, M., Nieuwenhuijzen, J.W., Meetsma, A.,

de Lange, B., Kaptein, B., Kellogg, R.M., and Broxterman, Q.B. (2004) Synthesis of enantiopure 1-aryl-1-butylamines and 1-aryl-3-butenylamines by diastereoselective addition of allylzinc bromide to imines derived from (*R*)-phenylglycine amide. *Eur. J. Org. Chem.*, **2004** (7), 1544–1557.

31. Onopchenko, A., Sabourin, E.T., and Selwitz, C.M. (1981) Selective nitration of benzophenone. *J. Org. Chem.*, **46**, 5014–5017.
32. Bennett, G.M. and Grove, J.F. (1945) By-products in aromatic nitration. Part II. Nitration of diphenyl, quinoline, and benzophenone. *J. Chem. Soc.*, 378–380.
33. Bosch, E. and Kochi, J.K. (1994) Direct nitrosation of aromatic hydrocarbons and ethers with the electrophilic nitrosonium cation. *J. Org. Chem.*, **59**, 5573–5586.
34. Potturi, H.K., Gurung, R.K., and Hou, Y. (2012) Nitromethane with IBX/TBAF as a nitrosating agent: synthesis of nitrosamines from secondary or tertiary amines under mild conditions. *J. Org. Chem.*, **77**, 626–631.
35. Liu, Y. and McWhorter, W.W. (2003) Synthesis of 8-desbromohinckdentine A. *J. Am. Chem. Soc.*, **125**, 4240–4252.
36. Atherton, J.H., Moodie, R.B., Noble, D.R., and O'Sullivan, B. (1997) Nitrosation of *m*-xylene, anisole, 4-nitrophenyl phenyl ether and toluene in trifluoroacetic acid or in acetic–sulfuric acid mixtures under nitric oxide. *J. Chem. Soc., Perkin Trans. 2*, 663–664.
37. Milligan, B. (1983) Some aspects of nitration of aromatics by lower oxidation states of nitrogen. *J. Org. Chem.*, **48**, 1495–1500.
38. Leoppky, R.N. and Tomasik, W. (1983) Stereoelectronic effects in tertiary amine nitrosation: nitrosative cleavage vs. aryl ring nitration. *J. Org. Chem.*, **48**, 2751–2757.
39. Zhang, L., Liu, Z., Li, H., Fang, G., Barry, B.-D., Belay, T.A., Bi, X., and Liu, Q. (2011) Copper-mediated chelation-assisted *ortho* nitration of (hetero)arenes. *Org. Lett.*, **13**, 6536–6539.
40. Liu, Y.-K., Lou, S.-J., Xu, D.-Q., and Xu, Z.-Y. (2010) Regiospecific synthesis of nitroarenes by palladium-catalyzed nitrogen-donor-directed aromatic C–H nitration. *Chem. Eur. J.*, **16**, 13590–13593.
41. Havlík, M., Král, V., and Dolenský, B. (2006) Overcoming regioselectivity issues inherent in bisTröger's base preparation. *Org. Lett.*, **8**, 4867–4870.
42. Olah, G.A., Laali, K., Farooq, O., and Olah, J.A. (1990) C-Nitration of 2,6-di- and 2,4,6-tri-*tert*-butylpyridine with nitronium tetrafluoroborate. *J. Org. Chem.*, **55**, 5179–5180.
43. Duffy, J.L. and Laali, K.K. (1991) Aprotic nitration ($NO_2^+BF_4^-$) of 2-halo- and 2,6-dihalopyridines and transfer-nitration chemistry of their *N*-nitropyridinium cations. *J. Org. Chem.*, **56**, 3006–3009.
44. Katritzky, A.R., Scriven, E.F.V., Majumder, S., Akhmedova, R.G., Vakulenko, A.V., Akhmedov, N.G., Muruganb, R., and Abboudc, K.A. (2005) Preparation of nitropyridines by nitration of pyridines with nitric acid. *Org. Biomol. Chem.*, **3**, 538–541.
45. Kavanagh, P. and Leech, D. (2004) Improved synthesis of 4,4′-diamino-2,2′-bipyridine from 4,4′-dinitro-2,2′-bipyridine-*N*,*N*′-dioxide. *Tetrahedron Lett.*, **45**, 121–123.
46. Newkome, G.R., Theriot, K.J., Majestic, V.K., Spruell, P.A., and Baker, G.R. (1990) Functionalization of 2-methyl- and 2,7-dimethy1-1,8-naphthyridine. *J. Org. Chem.*, **55**, 2838–2842.
47. Panke, G., Schwalbe, T., Stirner, W., Taghavi-Moghadam, S., and Wille, G. (2003) A practical approach of continuous processing to high energetic nitration reactions in microreactors. *Synthesis*, 2827–2830.
48. Aries, R.S. (1955) Process of making isocinchomeronic acid and decarboxylation of same to niacin. US Patent 2702802.
49. Frankel, M.B. and Karl, K. (1968) Preparation of nitramines. US Patent 3390183.
50. Girard, Y., Atkinson, J.G., Bélanger, P.C., Fuentes, J.J., Rokach, J., Rooney, C.S., Remy, D.C., and Hunt, C.A.

(1983) Synthesis, chemistry, and photochemical substitutions of 6,11-dihydro-5*H*-pyrrolo[2,1-*b*][3]benzazepin-11-ones. *J. Org. Chem.*, **48**, 3220–3234.

51. Olah, G.A. and Kuhn, S.J. (1973) 3,5-dinitro-*o*-tolunitrile. *Org. Synth.*, Coll. Vol. **5**, 480–485.

52. Gazzaeva, R.A., Fedotov, A.N., Trofimova, E.V., Popova, O.A., Mochalov, S.S., and Zefirov, N.S. (2006) Synthesis of *o*-nitrosoacylbenzenes from *o*-nitrobenzyl alcohols and their derivatives. *Russ. J. Org. Chem.*, **42**, 87–99.

53. Mochalov, S.S., Gazzaeva, R.A., Fedotov, A.N., Archegov, B.P., Trofimova, E.V., Shabarov, Y.S., and Zefirov, N.S. (2005) Transformations of *para*-substituted benzylcyclopropanes, allylbenzenes, and diphenylmethanes under nitration with nitric acid in acetic anhydride. *Russ. J. Org. Chem.*, **41**, 406–416.

54. Njoroge, F.G., Vibulbhan, B., Pinto, P., Chan, T.-M., Osterman, R., Remiszewski, S., Del Rosario, J., Doll, R., Girijavallabhan, V., and Ganguly, A.K. (1998) Highly regioselective nitration reactions provide a versatile method of functionalizing benzocycloheptapyridine tricyclic ring systems: application toward preparation of nanomolar inhibitors of farnesyl protein transferase. *J. Org. Chem.*, **63**, 445–451.

55. Lakshminarayana, N., Prasad, Y.R., Gharat, L., Thomas, A., Ravikumar, P., Narayanan, S., Srinivasan, C.V., and Gopalan, B. (2009) Synthesis and evaluation of some novel isochroman carboxylic acid derivatives as potential anti-diabetic agents. *Eur. J. Med. Chem.*, **44**, 3147–3157.

56. Bennett, G.M. and Youle, P.V. (1938) Hydroxy-by-products in aromatic nitration. *J. Chem. Soc.*, 1816–1818.

57. Feldman, K.S., McDermott, A., and Myhre, P.C. (1979) Ipso nitration. Preparation of 4-methyl-4-nitrocyclohexadienols and detection of intramolecular hydrogen migration (NIH shift) upon solvolytic rearomatization. *J. Am. Chem. Soc.*, **101**, 505–506.

58. Rodríguez González, R., Gambarotti, C., Liguori, L., and Bjørsvik, H.-R. (2006) Efficient and green telescoped process to 2-methoxy-3-methyl[1,4]benzoquinone. *J. Org. Chem.*, **71**, 1703–1706.

59. Fischer, A., Packer, J., Vaughan, J., and Wright, G.J. (1961) Acetoxylation accompanying nitration with nitric acid–acetic anhydride mixtures. *Proc. Chem. Soc.*, 369–370.

60. Smith, K., Musson, A., and DeBoos, G.A. (1998) A novel method for the nitration of simple aromatic compounds. *J. Org. Chem.*, **63**, 8448–8454.

61. Cordeiro, A., Shaw, J., O'Brien, J., Blanco, F., and Rozas, I. (2011) Synthesis of 6-nitro-1,2,3,4-tetrahydroquinoline: an experimental and theoretical study of regioselective nitration. *Eur. J. Org. Chem.*, **2011** (8), 1504–1513.

62. Nielsen, A.T., Henry, R.A., Norris, W.P., Atkins, R.L., Moore, D.W., Lepie, A.H., Coon, C.L., Spanggord, R.J., and Son, D.V.H. (1979) Synthetic routes to aminodinitrotoluenes. *J. Org. Chem.*, **44**, 2499–2504.

63. Boger, D.L. and Sakya, S.M. (1992) CC-1065 partial structures: enhancement of noncovalent affinity for DNA minor groove binding through introduction of stabilizing electrostatic interactions. *J. Org. Chem.*, **57**, 1277–1284.

64. Hester, J.B. (1964) A synthesis of 1,3,4,5-tetrahydropyrrolo[4,3,2-*de*]quinoline. *J. Org. Chem.*, **29**, 1158–1160.

65. Robinson, M.W., Overmeyer, J.H., Young, A.M., Erhardt, P.W., and Maltese, W.A. (2012) Synthesis and evaluation of indole-based chalcones as inducers of methuosis, a novel type of nonapoptotic cell death. *J. Med. Chem.*, **55**, 1940–1956.

66. Noland, W.E., Smith, L.R., and Rush, K.R. (1965) Nitration of indoles. III. Polynitration of 2-alkylindoles. *J. Org. Chem.*, **30**, 3457–3469.

67. Noland, W.E. and Rieke, R.D. (1962) New synthetic route to 6-nitroisatin via nitration of 3-indolealdehyde. *J. Org. Chem.*, **27**, 2250–2252.

68. Keawin, T., Rajviroongita, S., and Black, D.S. (2005) Reaction of some 4,6-dimethoxyindoles with nitric acid: nitration and oxidative dimerisation. *Tetrahedron*, **61**, 853–861.
69. Noland, W.E., Smith, L.R., and Johnson, D.C. (1963) Nitration of indoles. II. The mononitration of methylindoles. *J. Org. Chem.*, **28**, 2262–2266.
70. Fischer, A. and Sankararaman, S. (1987) Formation of 4-nitrocyclohexa-2,5-dienols by addition of organolithium reagents to 4-alkyl-4-nitrocyclohexa-2,5-dienones. *J. Org. Chem.*, **52**, 4464–4468.
71. Tobey, S.W. and Lourandos, M.Z. (1972) Direct nitration of alkylphenols with nitric acid. US Patent 3694513.
72. Matsuno, T., Matsukawa, T., Sakuma, Y., and Kunieda, T. (1989) A new nitration product, 3-nitro-4-acetamidophenol, obtained from acetaminophen with nitrous acid. *Chem. Pharm. Bull.*, **37**, 1422–1423.
73. Hodnett, E.M., Prakash, G., and Amirmoazzami, J. (1978) Nitrogen analogues of 1,4-benzoquinones. Activities against the ascitic sarcoma 180 of mice. *J. Med. Chem.*, **21**, 11–16.
74. Molander, G.A. and Cavalcanti, L.N. (2012) Nitrosation of aryl and heteroaryltrifluoroborates with nitrosonium tetrafluoroborate. *J. Org. Chem.*, **77**, 4402–4413.
75. Jose, J. and Burgess, K. (2006) Syntheses and properties of water-soluble Nile red derivatives. *J. Org. Chem.*, **71**, 7835–7839.
76. Grohmann, C., Wang, H., and Glorius, F. (2012) Rh[III]-catalyzed direct C–H amination using N-chloroamines at room temperature. *Org. Lett.*, **14**, 656–659.
77. Matsuda, N., Hirano, K., Satoh, T., and Miura, M. (2011) Copper-catalyzed direct amination of electron-deficient arenes with hydroxylamines. *Org. Lett.*, **13**, 2860–2863.
78. Ng, K.-H., Zhou, Z., and Yu, W.-Y. (2012) Rhodium(III)-catalyzed intermolecular direct amination of aromatic C–H bonds with N-chloroamines. *Org. Lett.*, **14**, 272–275.
79. Berman, A.M. and Johnson, J.S. (2009) Copper-catalyzed electrophilic amination of diorganozinc reagents: 4-phenylmorpholine. *Org. Synth.*, Coll. Vol. **11**, 684–690.
80. Zhao, H., Wang, M., Su, W., and Honga, M. (2010) Copper-catalyzed intermolecular amination of acidic aryl C–H bonds with primary aromatic amines. *Adv. Synth. Catal.*, **352**, 1301–1306.
81. John, A. and Nicholas, K.M. (2011) Copper-catalyzed amidation of 2-phenylpyridine with oxygen as the terminal oxidant. *J. Org. Chem.*, **76**, 4158–4162.
82. Verbeeck, S., Herrebout, W.A., Gulevskaya, A.V., van der Veken, B.J., and Maes, B.U.W. (2010) ONSH: optimization of oxidative alkylamination reactions through study of the reaction mechanism. *J. Org. Chem.*, **75**, 5126–5133.
83. Gulevskaya, A.V., Verbeeck, S., Burov, O.N., Meyers, C., Korbukova, I.N., Herrebout, W., and Maes, B.U.W. (2009) Synthesis of (alkylamino)nitroarenes by oxidative alkylamination of nitroarenes. *Eur. J. Org. Chem.*, **2009**, 564–574.
84. Masters, K.-S., Rauws, T.R.M., Yadav, A.K., Herrebout, W.A., Van der Veken, B., and Maes, B.U.W. (2011) On the importance of an acid additive in the synthesis of pyrido[1,2-a]benzimidazoles by direct copper-catalyzed amination. *Chem. Eur. J.*, **17**, 6315–6320.
85. Lipilin, D.L., Churakov, A.M., Ioffe, S.L., Strelenko, Y.A., and Tartakovsky, V.A. (1999) Nucleophilic aromatic substitution of hydrogen in the reaction of tert-alkylamines with nitrosobenzenes – synthesis and NMR study of N-(tert-alkyl)-ortho-nitrosoanilines. *Eur. J. Org. Chem.*, **1999**, 29–35.
86. Yoshifuji, M., Tanaka, S., and Inamoto, N. (1975) Friedel–Crafts reaction of 1,3,5-trialkylbenzenes with sulfur monochloride and N-chlorodimethylamine. *Bull. Chem. Soc. Jpn.*, **48**, 2607–2608.

87. Wu, W.-B. and Huang, J.-M. (2012) Highly regioselective C–N bond formation through C–H azolation of indoles promoted by iodine in aqueous media. *Org. Lett.*, **14**, 5832–5835.
88. Li, Y.-X., Ji, K.-G., Wang, H.-X., Ali, S., and Liang, Y.-M. (2011) Iodine-induced regioselective C–C and C–N bonds formation of N-protected indoles. *J. Org. Chem.*, **76**, 744–747.
89. Liégault, B., Lee, D., Huestis, M.P., Stuart, D.R., and Fagnou, K. (2008) Intramolecular Pd(II)-catalyzed oxidative biaryl synthesis under air: reaction development and scope. *J. Org. Chem.*, **73**, 5022–5028.
90. Kumar, R.K., Ali, M.A., and Punniyamurthy, T. (2011) Pd-Catalyzed C–H activation/C–N bond formation: a new route to 1-aryl-1H-benzotriazoles. *Org. Lett.*, **13**, 2102–2105.
91. He, H.-F., Wang, Z.-J., and Baoa, W. (2010) Copper(II) acetate/oxygen-mediated nucleophilic addition and intramolecular C–H activation/C–N or C–C bond formation: one-pot synthesis of benzimidazoles or quinazolines. *Adv. Synth. Catal.*, **352**, 2905–2912.
92. Guru, M.M., Ali, M.A., and Punniyamurthy, T. (2011) Copper-mediated synthesis of substituted 2-aryl-N-benzylbenzimidazoles and 2-arylbenzoxazoles via C–H functionalization/C–N/C–O bond formation. *J. Org. Chem.*, **76**, 5295–5308.
93. Xiao, B., Gong, T.-J., Xu, J., Liu, Z.-J., and Liu, L. (2011) Palladium-catalyzed intermolecular directed C–H amidation of aromatic ketones. *J. Am. Chem. Soc.*, **133**, 1466–1474.
94. (a) Mei, T.-S., Wang, X., and Yu, J.-Q. (2009) Pd(II)-Catalyzed amination of C–H bonds using single-electron or two-electron oxidants. *J. Am. Chem. Soc.*, **131**, 10806–10807; (b) Li, J.-J., Mei, T.-S., and Yu, J.-Q. (2008) Synthesis of indolines and tetrahydroisoquinolines from arylethylamines by Pd[II]-catalyzed C–H activation reactions. *Angew. Chem. Int. Ed.*, **47**, 6452–6455.
95. De Rosa, M., Cabrera Nieto, G., and Ferrer Gago, F. (1989) Effect of N-chloro structure and 1-substituent on σ-substitution (addition–elimination) in pyrroles. *J. Org. Chem.*, **54**, 5347–5350.
96. Wang, Q. and Schreiber, S.L. (2009) Copper-mediated amidation of heterocyclic and aromatic C–H bonds. *Org. Lett.*, **11**, 5178–5180.
97. Samanta, R., Bauer, J.O., Strohmann, C., and Antonchick, A.P. (2012) Organocatalytic, oxidative, intermolecular amination and hydrazination of simple arenes at ambient temperature. *Org. Lett.*, **14**, 5518–5521.
98. Gu, L., Neo, B.S., and Zhang, Y. (2011) Gold-catalyzed direct amination of arenes with azodicarboxylates. *Org. Lett.*, **13**, 1872–1874.
99. Lenaršič, R., Kočevar, M., and Polanc, S. (1999) $ZrCl_4$-Mediated regioselective electrophilic amination of activated arenes with new alkyl arylaminocarbonyldiazenecarboxylates: intermolecular and intramolecular reactions. *J. Org. Chem.*, **64**, 2558–2563.
100. Dabbagh, H.A. and Lwowski, W. (1989) Functionalization of phenyl rings by imidoylnitrenes. *J. Org. Chem.*, **54**, 3952–3957.
101. Yamamoto, Y., Mizuno, H., Tsuritani, T., and Mase, T. (2009) Synthesis of α-chloroaldoxime O-methanesulfonates and their use in the synthesis of functionalized benzimidazoles. *J. Org. Chem.*, **74**, 1394–1396.
102. Venkatesh, C., Singh, B., Mahata, P.K., Ila, H., and Junjappa, H. (2005) Heteroannulation of nitroketene N,S-arylaminoacetals with $POCl_3$: a novel highly regioselective synthesis of unsymmetrical 2,3-substituted quinoxalines. *Org. Lett.*, **7**, 2169–2172.
103. Liu, X.-Y., Gao, P., Shen, Y.-W., and Liang, Y.-M. (2011) Palladium-/copper-catalyzed regioselective amination and chloroamination of indoles. *Org. Lett.*, **13**, 4196–4199.
104. Togo, H., Hoshina, Y., Muraki, T., Nakayama, H., and Yokoyama, M. (1998) Study on radical amidation onto aromatic rings with (diacyloxyiodo)arenes. *J. Org. Chem.*, **63**, 5193–5200.
105. He, G., Lu, C., Zhao, Y., Nack, W.A., and Chen, G. (2012) Improved protocol

106. Lubriks, D., Sokolovs, I., and Suna, E. (2012) Indirect C–H azidation of heterocycles via copper-catalyzed regioselective fragmentation of unsymmetrical λ^3-iodanes. *J. Am. Chem. Soc.*, **134**, 15436–15442.
107. Chen, Y., Ju, T., Wang, J., Yu, W., Du, Y., and Zhao, K. (2010) Concurrent α-iodination and *N*-arylation of cyclic β-enaminones. *Synlett*, 231–234.
108. Itoh, N., Sakamoto, T., Miyazawa, E., and Kikugawa, Y. (2002) Introduction of a hydroxy group at the *para* position and *N*-iodophenylation of *N*-arylamides using phenyliodine(III) bis(trifluoroacetate). *J. Org. Chem.*, **67**, 7424–7428.
109. Kim, H.J., Kim, J., Cho, S.H., and Chang, S. (2011) Intermolecular oxidative C–N bond formation under metal-free conditions: control of chemoselectivity between aryl sp^2 and benzylic sp^3 C–H bond imidation. *J. Am. Chem. Soc.*, **133**, 16382–16385.
110. Kantak, A.A., Potavathri, S., Barham, R.A., Romano, K.M., and DeBoef, B. (2011) Metal-free intermolecular oxidative C–N bond formation via tandem C–H and N–H bond functionalization. *J. Am. Chem. Soc.*, **133**, 19960–19965.
111. Wang, Z., Zhang, Y., Fu, H., Jiang, Y., and Zhao, Y. (2008) Efficient intermolecular iron-catalyzed amidation of C–H bonds in the presence of *N*-bromosuccinimide. *Org. Lett.*, **10**, 1863–1866.
112. Neumann, J.J., Rakshit, S., Dröge, T., and Glorius, F. (2009) Palladium-catalyzed amidation of unactivated C(sp^3)–H bonds: from anilines to indolines. *Angew. Chem. Int. Ed.*, **48**, 6892–6895.
113. Youn, S.W., Bihn, J.H., and Kim, B.S. (2011) Pd-Catalyzed intramolecular oxidative C–H amination: synthesis of carbazoles. *Org. Lett.*, **13**, 3738–3741.
114. Cho, S.H., Yoon, J., and Chang, S. (2011) Intramolecular oxidative C–N bond formation for the synthesis of carbazoles: comparison of reactivity between the copper-catalyzed and metal-free conditions. *J. Am. Chem. Soc.*, **133**, 5996–6005.

7 通过亲电反应形成芳烃 C—S 键

7.1 磺酰化反应

7.1.1 概述

从芳烃制备芳基磺酸、芳基磺酰氯和砜是一种广泛使用的通用反应(图式7.1)。由于反应产物不活泼并且仅仅能缓慢磺化过量的芳烃,所以该反应收率一般很高。尽管如此,少量的砜常常能够作为副产物在芳族磺酸或磺酰卤的合成中得到[1]。

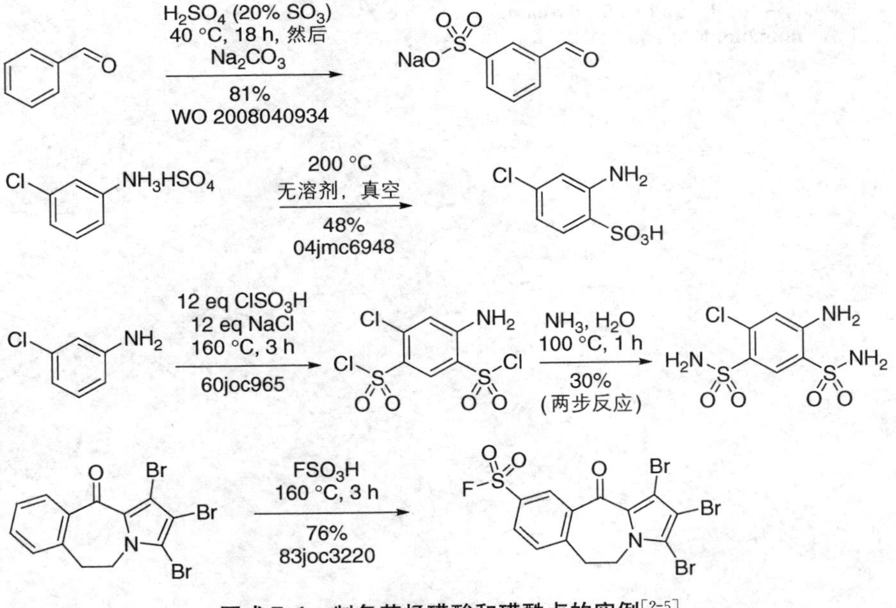

图式 7.1 制备芳烃磺酸和磺酰卤的实例[2-5]

对于简单反应底物而言,三氯化铝引发的芳烃磺酰氯转化成二芳基砜的反应能够顺利进行(图式 7.2)。对于烷基磺酰卤则需要特殊的催化剂和反应

7 通过亲电反应形成芳烃 C—S 键

条件[6]。磺酸酐磺化芳烃比磺酰卤纯净[7,8]，因为后者往往可用作卤代剂或氧化剂。

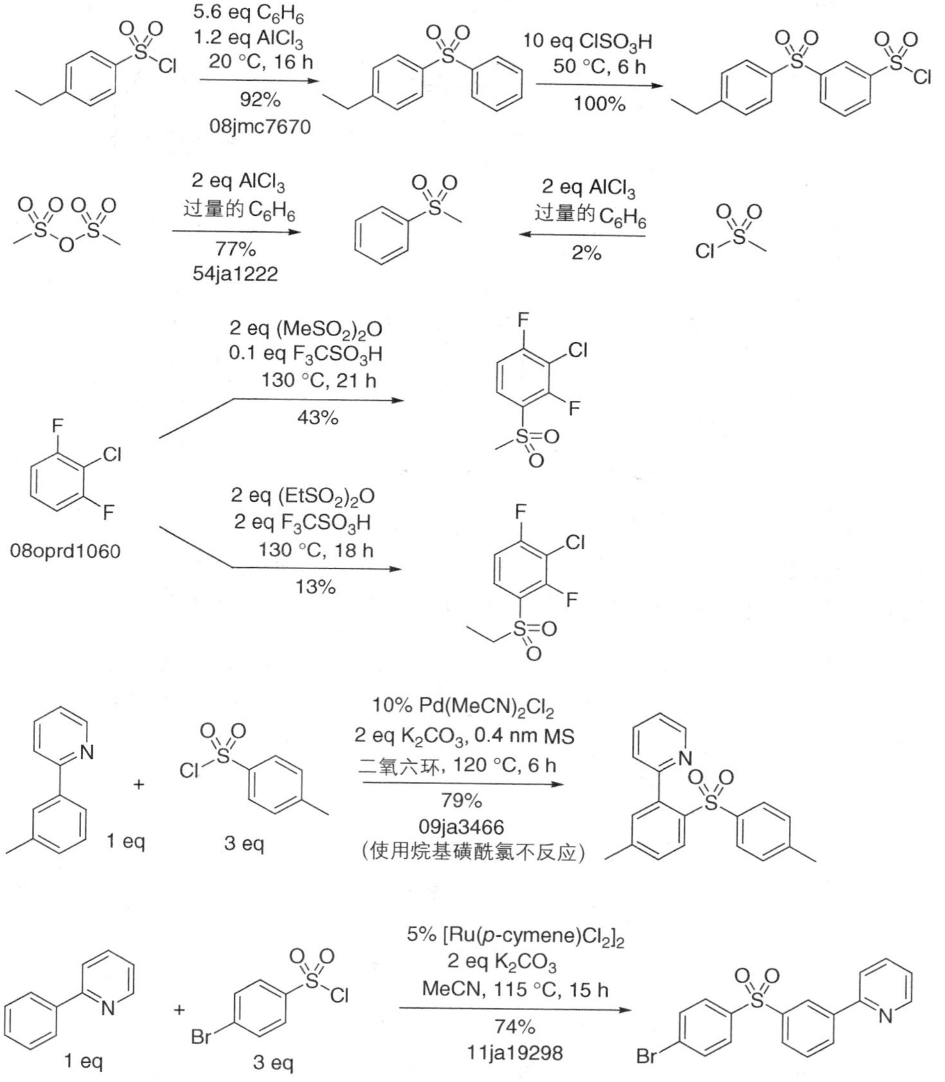

图式 7.2 芳基磺酰化的实例[7-11]（更多实例：[6,12,13]）

7.1.2 典型的副反应

用磺酰卤进行磺酰化收率低的一个原因是磺酰卤经常用作卤化剂而非磺酰化试剂（图式 7.3）。与酰基负离子相比，亚磺酸负离子（RSO_2^-）是稳定、良好的离去基团，并且经常用作离去基团（例如，作为 β-消除的离去基团，RSO_2X 作为 X$^+$ 的等价体）。甚至在无水条件下从磺酰氯制备磺胺类药物的收率也很低。磺胺类药物的高收率通常是在西奥特-鲍曼反应条件(在含水碱存在时)下获得的。因为磺酰卤,特别是芳基磺酰氯,不容易与水或含水碱进行反应,磺酰卤的水解很少会成为问题。

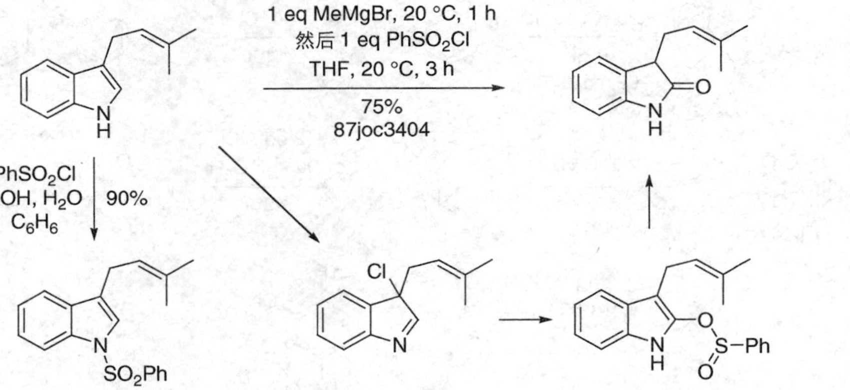

图式 7.3 用苯磺酰氯氧化吲哚[14]

磺酸衍生物的酸催化芳烃磺酰化反应是可逆反应,因而可以通过异构化成为更稳定的异构体——磺酰基迁移产物,或者脱磺酰基产物(图式 7.4)。这些副反应可以通过降低反应温度和酸的用量被抑制。

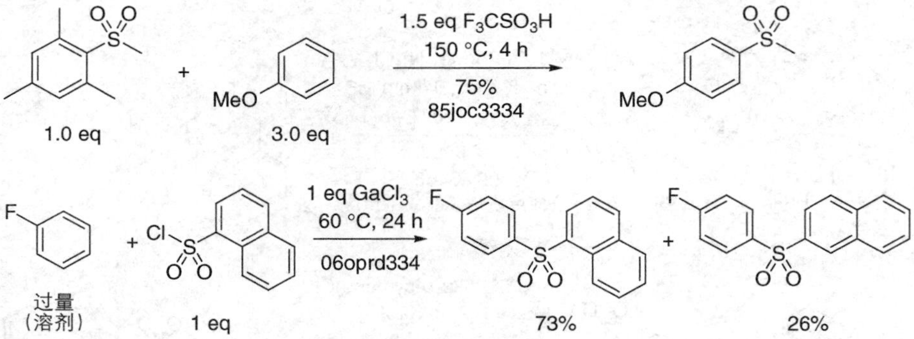

图式 7.4 砜的异构化[15,16]

磺酸衍生物能作为氧化剂,将富电子芳烃转化成为自由基阳离子。联芳基是从这种自由基中间体形成的典型副产物(图式 7.5)。在过渡金属存在下,芳基磺酰卤和其他芳基磺酸衍生物也可作为芳基化试剂[17]。

氯化亚砜是比芳基磺酰氯更强的氯化剂,因而不方便用于制备磺酰氯或砜(图式 7.6)。只有当金属化芳烃要在无水条件下进行氯磺酰化的时候,氯化亚砜才可用作亲电试剂,但收率往往较低。用卤代磺酸($HalSO_3H$)转化芳烃生成芳基磺酰卤的反应最佳。但是在碘的存在下,即使氯磺酸也能作为氯代试剂[19]。

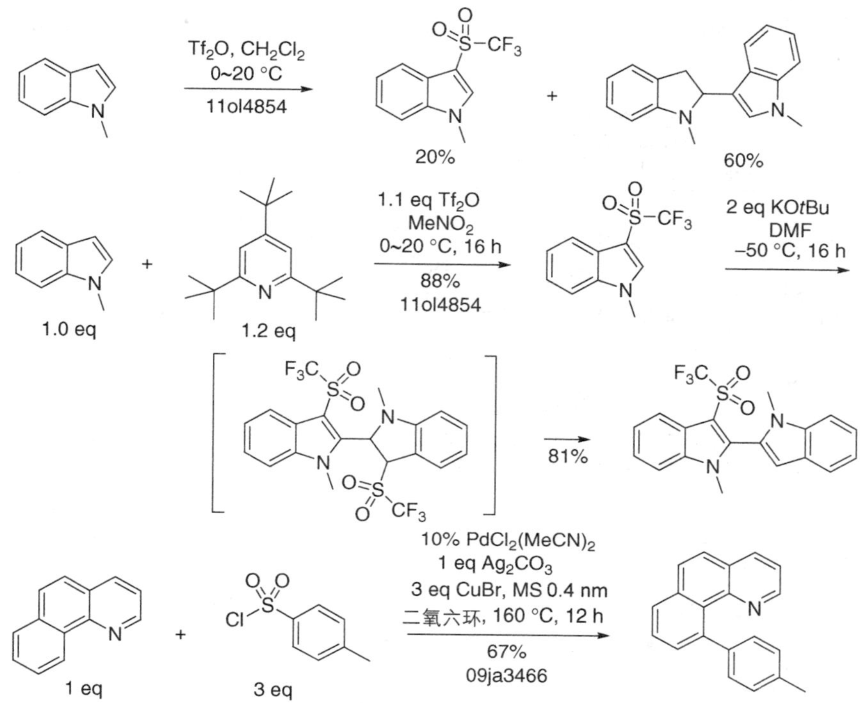

图式 7.5 与磺酸衍生物的氧化反应和芳基化反应[10,18]

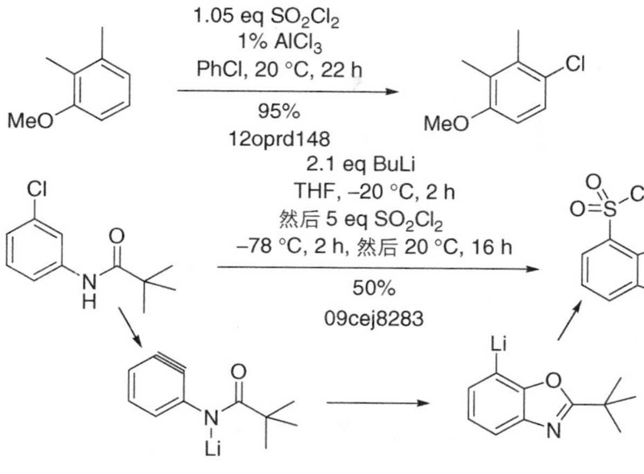

图式 7.6 芳烃与硫酰氯的反应[20,21]

7.2 亚磺酰化反应

7.2.1 概述

亚砜能通过无取代芳烃与亚磺酸衍生物进行直接亚磺酰化反应而制备

（图式 7.7），但副产物众多。直接亚磺酰（或磺酰）化的一种替代方法是硫醚化随后氧化[8,22]。

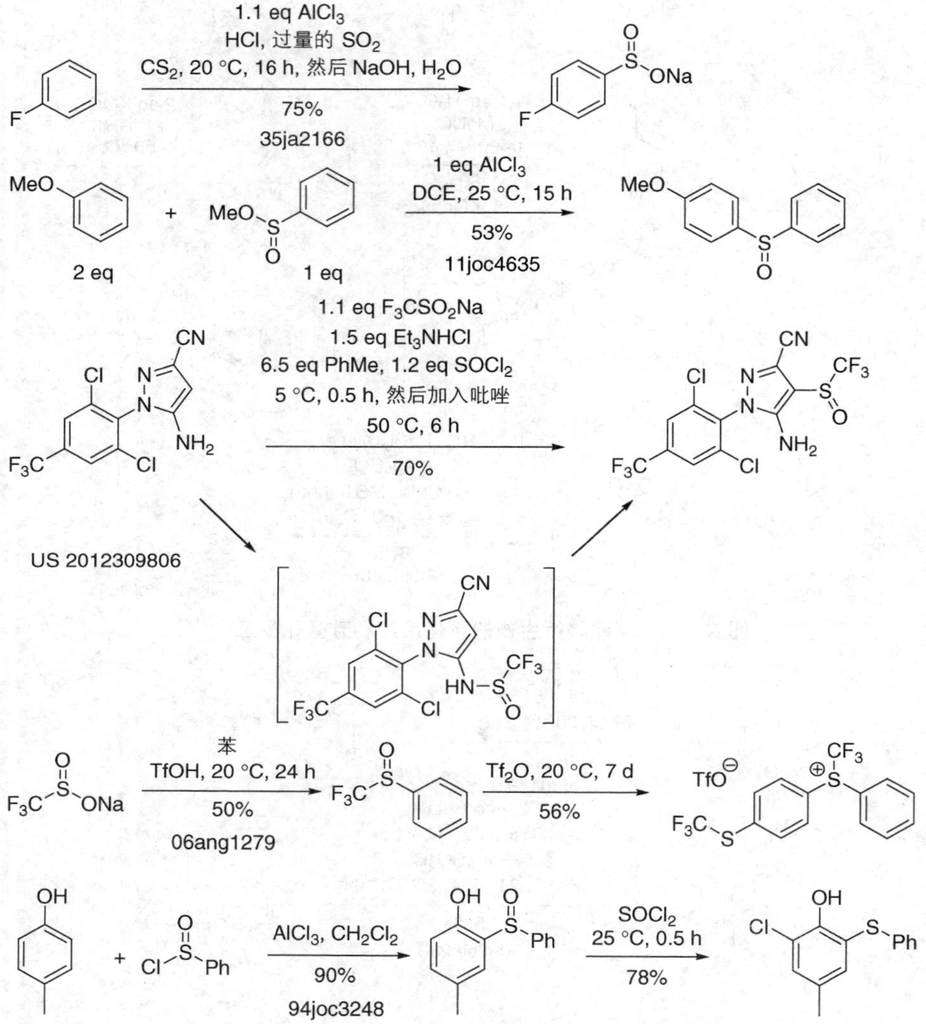

图式 7.7 芳基亚磺酰化的实例[23-27]

7.2.2 典型的副反应

如图式 7.7 和图式 7.8 所示，亚磺酸和亚砜在其制备条件下能作为亲电试剂导致了亚磺酰化的问题。即便是弱亲核试剂（如卤代物），也能够对从亚砜和亲电试剂产生的阳离子中间体进行加成。亲电试剂存在时，亚砜通常能够将脂肪醇氧化成醛或酮［斯文（Swern）和普菲茨纳-莫法特（Pfitzner-Moffatt）氧化］。

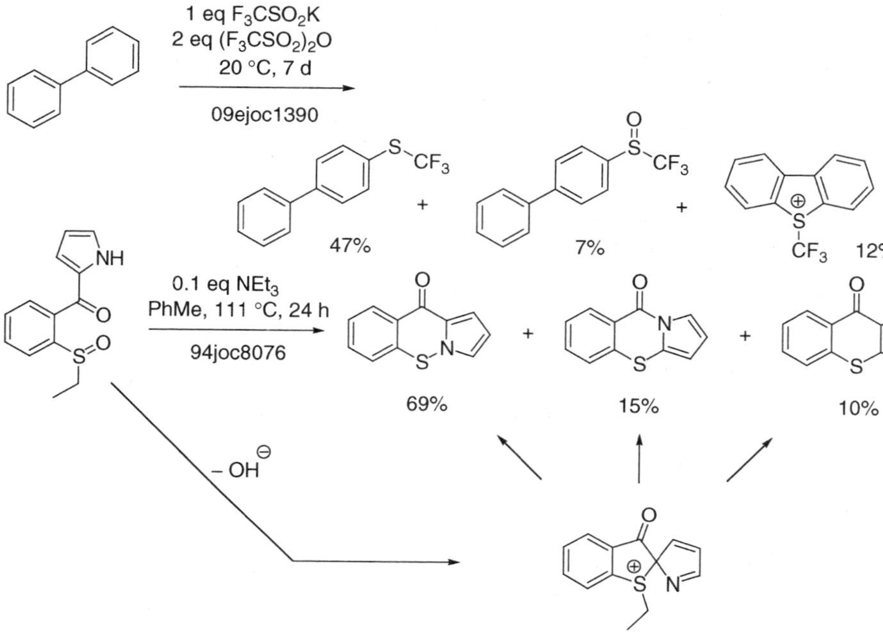

图式 7.8　亚砜分解反应[28,29]

类似于磺酰化,亚磺酰化也是可逆反应。用酸或其他催化剂处理芳基亚砜可以导致重排成更稳定的异构体(图式 7.9)。此外,有的酸,如三氯化铝,能导致亚砜歧化成为砜和硫醚。强碱[30]、四氧化锇[31]或亲电试剂[如普梅瑞尔(Pummerer)重排反应里]也能诱导亚砜的歧化或 $ArSO^-$ 的 β-消除[32]。由于亚砜的高反应活性,二甲基亚砜(DMSO)是一种有害溶剂,不适合于大多数大规模应用。

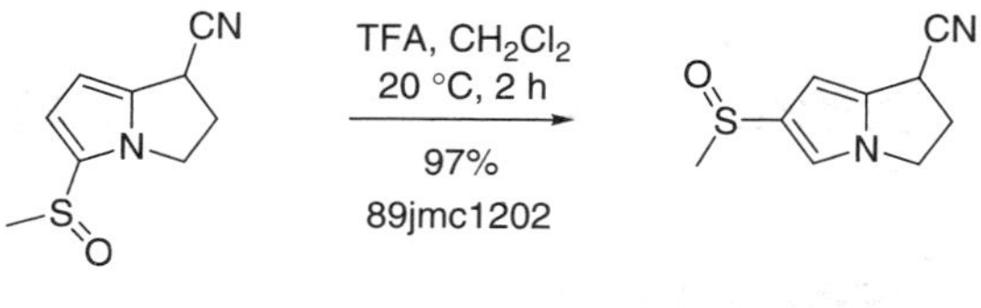

图式 7.9　酸引发的亚砜重排[33]（更多实例：[34]）

从氯化亚砜制备亚硫酰氯或亚砜的实例已有报道[35],但必须谨慎操作。除了上面提到的问题,各种官能团可以被氯化亚砜氯代、亚磺酰化、脱水乃至氧化,这可能会形成意想不到的产物(图式 7.10)。由于亲电亚磺酰化存在许多潜在的副反应,亚磺酸通常通过磺酰氯还原进行制备,而不是通过氯化亚砜或二氧化硫的亲电亚磺酰化反应制备。

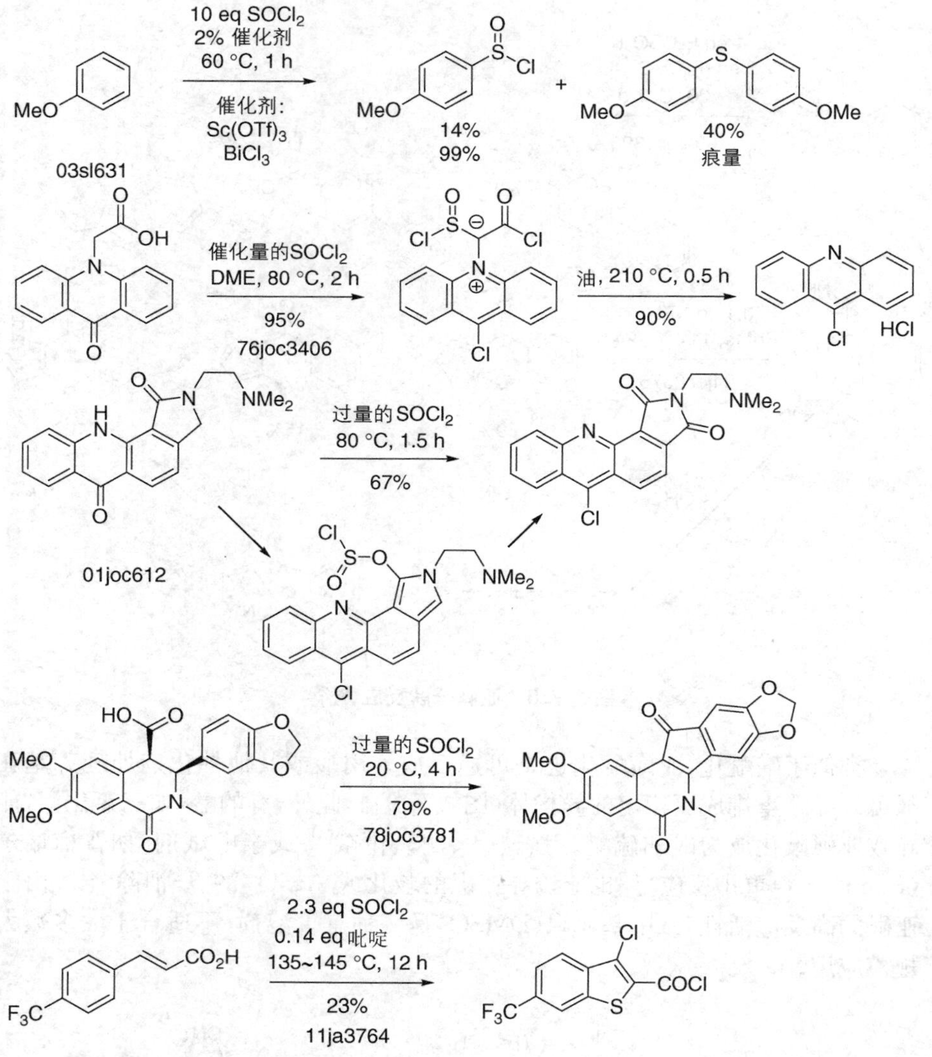

图式 7.10 与氯化亚砜反应中形成的副产物[36-40]（更多实例：[41]）

7.3 硫醚化（次磺酰化）反应

7.3.1 概述

芳基硫醚可以由富电子芳烃和二硫醚[33]、硫醚、次磺酰卤（RSX）、次磺酰胺（RS-NR$_2$）[43]或次磺酸酯（RS-OR）来制备（图式 7.11）。这些反应可以被酸、氧化剂[44]和铜或钯的络合物催化。在强酸存在下，芳烃与亚砜反应生成锍盐，该化合物可与各种亲核试剂进行脱烷基化反应得到硫醚[29,45]。

图式 7.11 芳烃的亲电硫醚化反应[42,46-51]（更多实例：[52]）

7.3.2 典型的副反应

使用二硫醚进行硫醚化反应能够形成硫醇副产物。这些化合物具有强亲核性，会导致不希望的取代反应，例如硫醚的脱烷基化。其他副反应包括多硫醚化和芳烃的卤代（图式7.12）。

图式7.12 芳烃亲电硫醚化反应中的副反应[53,54]

参考文献

1. Tanaka, M. and Souma, Y. (1992) Sulfonation of aromatic compounds in HSO_3F–SbF_5. *J. Org. Chem.*, **57**, 3738–3740.

2. Davidson, A., Moffat, D.F.C., Day, F.A., and Donald, A.D.G. (2008) HDAC inhibitors. WO Patent 2008040934.

3. Valgeirsson, J., Nielsen, E.Ø., Peters, D., Mathiesen, C., Kristensen, A.S., and Madsen, U. (2004) Bioisosteric modifications of 2-arylureidobenzoic acids: selective noncompetitive antagonists for the homomeric kainate receptor subtype GluR5. *J. Med. Chem.*, **47**, 6948–6957.

4. Novello, F.C., Bell, S.C., Abrams, E.L.A., Ziegler, C., and Sprague, J.M. (1960) Diuretics: aminobenzenedisulfonamides. *J. Org. Chem.*, **25**, 965–970.

5. Girard, Y., Atkinson, J.G., Bélanger, P.C., Fuentes, J.J., Rokach, J., Rooney, C.S., Remy, D.C., and Hunt, C.A. (1983) Synthesis, chemistry, and photochemical substitutions of 6,11-dihydro-5*H*-pyrrolo[2,1-*b*][3]benzazepin-11-ones. *J. Org. Chem.*, **48**, 3220–3234.

6. Peyronneau, M., Boisdon, M.-T., Roques, N., Mazières, S., and Le Roux, C. (2004) Total synergistic effect between triflic acid and bismuth(III) or antimony(III) chlorides in catalysis of the methanesulfonylation of arenes. *Eur. J. Org. Chem.*, **2004**, 4636–4640.

7. Field, L. and Settlage, P.H. (1954) Alkanesulfonic acid anhydrides. *J. Am. Chem. Soc.*, **76**, 1222–1225.

8. Parker, J.S., Bower, J.F., Murray, P.M., Patel, B., and Talavera, P. (2008) Kepner–Tregoe decision analysis as a tool to aid route selection. Part 3. Application to a back-up series of compounds in the PDK project. *Org. Process Res. Dev.*, **12**, 1060–1077.

9. Gopalsamy, A., Shi, M., Stauffer, B., Bahat, R., Billiard, J., Ponce-de-Leon, H., Seestaller-Wehr, L., Fukayama, S., Mangine, A., Moran, R., Krishnamurthy, G., and Bodine, P. (2008) Identification of diarylsulfone sulfonamides as secreted frizzled related protein-1 (sFRP-1) antagonist. *J. Med. Chem.*, **51**, 7670–7672.

10. Zhao, X., Dimitrijević, E., and Dong, V.M. (2009) Palladium-catalyzed C–H bond functionalization with arylsulfonyl chlorides. *J. Am. Chem. Soc.*, **131**, 3466–3467.

11. Saidi, O., Marafie, J., Ledger, A.E.W., Liu, P.M., Mahon, M.F., Kociok-Köhn, G., Whittlesey, M.K., and Frost, C.G. (2011) Ruthenium-catalyzed *meta* sulfonation of 2-phenylpyridines. *J. Am. Chem. Soc.*, **133**, 19298–19301.

12. Singh, R.P., Kamble, R.M., Chandra, K.L., Saravanan, P., and Singh, V.K. (2001) An efficient method for aromatic

Friedel–Crafts alkylation, acylation, benzoylation, and sulfonylation reactions. *Tetrahedron*, **57**, 241–247.
13. Shirley, D.A. and Lehto, E.A. (1957) The metalation of 4-*t*-butyldiphenyl sulfone with n-butyllithium. *J. Am. Chem. Soc.*, **79**, 3481–3485.
14. Wenkert, E., Moeller, P.D.R., Piettre, S.R., and McPhail, A.T. (1987) Unusual reactions of magnesium indolates with benzenesulfonyl chloride. *J. Org. Chem.*, **52**, 3404–3409.
15. El-Khawagat, A.M. and Roberts, R.M. (1985) Transsulfonylations between aromatic sulfones and arenes. *J. Org. Chem.*, **50**, 3334–3336.
16. Fleck, T.J., Chen, J.J., Lu, C.V., and Hanson, K.J. (2006) Isomerization-free sulfonylation and its application in the synthesis of PHA-565272A. *Org. Process Res. Dev.*, **10**, 334–338.
17. Yu, X., Li, X., and Wan, B. (2012) Palladium-catalyzed desulfitative arylation of azoles with arylsulfonyl hydrazides. *Org. Biomol. Chem.*, **10**, 7479–7482.
18. Xu, X.-H., Liu, G.-K., Azuma, A., Tokunaga, E., and Shibata, N. (2011) Synthesis of indole and biindolyl triflones: trifluoromethanesulfonylation of indoles with Tf$_2$O/TTBP (2,4,6-tri-*tert*-butylpyridine) system. *Org. Lett.*, **13**, 4854–4857.
19. Cremlyn, R.J.W. and Cronje, T. (1979) Chlorination of aromatic halides with chlorosulfonic acid. *Phosphorus Sulfur Rel. Elem.*, **6**, 495–504.
20. Allwein, S.P., Roemmele, R.C., Haley, J.J., Mowrey, D.R., Petrillo, D.E., Reif, J.J., Gingrich, D.E., and Bakale, R.P. (2012) Development and scale-up of an optimized route to the ALK inhibitor CEP-28122. *Org. Process Res. Dev.*, **16**, 148–155.
21. Bruyneel, F., Payen, O., Rescigno, A., Tinant, B., and Marchand-Brynaert, J. (2009) Laccase-mediated synthesis of novel substituted phenoxazine chromophores featuring tuneable water solubility. *Chem. Eur. J.*, **15**, 8283–8295.
22. Finikova, O.S., Cheprakov, A.V., and Vinogradov, S.A. (2005) Synthesis and luminescence of soluble *meso*-unsubstituted tetrabenzo- and tetranaphtho[2,3]porphyrins. *J. Org. Chem.*, **70**, 9562–9572.
23. Hann, R.M. (1935) Some derivatives of *p*-fluorophenyl sulfinic acid. *J. Am. Chem. Soc.*, **57**, 2166–2167.
24. Yuste, F., Hernández Linares, A., Mastranzo, V.M., Ortiz, B., Sánchez-Obregón, R., Fraile, A., and García Ruano, J.L. (2011) Methyl sulfinates as electrophiles in Friedel–Crafts reactions. Synthesis of aryl sulfoxides. *J. Org. Chem.*, **76**, 4635–4644.
25. Sukopp, M., Kuhn, O., Gröning, C., Keil, M., and Longlet, J.J. (2012) Process for the sulfinylation of a pyrazole derivative. US Patent 2012309806.
26. Magnier, E., Blazejewski, J.-C., Tordeux, M., and Wakselman, C. (2006) Straightforward one-pot synthesis of trifluoromethyl sulfonium salts. *Angew. Chem. Int. Ed.*, **45**, 1279–1282.
27. Jung, M.E., Kim, C., and von dem Bussche, L. (1994) Vicarious nucleophilic aromatic substitution via trapping of an α-ketosulfonium ion generated by Pummerer-type rearrangement of 2-(phenylsulfinyl)phenols: preparation of biaryls. *J. Org. Chem.*, **59**, 3248–3249.
28. Macé, Y., Raymondeau, B., Pradet, C., Blazejewski, J.-C., and Magnier, E. (2009) Benchmark and solvent-free preparation of sulfonium salt based electrophilic trifluoromethylating reagents. *Eur. J. Org. Chem.*, **2009**, 1390–1397.
29. Bates, D.K. and Tafel, K.A. (1994) Sulfenylation using sulfoxides. Intramolecular cyclization of 2- and 3-acylpyrroles. *J. Org. Chem.*, **59**, 8076–8080.
30. Furukawa, N., Ogawa, S., Matsumura, K., and Fujihara, H. (1991) Extremely facile ligand-exchange and disproportionation reactions of diaryl sulfoxides, selenoxides, and triarylphosphine oxides with organolithium and Grignard reagents. *J. Org. Chem.*, **56**, 6341–6348.
31. Davis, H.R. and Sorensen, D.P. (1959) Disproportionation of organic sulfoxides. US Patent 2870215.
32. Forbes, D.C., Bettigeri, S.V., Al-Azzeh, N.N., Finnigan, B.P., and Kundukulama, J.A. (2009) Sulfenylation chemistry using polymer-supported sulfides. *Tetrahedron Lett.*, **50**, 1855–1857.

33. Muchowski, J.M., Galeazzi, E., Greenhouse, R., Guzmán, A., Peréz, V., Ackerman, N., Ballaron, S.A., Rovito, J.R., Tomolonis, A.J., Young, J.M., and Rooks, W.H. (1989) Synthesis and anti-inflammatory and analgesic activity of 5-aroyl-6-(methylthio)-1,2-dihydro-3*H*-pyrrolo[1,2-*a*]pyrrole-1-carboxylic acids and 1-methyl-4-(methylthio)-5-aroylpyrrole-2-acetic acids. *J. Med. Chem.*, **32**, 1202–1207.

34. Carmona, O., Greenhouse, R., Landeros, R., and Muchowski, J.M. (1980) Synthesis and rearrangement of 2-(arylsulfinyl)- and 2-(alkylsulfinyl)pyrroles. *J. Org. Chem.*, **45**, 5336–5339.

35. Reddy, B.M., Sreekanth, P.M., and Lakshmanan, P. (2005) Sulfated zirconia as an efficient catalyst for organic synthesis and transformation reactions. *J. Mol. Catal. A*, **237**, 93–100.

36. Peyronneau, M., Roques, N., Mazières, S., and Le Roux, C. (2003) Catalytic Lewis acid activation of thionyl chloride: application to the synthesis of aryl sulfinyl chlorides catalyzed by bismuth(III) salts. *Synlett*, 631–634.

37. Ning, R.Y., Madan, P.B., Blount, J.F., and Fryer, R.I. (1976) Structure and reactions of an unusual thionyl chloride oxidation product. 9-Chloroacridinium 2-chloro-l-(chlorosulfinyl)-2-oxoethylide. *J. Org. Chem.*, **41**, 3406–3409.

38. Feng, S., Panetta, C.A., and Graves, D.E. (2001) An unusual oxidation of a benzylic methylene group by thionyl chloride: a synthesis of 1,3-dihydro-2-[2-(dimethylamino)ethyl]-1,3-dioxopyrrolo[3,4-*c*]acridine derivatives. *J. Org. Chem.*, **66**, 612–616.

39. Cushman, M. and Cheng, L. (1978) Stereoselective oxidation by thionyl chloride leading to the indeno[1,2-*c*]isoquinoline system. *J. Org. Chem.*, **43**, 3781–3783.

40. Cheng, K., Wang, X., and Yin, H. (2011) Small-molecule inhibitors of the TLR3/dsRNA complex. *J. Am. Chem. Soc.*, **133**, 3764–3767.

41. Higa, T. and Krubsack, A.J. (1975) Oxidations by thionyl chloride. VI. Mechanism of the reaction with cinnamic acids. *J. Org. Chem.*, **40**, 3037–3045.

42. Gillis, H.M., Greene, L., and Thompson, A. (2009) Preparation of sulfenyl pyrroles. *Synlett*, 112–116.

43. Ranken, P.F. and McKinnie, B.G. (1989) Alkylthio aromatic amines. *J. Org. Chem.*, **54**, 2985–2988.

44. Prasad, C.D., Balkrishna, S.J., Kumar, A., Bhakuni, B.S., Shrimali, K., Biswas, S., and Kumar, S. (2013) Transition-metal-free synthesis of unsymmetrical diaryl chalcogenides from arenes and diaryl dichalcogenides. *J. Org. Chem.*, **78**, 1434–1443.

45. Ukai, S. and Hirose, K. (1968) Die Reaktion der Phenolderivate mit Sulfoxiden. IV. (einschließlich der von Sulfiden und Wasserstoffperoxid ausgehenden Reaktion). Die Synthese von 4-thiosubstituierten 1,2-Naphthochinonderivaten. *Chem. Pharm. Bull.*, **16**, 606–612.

46. Delprato, I. and Bertoldi, M. (1997) Process for preparation of 4-mercapto-1-naphthol compounds. EP Patent 0763526.

47. Schlosser, K.M., Krasutsky, A.P., Hamilton, H.W., Reed, J.E., and Sexton, K. (2004) A highly efficient procedure for 3-sulfenylation of indole-2-carboxylates. *Org. Lett.*, **6**, 819–821.

48. Yang, Y., Jiang, X., and Qing, F.-L. (2012) Sequential electrophilic trifluoromethanesulfanylation-cyclization of tryptamine derivatives: synthesis of C(3)-trifluoromethanesulfanylated hexahydropyrrolo[2,3-*b*]indoles. *J. Org. Chem.*, **77**, 7538–7547.

49. Chen, X., Hao, X.-S., Goodhue, C.E., and Yu, J.-Q. (2006) Cu(II)-catalyzed functionalizations of aryl C–H bonds using O_2 as an oxidant. *J. Am. Chem. Soc.*, **128**, 6790–6791.

50. Zhang, S., Qian, P., Zhang, M., Hu, M., and Cheng, J. (2010) Copper-catalyzed thiolation of the di- or trimethoxybenzene arene C–H bond with disulfides. *J. Org. Chem.*, **75**, 6732–6735.

51. Chu, L., Yue, X., and Qing, F.-L. (2010) Cu(II)-mediated methylthiolation of aryl C–H bonds with DMSO. *Org. Lett.*, **12**, 1644–1647.

52. Tran, L.D., Popov, I., and Daugulis, O. (2012) Copper-promoted sulfenylation of

sp² C–H bonds. *J. Am. Chem. Soc.*, **134**, 18237–18240.

53. Yu, C., Zhang, C., and Shi, X. (2012) Copper-catalyzed direct thiolation of pentafluorobenzene with diaryl disulfides or aryl thiols by C–H and C–F bond activation. *Eur. J. Org. Chem.*, **2012**, 1953–1959.

54. Klečka, M., Pohl, R., Čejka, J., and Hocek, M. (2013) Direct C–H sulfenylation of purines and deazapurines. *Org. Biomol. Chem.*, **11**, 5189–5193.

8 芳烃的亲核取代反应

8.1 概述

虽然不如芳烃亲电取代重要,但芳烃的亲核取代反应在有机合成中也很有价值。许多药物、农用化学品和其他结构复杂的化合物在其工业合成路线中都涉及芳环的亲核取代反应。许多综述已经被发表[1-4]。

8.1.1 机理

芳烃的亲核取代反应能够通过加成-消除(S_NAr)、攫取-加成(S_N1)、自由基($S_{RN}1$)机理或形成芳炔中间体进行(图式8.1)。第一种机理主要见于具有离域负电荷到电负性原子能力的缺电子芳烃或杂芳烃。在离去基团的邻位或对位有吸电子基团,可以提高芳香亲核取代反应的速率,而给电子基团则抑制该反应。能大大增加反应速率的基团有重氮、亚胺、亚硝基和硝基。

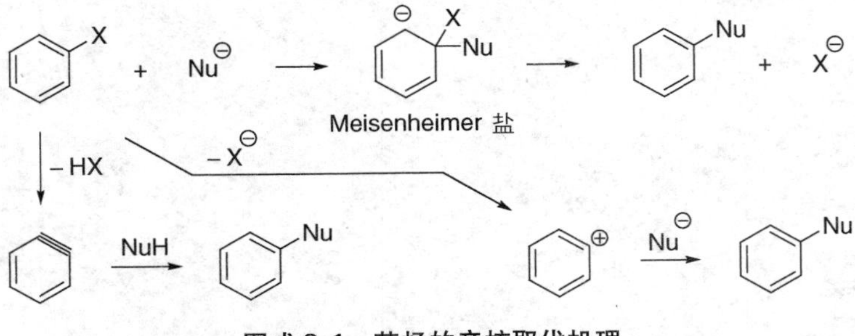

图式8.1 芳烃的亲核取代机理

由于萘去芳构化需要的能量比单环芳烃低(萘的两个环中仅一个是完全芳香性的)[5,6],萘的衍生物尤其容易进行芳香亲核取代反应(图式8.2)。例如,2-硝基萘的化学性质更像硝基烯烃而不是硝基芳烃[7]。

8.1.2 区域选择性

由于亲核试剂对缺电子芳烃的加成是可逆反应,故只有在芳烃上存在优

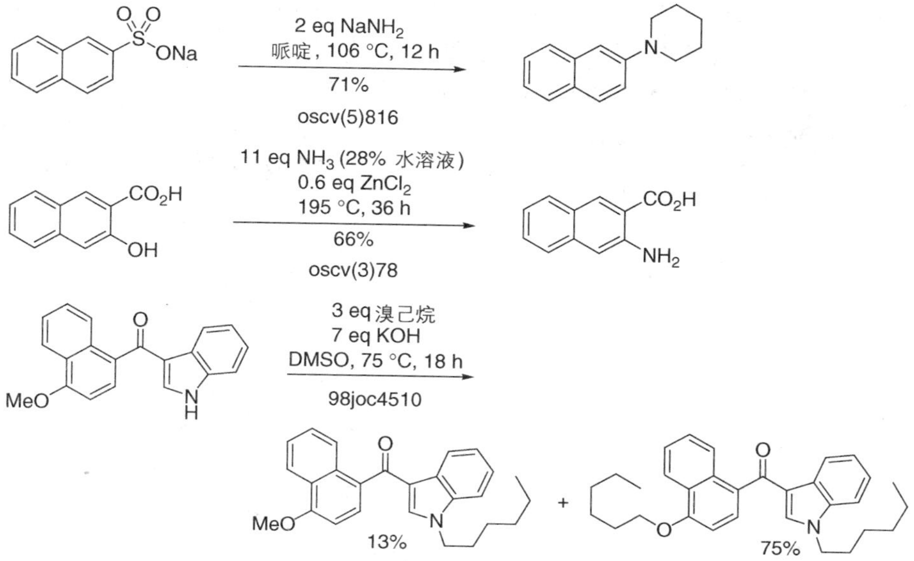

图式 8.2 萘的芳香亲核取代反应[6,8,9]

良的离去基团时，取代反应可以顺利进行。对于芳香亲核取代反应，优良的离去基团包括氟、亚硝酸盐、磺酸盐、其他卤素原子和亚磺酸酯。在氧化剂的存在下，氢原子也可以作为离去基团。离去基团的精准排序也取决于溶剂[亲核试剂、迈森海默（Meisenheimer）复合物和离去基团的溶剂化]和亲核试剂（迈森海默复合物里的氢键）[10-12]。例如，氨基在非质子溶剂里是不良离去基团，但是用水作溶剂时要比氯原子更容易被羟基取代（图式 8.3）。6-（烷基磺酰基）嘌呤与 MeOH/DBU（1,8-二氮杂双环[5.4.0]十一碳-5-烯）的反应速度比 6-碘代嘌呤快，但是与苯胺的反应速度却比 6-碘代嘌呤慢了许多[13]。芳烃亲核取代反应中更多溶剂效应实例如图式 8.4 所示。

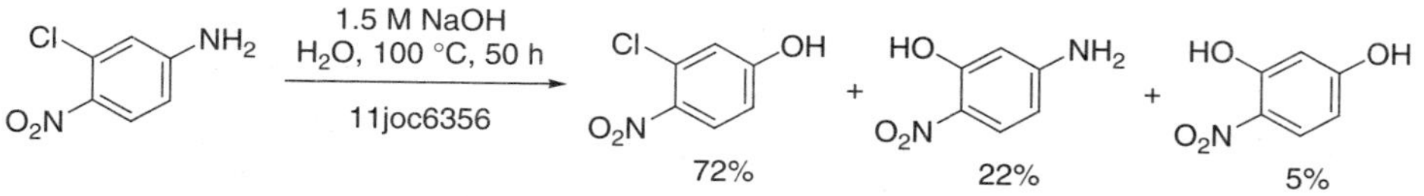

图式 8.3 在水相的氨基的取代反应[11]

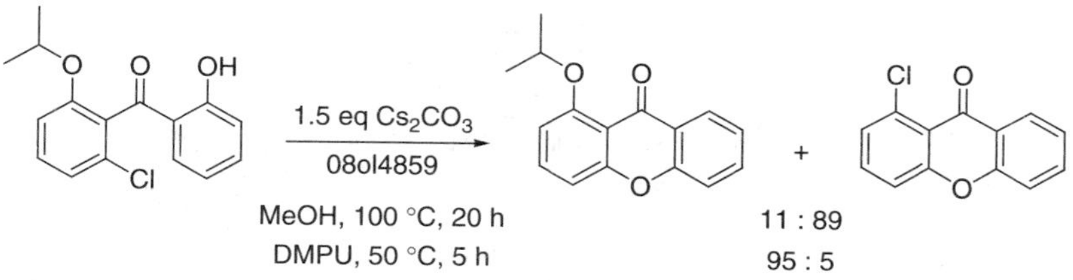

图式 8.4 芳烃亲核取代的溶剂效应[14]

邻近和远程取代反应,即亲核试剂进攻离去基团邻近的或更远的远程位置,在芳烃的亲核取代反应中很少见[15](如图式 8.5、图式 8.6 及图式 3.3 所示)。在大多数情况下,亲电试剂是高度缺电子的低芳香性芳烃或杂芳烃,对于发生质子转移和双键迁移反应,非芳香性反应中间体有足够长的半衰期。质子溶剂常常促进这些让人惊奇的反应,使用过渡金属催化剂是避免这些反应的一种策略。

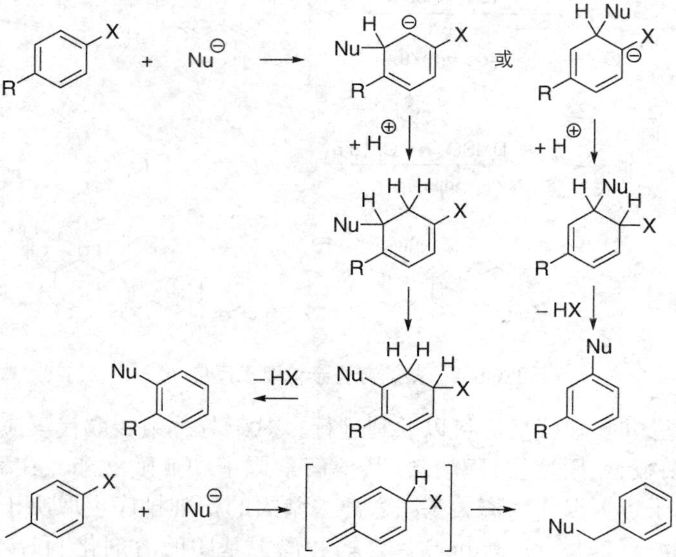

图式 8.5 邻近和远程芳烃亲核取代反应的机理

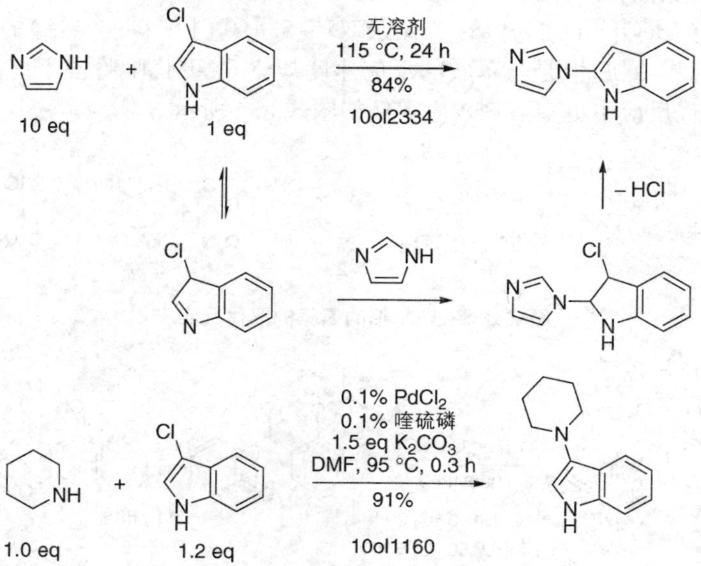

图式 8.6 芳烃亲核取代中的邻近取代反应[16-22] (更多实例:[23-26])

续图式 8.6

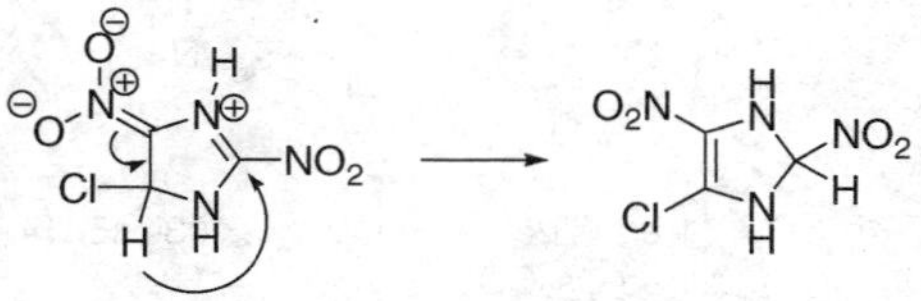

续图式 8.6

210 芳烃亲核取代反应产物也可以重排成更稳定的异构体。例如,多卤代芳烃在强碱存在下很容易异构化(图式 8.7)。在酸的存在下,吲哚硫醚有时可在非常温和的反应条件下发生异构化(图式 8.7)。

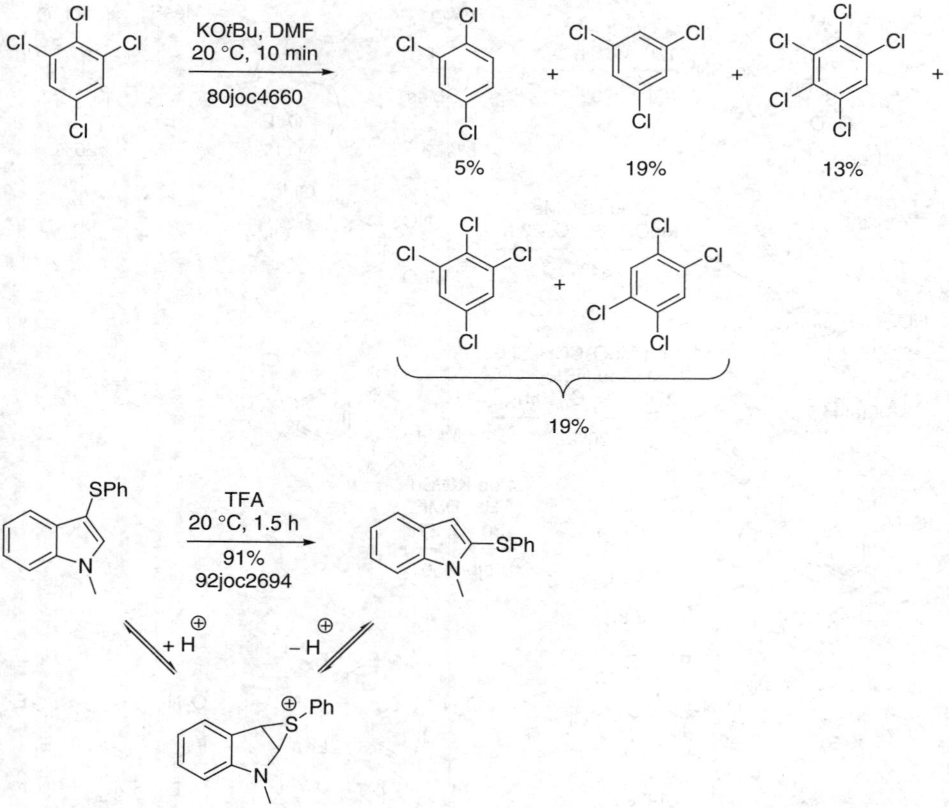

图式 8.7 多卤代芳烃和吲哚硫醚的异构化[27,28]

卤代烷在芳香亲核取代反应过程中可能会与芳香亲电试剂竞争。通常情况下,非催化的芳香亲核取代反应要比活泼的卤代烷的亲核取代反应慢(图 8.8)。但是,确切的反应性顺序与反应物的结构、溶剂及具体的反应条件有关[29]。因为卤代芳烃与过渡金属络合物的反应往往比卤代烷快,过渡金属催化可以加速芳香取代反应。

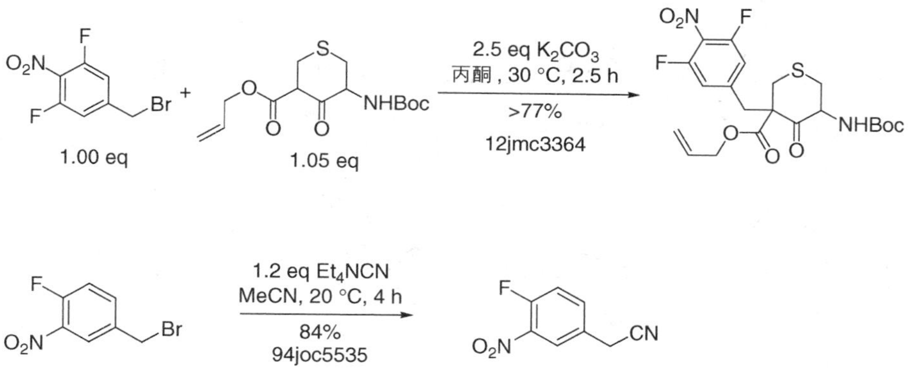

图式 8.8 脂肪族亲核取代反应可能与芳香亲核取代反应竞争[30,31]

苄基卤与亲核试剂反应,可以是直接的脂肪链上的亲核取代反应,也可以亲核进攻芳环。缺电子芳烃或低芳香性的芳烃(例如萘)倾向于发生后一种类型的取代,当苄基 C—H 受到空间位阻屏蔽时也有利于亲核试剂直接进攻芳烃,但也有令人惊讶的意外结果(图式 8.9)。

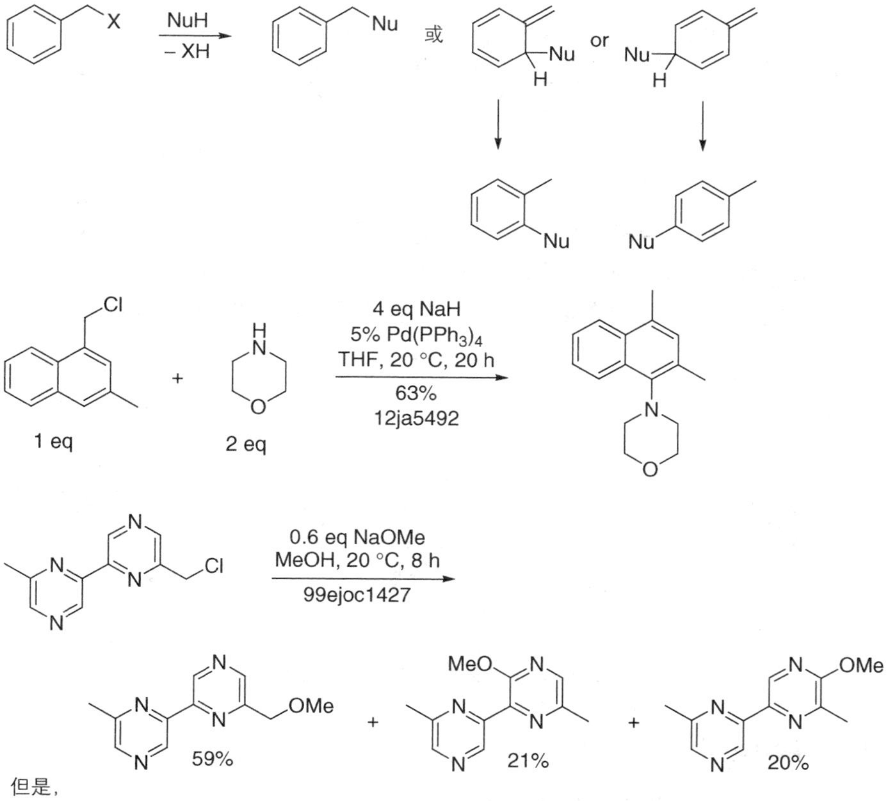

但是,

图式 8.9 苄基卤和亲核试剂反应[32-36]

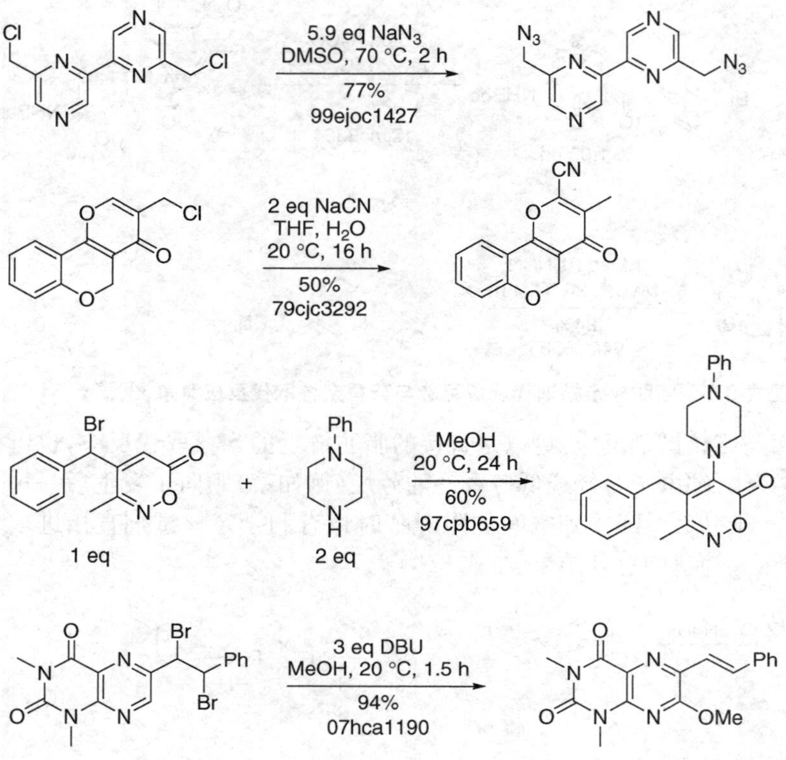

续图式 8.9

通过螯合方式，芳烃和烷烃可在意想不到的位置进行金属化。已有一些例子使用这种策略对烷基进行芳基化反应（图式 8.10）。由于烷烃和芳基卤皆可以金属化，所以自偶联产物是在这些有趣的反应中能够预料到的副产物。

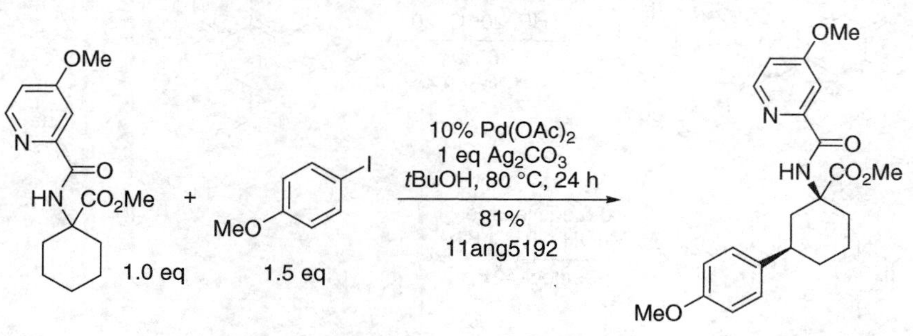

图式 8.10　环己基通过螯合进行芳基化[37]

8.1.3　酸/碱催化

芳烃的亲核取代反应无论由酸催化还是由碱催化均取决于亲核试剂和亲电试剂的碱性。质子化通常可以增加亲电试剂的亲电性，但同时降低亲核试剂的亲核性。

缺电子的杂芳烃,如吡啶、嘧啶或吡嗪,质子化后与亲核试剂反应的活性更高。因此,这些杂芳烃的芳香取代反应可以通过酸催化或自催化(没有额外的酸或碱加入)(图式8.11)。无论如何,随着亲核试剂的碱性增加会减弱这种催化的效果。

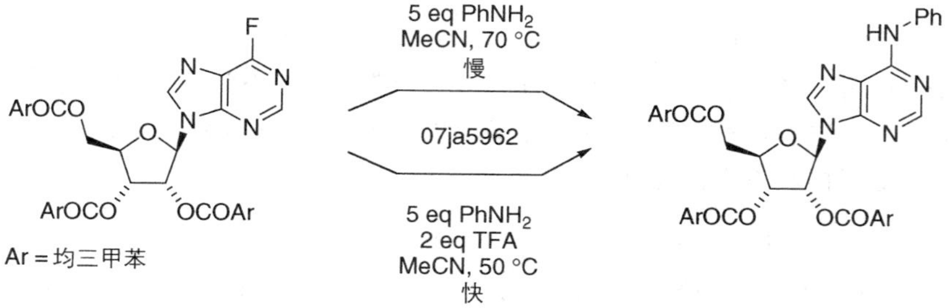

图 8.11　酸催化的 6-氟嘌呤的芳香亲核取代反应[13]

8.1.4　过渡族金属催化

芳香亲核取代也可以通过过渡金属催化进行。钯、镍和铜的络合物能强烈促进取代反应,在没有催化剂存在下,这些反应需要高温才可进行。例如,钯催化可以使芳基溴代物转化成为苯胺而没有发生烷基氯的位移或消除(图式 8.12 中第一个反应式)。

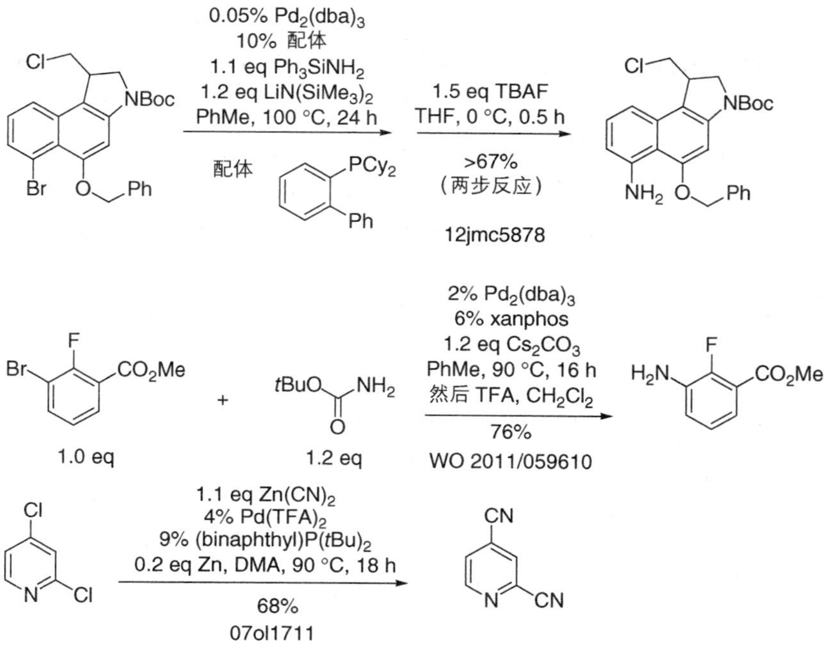

图式 8.12　钯和铜催化的芳烃亲核取代反应[38-41]

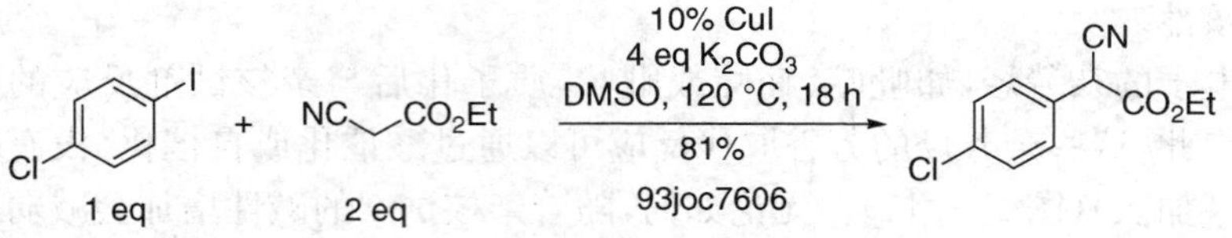

[图式反应: 4-氯碘苯 + NC-CH₂-CO₂Et, 10% CuI, 4 eq K₂CO₃, DMSO, 120 °C, 18 h, 81%, 93joc7606, 产物为 4-氯苯基-CH(CN)-CO₂Et]
1 eq + 2 eq

续图式 8.12

一价铜催化的反应被认为是按照图式 8.13 给出的方式进行。醇或胺与一价铜盐络合增强了它们的酸性,进一步在碱的存在下形成铜醇化合物或铜胺化合物。这些物种与芳基碘或芳基溴发生氧化加成得到三价铜中间体,而后释放出产物及再生催化活性的一价铜络合物。

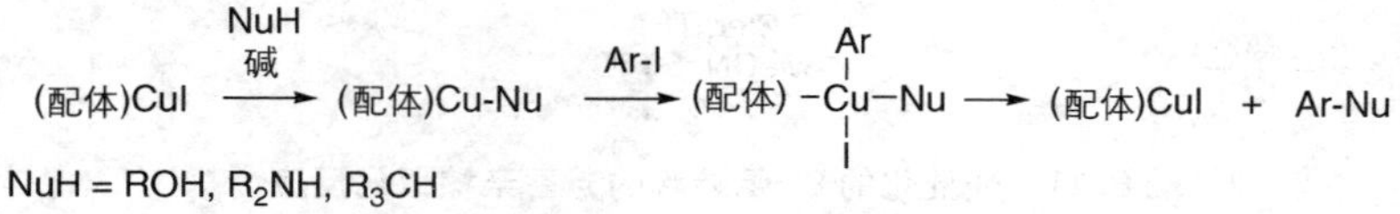

[图式: (配体)CuI →(NuH,碱)→ (配体)Cu-Nu →(Ar-I)→ (配体)Ar-Cu(-Nu) → (配体)CuI + Ar-Nu]
NuH = ROH, R₂NH, R₃CH

图式 8.13 一价铜催化芳烃亲核取代反应的机理[42,43]

钯络合物催化类似于铜催化,主要区别在于芳基卤的氧化加成发生在与亲核试剂反应之前。此外,钯催化剂是更适合于不活泼的亲电试剂,如芳基氯、磺酸酯或氨基甲酸酯。

对于这两种金属而言,强配体能促进催化活性络合物的再生,并防止其与亲核试剂形成稳定的无催化反应活性的络合物。因此,使用螯合型配体(1,3-二酮、双膦、乙二胺、邻二氮杂菲、联吡啶、氨基酸等)显著地拓宽铜和钯催化的芳香取代的使用范围。使用双齿膦配体甚至可以实现用量低至 0.01% 钯催化硫醇的芳香亲核取代反应[44]。

由于过渡金属催化反应需通过芳基卤对金属的氧化加成进行,芳基卤对亲核试剂的反应活性不同于其在非催化芳香亲核取代反应里的活性(图式 8.14)。铜催化对芳基碘代物和溴代物最为有效,而钯催化剂适用于芳基氯代物、溴代

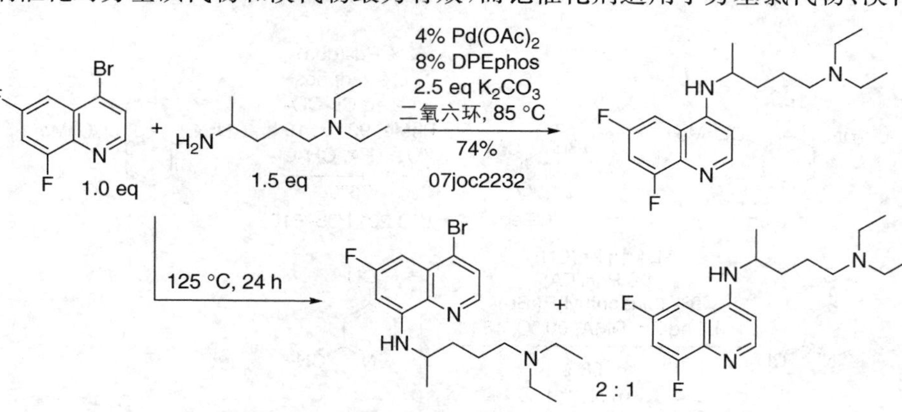

[图式: 4-溴-6,8-二氟喹啉 + H₂N-CH(CH₃)CH₂CH₂CH₂-N(Et)₂, 4% Pd(OAc)₂, 8% DPEphos, 2.5 eq K₂CO₃, 二氧六环, 85 °C, 74%, 07joc2232]
1.0 eq + 1.5 eq

125 °C, 24 h → 两种产物 2:1

图式 8.14 钯催化和未催化的多卤代喹啉取代反应的选择性[45]

物、碘代物、磺酸盐、氨基甲酸盐和重氮盐。芳基氟化物、苯胺、硝基芳烃和芳基砜对铜或钯络合物通常是惰性的。

过渡金属催化的芳香亲核取代反应一个潜在的副反应是联芳基的形成（沃尔曼反应）。低浓度亲核试剂、高温和氧化剂（例如空气）的存在能够促进这一反应。苯胺和苯酚特别容易进行氧化二聚，而二芳基胺和二芳基醚通过氧化脱氢能够生成咔唑和二苯并呋喃。氧化剂与过渡金属催化剂、高温组合也可能导致许多不需要的氧化反应，例如苄胺[46]、苄醇或其他官能团的脱氢。如果二甲基亚砜（DMSO）被用作铜催化芳香亲核取代反应的溶剂，可能产生甲基硫醚副产物[47]（图式 7.11）。有 N,N-二甲基甲酰胺（DMF）的存在时，芳基碘代物可以被钯催化转化成为 N,N-二甲基苯甲酰胺[48]。

在各种还原剂如锌[49]、葡萄糖[50]、羟胺、异丙醇、氢醌、DMF、胺[51] 或膦[52] 存在下，芳基钯和芳基镍络合物容易发生二聚。还原剂也可以促进无取代芳烃的形成（脱卤反应）。

8.2 亲电试剂的问题

8.2.1 不兼容的官能团

由于芳香亲核取代反应通常需要强亲核试剂，而其他官能团也有可能与亲核试剂反应。醛特别敏感，可以发生卡尼扎罗（Cannizzaro）型氧化还原反应、脱甲酰化或与亲核试剂发生不可逆的加成反应（例如得到重亚硫酸加成产物、氰醇或硫缩醛）。然而，也有许多成功使用卤代苯甲醛进行芳香亲核取代反应的报道（图式 8.15）。特别稳定的（和不活泼的）是 4-烷氧基、4-羟基和 4-氨基苯甲醛。

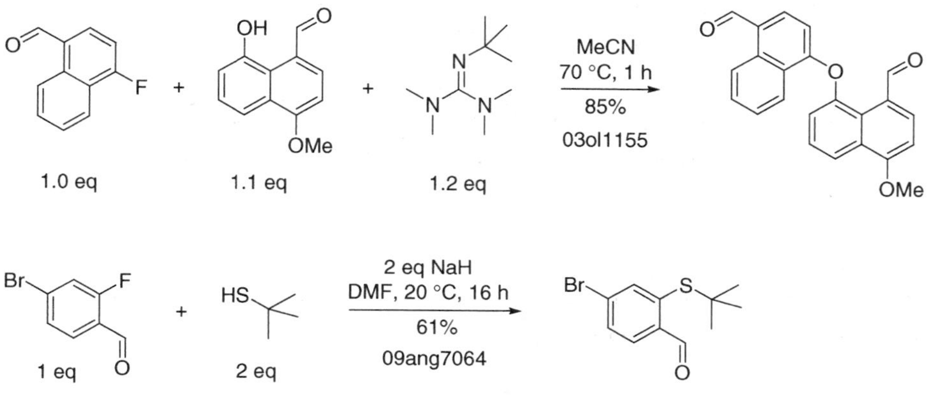

图式 8.15　卤代苯甲醛的芳香亲核取代反应[53,54]

一些亲核试剂,如硫醇、硫化物[55]、亚硫酸盐、亚磺酸、醇和氰化物,也能够还原亲电试剂中的官能团。硝基、亚硝基、甲酰基和重氮基团特别容易被还原。如果亲电试剂是强缺电子并且在空间上不能接近,则可能发生单电子转移(SET)的还原反应。

8.2.2 非活化芳烃

在没有过渡金属络合物存在的情况下,非缺电子的芳基卤仅仅在剧烈反应条件(图式 8.16)或光化学条件[56]下进行 $S_N Ar$ 反应。如果亲核试剂具有强碱性,反应可能经过芳基炔中间体形成区域异构体的混合物。碱性的亲核试剂和高反应温度都能够导致许多副反应,例如醛的卡尼扎罗反应和酯、酰胺、碳酸酯、氨基甲酸酯、酮(哈勒-鲍尔反应)、砜及亚砜的裂解[57]。在氧气或其他氧化剂的存在下,强碱可使许多有机化合物形成过氧化物、羟基化或脱氢。剧烈反应条件,尤其在痕量过渡金属存在下,可能导致一些通常不认为是离去基团的取代基发生取代反应。

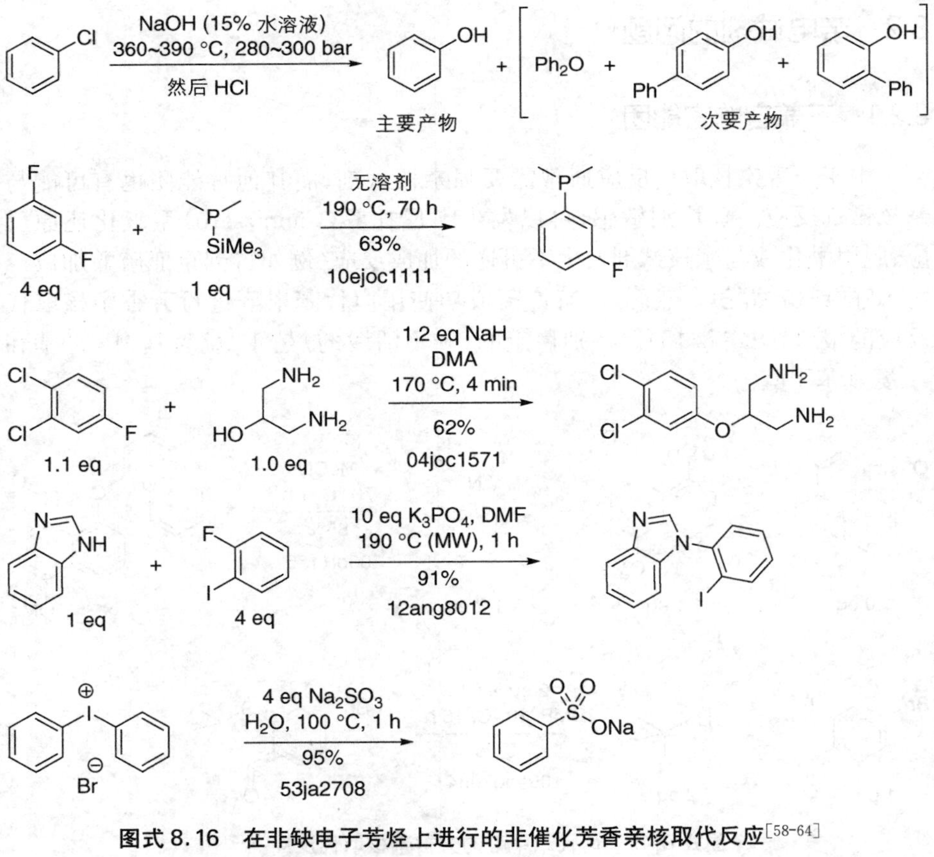

图式 8.16 在非缺电子芳烃上进行的非催化芳香亲核取代反应[58-64]

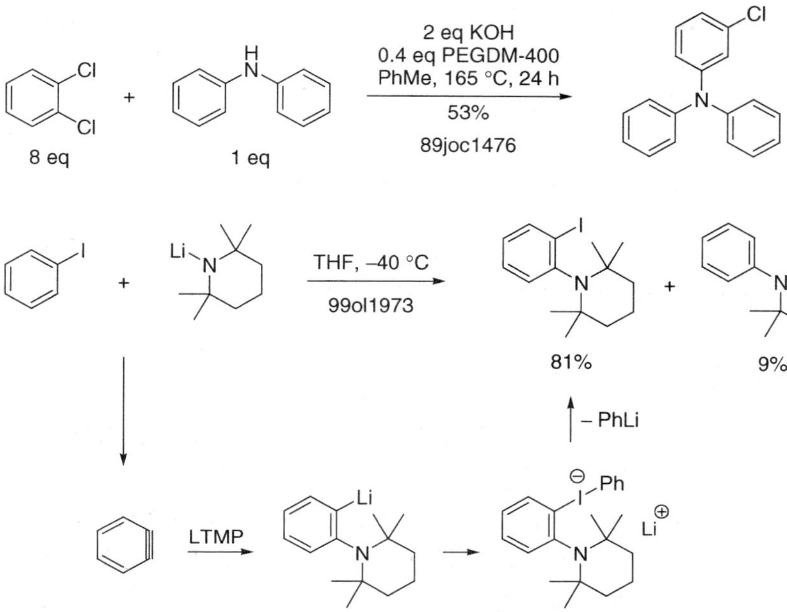

续图式 8.16

芳基磺酸与氢氧化钠或氢氧化钾在 250～300℃熔融是一个常用的合成苯酚的方法[65,66]。该反应通常只得到取代反应产物，没有区域异构体产生。1926 年，在米德兰的陶氏化学通过将起始原料泵入一个长约 1 英里①、温度达 300℃的管道[停留时间为 10～30 min，哈勒-布雷顿(Hale-Britton)工艺，图式 8.16 中第一个反应]的方式从氯苯和氢氧化钠生产苯酚[67]。

氟和其他卤素可以增强芳烃 C—H 的酸性。因此多卤代芳烃可以用作亲电卤代试剂(卤素阳离子转移到亲核试剂)，或在强碱存在下进行异构化("卤素舞蹈")(图式 8.7)。

活化非缺电子卤代芳烃的一个实验室秘诀是与三羰基铬络合[68](图式 8.17)。不幸的是，对于大多数工业应用而言，此方法会产生太多的废物并且过于昂贵。

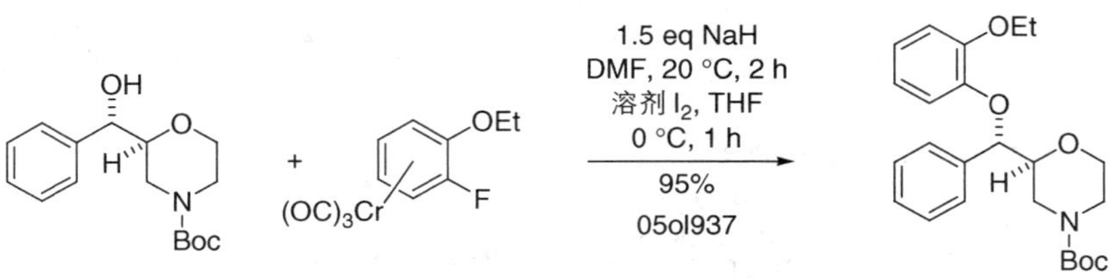

图式 8.17 芳烃铬三羰基络合物的芳香亲核取代反应[69]

---

① 1 英里=1.609 km。

空间位阻的需求似乎并不能阻止 S_NAr 反应的发生,已有 2,6-二烷基芳基卤代物成功取代的例子被报道[70](图式 8.18)。然而,在这些实例中的亲核试剂体积相当小(氰化物、卤素、叠氮化物、亚硝酸盐等)。

图式 8.18　空间位阻要求苛刻的芳基溴化物的 S_NAr 反应[71]

8.2.3　硝基芳烃

由于其强的吸电子作用,硝基是芳香亲核取代反应首选的活化基团。1-氯或 1-氟-2,4-二硝基苯在室温下即可与许多亲核试剂反应。但需要注意的是,硝基也是一个很好的离去基团,有时很难预测哪一个基团会被取代。此外,有时硝基芳烃的芳香性很低,反应更像硝基烯烃。例如,硝基芳烃可与甲亚胺叶立德进行 1,3-偶极环加成反应[72]。硝基也容易被富电子的亲核试剂还原,从而导致另一类型的副产物(图式 8.19)。

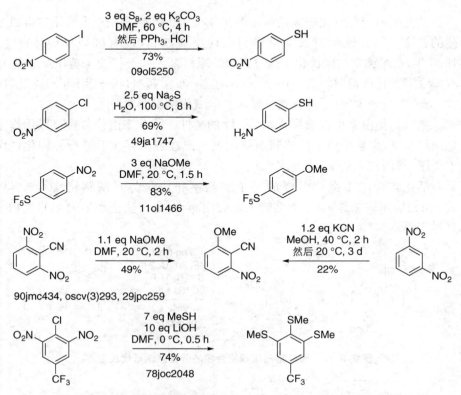

图式 8.19　硝基苯的芳香亲核取代反应[73-80](更多实例:[57,81])

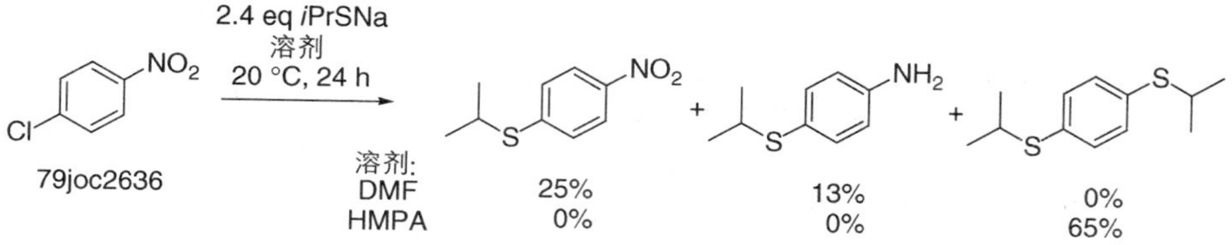

溶剂:			
DMF	25%	13%	0%
HMPA	0%	0%	65%

续图式 8.19

由于其具有高反应活性,硝基苯是一种危险的溶剂,在过度加热或在用酸或碱处理时会猛烈分解。已经有无数关于硝基苯爆炸的报道。

硝基芳烃及一些缺电子杂芳烃可通过加成消除机理(替代亲核取代)进行烷基化(图式 8.20)。进攻氢取代的位置通常比加成到卤素取代的位置更快,使得在氧化剂(硝基芳烃、DMSO)的存在下,2-(或 4-)卤代硝基苯氢被取代的速度要比卤素被取代的速度更快[82-85]。也有报道伯胺或仲胺对硝基芳烃进行加成,随后脱氢[86,87]。

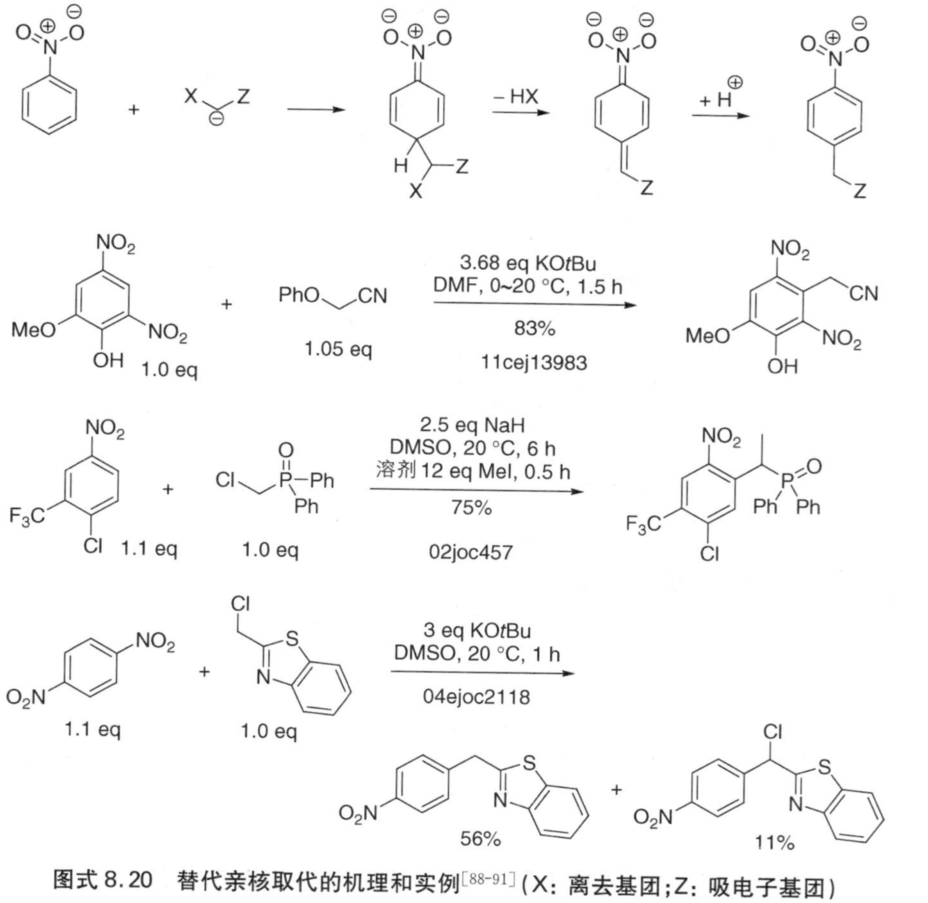

图式 8.20 替代亲核取代的机理和实例[88-91] (X: 离去基团;Z: 吸电子基团)

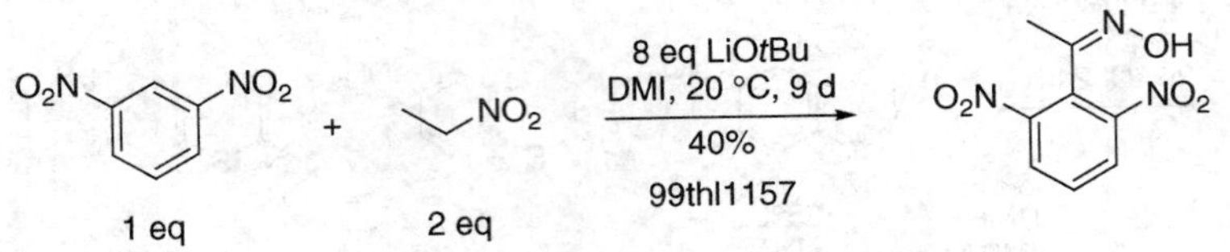

续图式 8.20

适合于进行替代亲核取代的烷基化试剂有 α-卤代砜和具有 C—H 酸性且酸性 C—H 位置带离去基团的化合物。不幸的是，此反应仅仅适用于硝基芳香化合物，其他缺电子芳烃进行该反应的报道几乎没有[92]。

与用碳亲核试剂进行替代亲核取代密切相关的是硝基芳烃与过氧化物（图式 8.21）、肼（例如，$H_2N-NMe_3 I$[93]）、叠氮化物（第 8.3.6 节）或者羟胺反应。由于在硝基芳烃未取代的位置也可以发生反应，故这些反应的区域选择性很难预测。

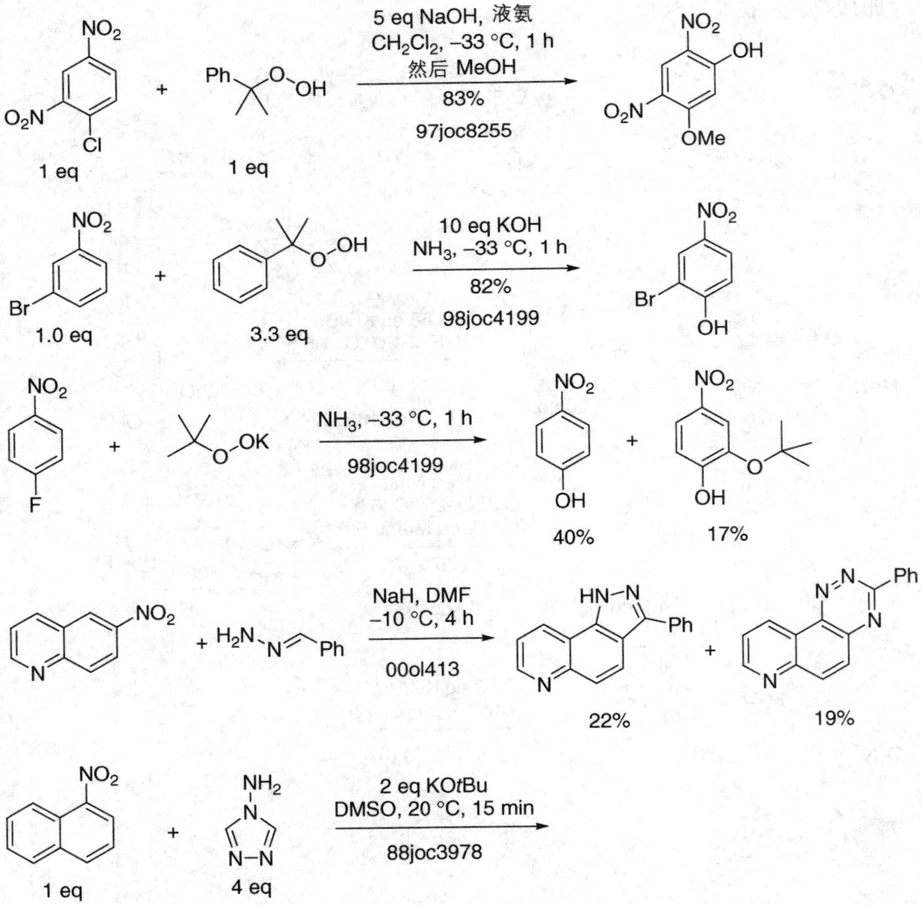

图式 8.21 硝基芳烃的氨化和羟化[94-97]（更多实例：[7]）

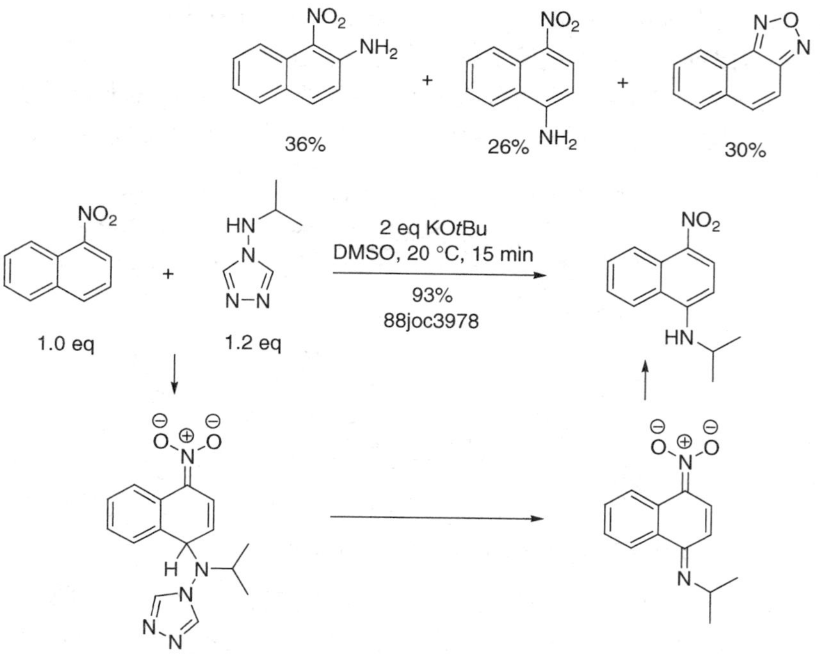

续图式 8.21

硝基苯的一个著名的但意外的反应是冯·李希特(von Richter)重排(图式 8.22)。氰化物在含水乙醇中回流处理,将 1-硝基芳烃转化成 2-羧酸芳烃。此反应比溴或氯的芳香亲核取代速度更快,因为亲核试剂对芳烃上无取代基位置的加成通常比有取代基的位置的取代更快[84]。

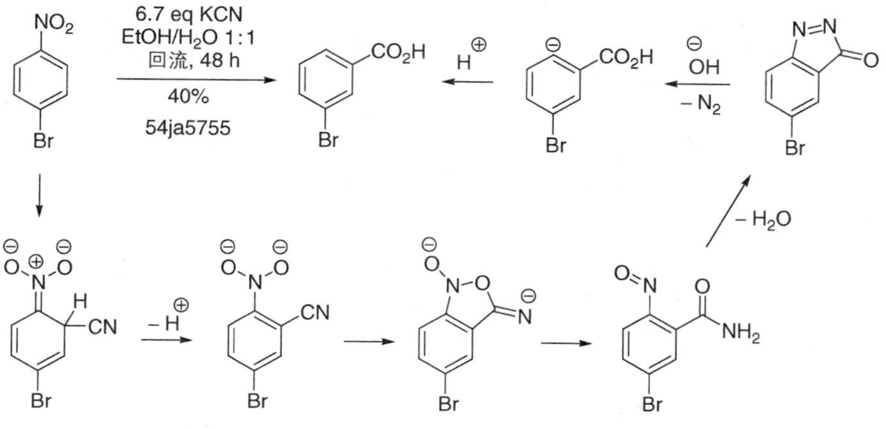

图式 8.22 冯·李希特重排的机理[98]

以铁氰化钾(II)粉末(该试剂不引发冯·李希特重排[98])或氰化锌作为氰基来源,用镍[99]、铜或钯[100-102]催化能够将溴代硝基苯转化为硝基苯腈(图式 8.23)。在这些反应中很重要的一点是,保持氰化物浓度尽可能低,以防止形成稳定的无催化活性的过渡金属氰化物。通过加入还原剂(锌或异丙醇[103])可以从无活性的氰化钯再生具有催化活性零价钯,通过在严格的非质子碱性条件下操作,可以防止氰化氢的形成,氰化氢与钯络合物的反应速度比氰化物更快[104]。

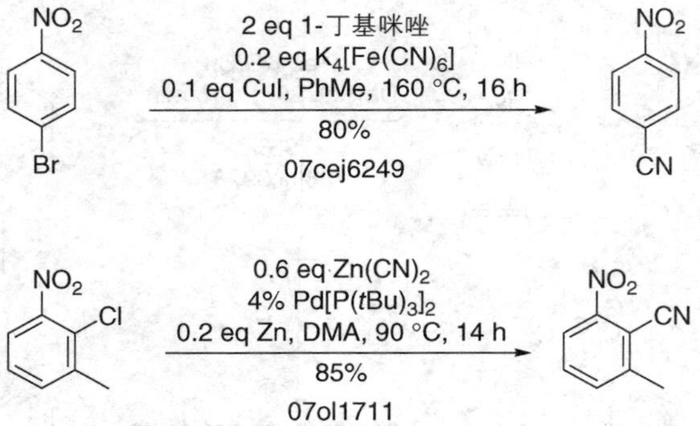

图式 8.23 硝基苯卤化物与氰化物的催化芳香亲核取代反应[40,105]

硝基芳烃也可以进行氰化物加成,然后被氧化成硝基苯腈(图式 8.19),或者消除水分子得到 2-亚硝基苯腈(图式 8.25)。这些反应的低收率是硝基芳烃既可作为氧化剂又是产物前体的双重身份引起的。

硝基芳烃容易还原成亚硝基芳烃或苯胺,或者通过单电子转移形成芳基自由基。可以还原硝基芳烃的典型亲核试剂包括硫化物、亚硫酸盐和醇化物(图式 8.24)。这些还原反应有时可以通过降低亲核试剂用量或反应温度避免[79]。

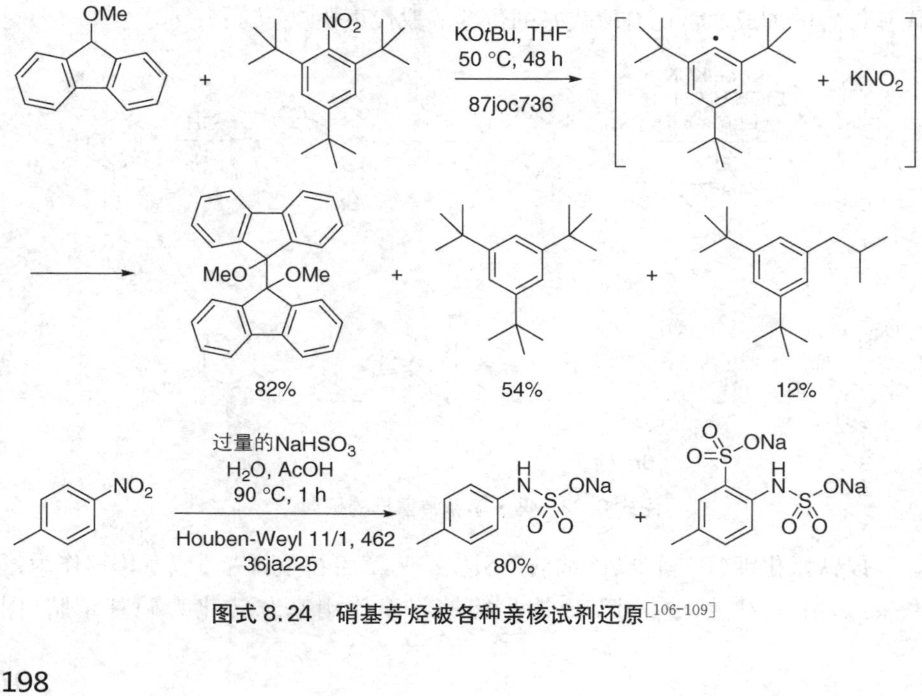

图式 8.24 硝基芳烃被各种亲核试剂还原[106-109]

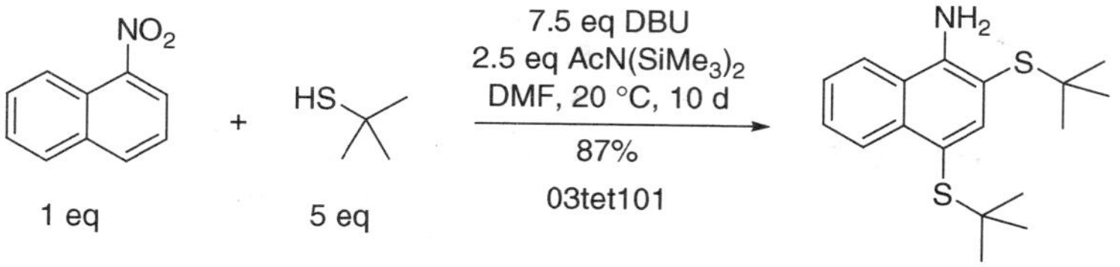

续图式 8.24

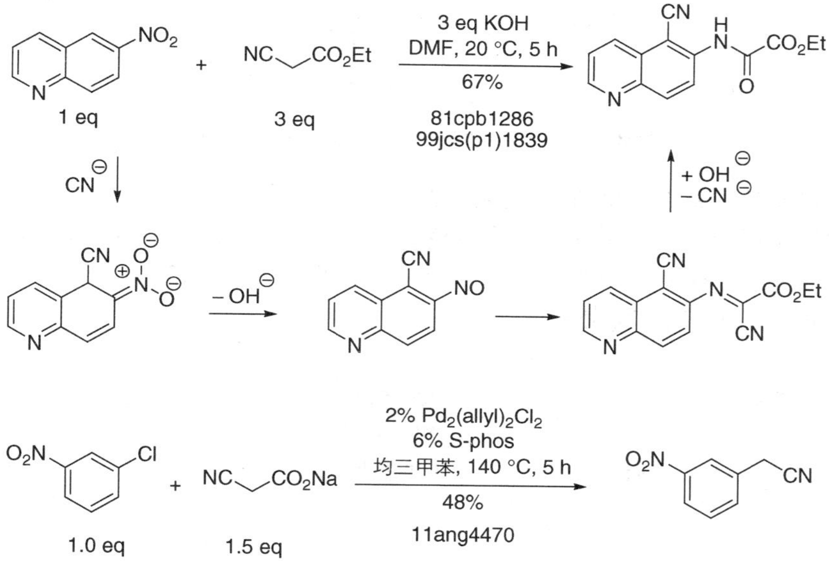

图式 8.25 氰基乙酸衍生物与硝基芳烃的反应[110-112]（更多实例：[113]）

当用富电子亲核试剂进行反应时，空间位阻大的硝基芳烃能够作为芳基自由基的前体（图式 8.24）。因其空间位阻大，使得产生的自由基大多会从其他反应底物攫氢或发生重排反应。

硝基芳烃能被亚硫酸盐水溶液还原。亚硫酸盐容易对亚硝基芳烃中间体加成，通常会得到氨基苯磺酸或（磺酰氨基）苯磺酸［皮瑞尔（Piria）反应］。酸性反应条件下，N-芳基羟基胺中间体可以重排成氨基酚。

C—H 酸性化合物，如氰基乙酸酯、丙二酸酯或 β 酮酯，可与硝基芳烃缩合。在碱存在下，氰基乙酸酯与缺电子的硝基芳烃反应形成 2-氨基苯腈的草酰胺（图式 8.25），反应经 2-亚硝基苯腈进行[110]。该反应可以通过过渡金属络合物催化卤代苯腈的 $S_N Ar$ 反应被阻止。

图式 8.25 中的第一个反应所需的氰化物来自氰基乙酰胺在碱引发下的降解。事实上，在碱的存在下，氰基乙酸衍生物和丙二腈可以都作为氰化物来源（图式 8.26）。因此，与氰基乙酸衍生物的芳香亲核取代反应的副反应之一是形成苄腈或苯甲酸。后者也可能源于芳基化氰基乙酸衍生物的氧化降解[114]。

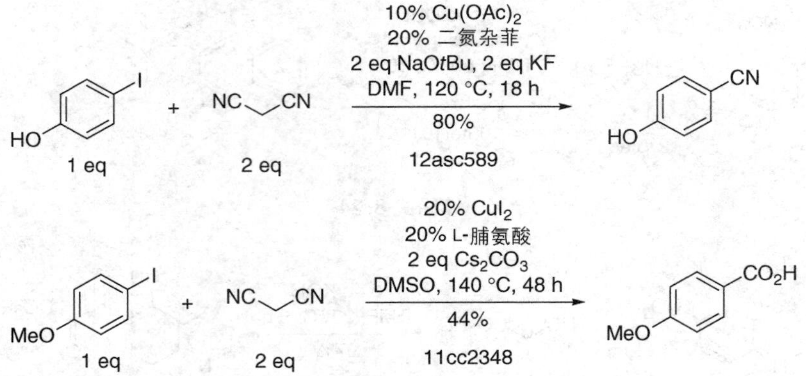

图式 8.26 丙二腈的芳香亲核取代反应[114,115]

硝基芳烃是强氧化剂,并且可以与醇、胺或其他潜在氢供体反应,尤其是在过渡金属的存在下。有时芳香亲核取代反应需要高的反应温度,残留的催化剂可能导致硝基芳烃的还原和各种副产物的形成。例如,硝基芳烃与伯醇反应可以产生喹啉(图式 8.27)。

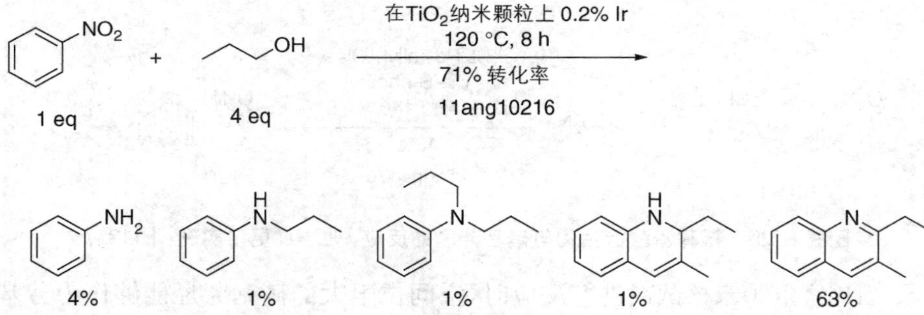

图式 8.27 铱催化的丙醇与硝基苯的反应[116]

一些杂芳烃在用亲核试剂处理时会发生开环反应,例如,3-硝基苯硫酚被仲胺裂解(图式 8.28)。

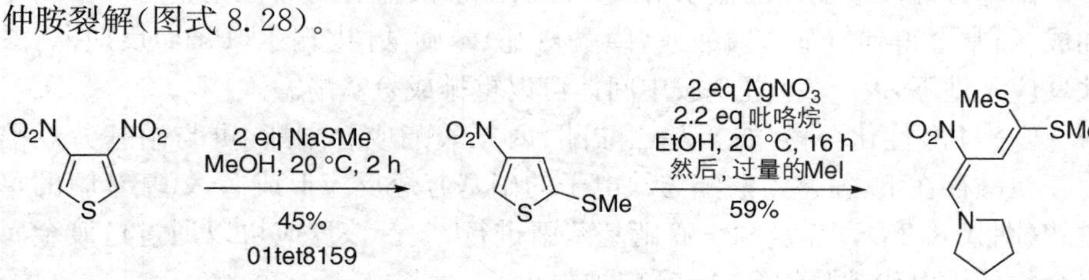

图式 8.28 硝基噻吩的环裂变[117]

8.2.4 重氮盐

尽管氮气是一个很好且环保的离去基团,但芳基重氮盐并不总是芳香亲核取代合适的亲电试剂。碱性反应条件下芳基重氮盐可以在氮原子上被亲核试剂加成，是强单电子氧化剂。需要大量化学计量铜盐的经典桑德迈尔(Sandmeyer)反应和相关的无铜取代反应可以将芳基重氮盐转化为芳基卤、芳基叠氮化物、硝基芳烃、酚、苯甲酸、苯腈、苯甲醛、苯乙酮、苯磺酰氯、芳基硫醚以及芳基硫酯[118]。通过梅尔外因(Meerwein)反应和刚伯格-巴赫姆反应可实现芳基重氮盐的烯基化和芳基化。然而，没有一个高收率且具普遍适用于胺和醇取代重氮基的工艺。在酸性条件下用醇处理芳基重氮盐会得到醚和还原芳烃的混合物，其比例取决于芳基重氮盐发生的取代模式[119]。只有烷基或烷氧基取代，或者邻位有能够稳定芳基阳离子的取代基的重氮盐，有时候能够在酸存在下纯净地进行 S_NAr 反应[120]（图式 8.29 中第一个例子）。而在碱性条件下，胺和醇加成到重氮基团上得到三氮烯和重氮醚，其通过芳基自由基的热分解得到芳烃还原产物和其他产物。胺和醇在与芳基自由基的反应中主要作为氢原子供体（从 H—C—N/O 攫取氢）。在碱性条件下，醇化物也可以通过氢转移还原重氮盐[121]。

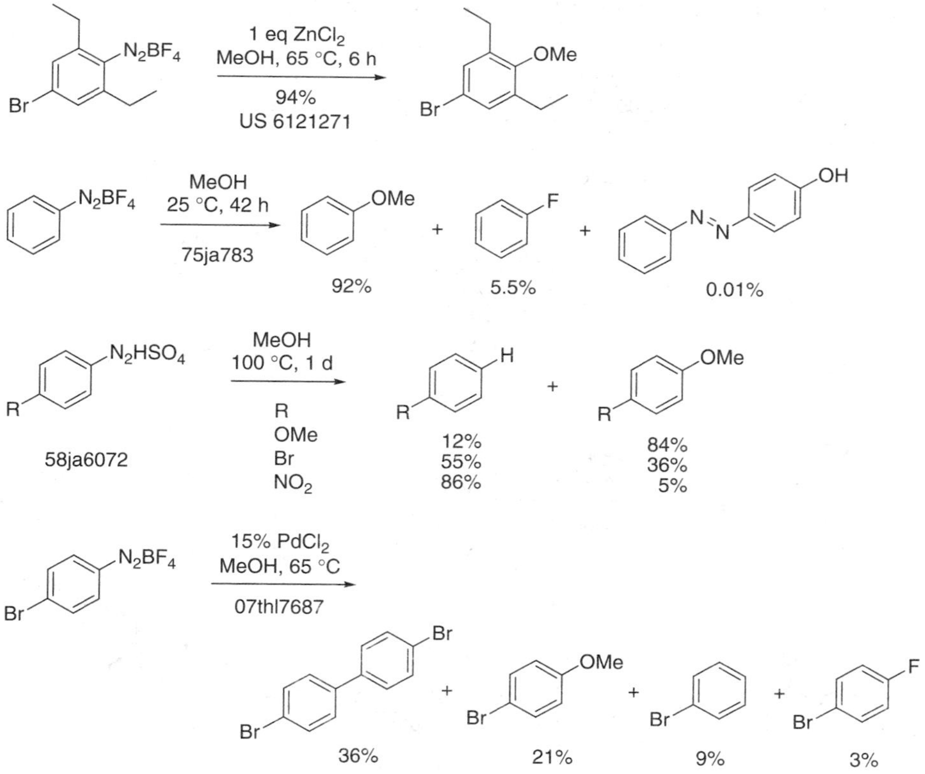

图式 8.29　芳基重氮盐与醇的反应[119,122-124]

重氮盐基团强烈活化芳烃有利于芳香亲核取代反应。特别是缺电子芳烃重氮盐的形成,需要强酸和较高的反应温度(因为最后的脱水步骤缓慢),其邻位的卤素或硝基取代基容易进行芳香亲核取代[125](图式 8.30)。这样的反应可以通过排除反应体系中可能的亲核试剂被避免,例如,使用四氟硼酸代替盐酸或硫酸进行重氮化。

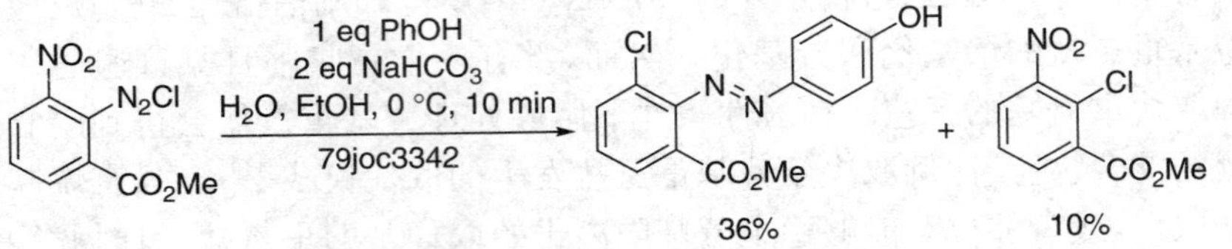

图式 8.30 芳基重氮盐的芳香亲核取代反应[126]

8.2.5 酚类

一些酚,特别是多羟基苯和萘酚,偶尔表现出酮类(环己二烯酮)的性质,可以与氨或其他亲核试剂发生羟基的亲核替代反应。至于与酮类(例如,形成缩醛或烯胺),这样的反应优先在酸存在下发生(图式 8.31)。

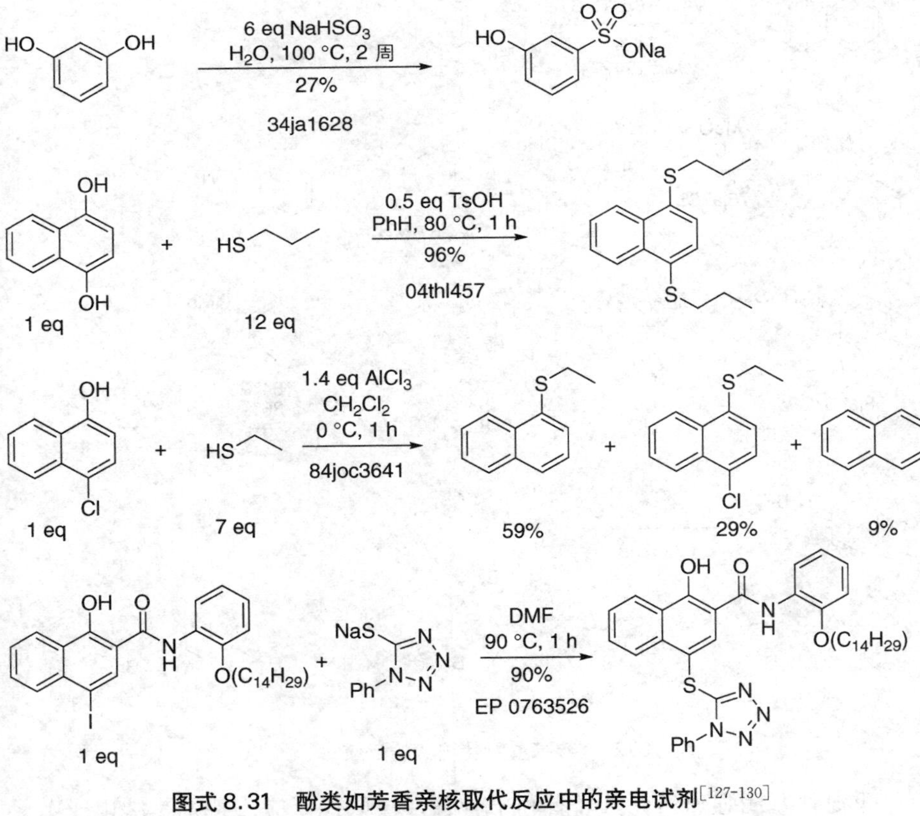

图式 8.31 酚类如芳香亲核取代反应中的亲电试剂[127-130]

8.2.6 芳基醚和芳基硫醚

烷氧基苯是合适的芳香亲核取代反应亲电试剂(图式 8.32)。已有许多 2-甲氧基苯甲酸衍生物的甲氧基被亲核试剂取代的例子被报道[131,132]。但是芳基醚也可能作为烷基化试剂反应[133,134]。

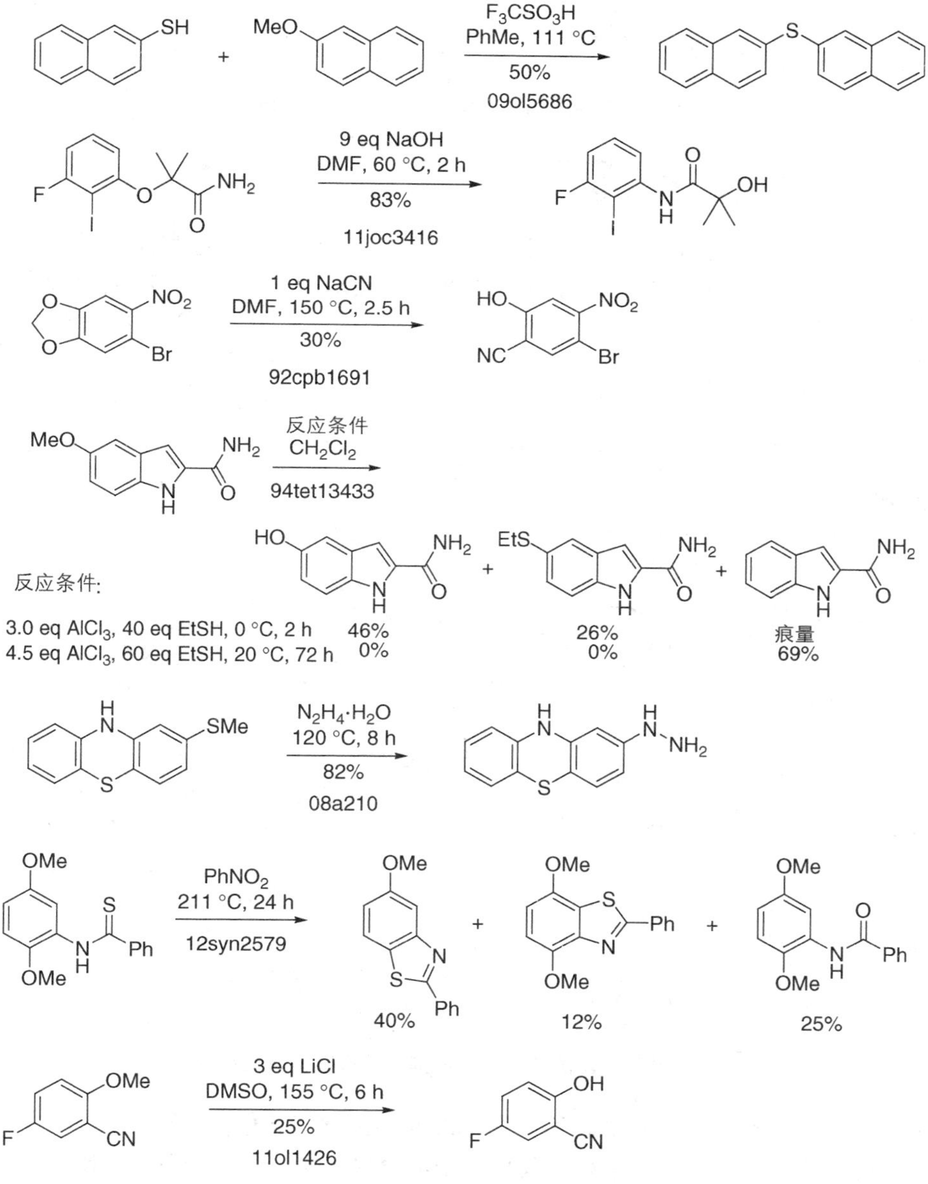

图式 8.32 芳基醚与亲核试剂的反应[133,135-140] (更多实例: [141])

芳基硫醚也被用作芳基化的亲核试剂(例如,格氏试剂[142]),但也可以作为烷基化试剂(图式 8.33)。例如,用硫醇处理芳基卤代物能够通过硫醇引发的芳基硫醚中间体脱烷基化形成苯硫酚。

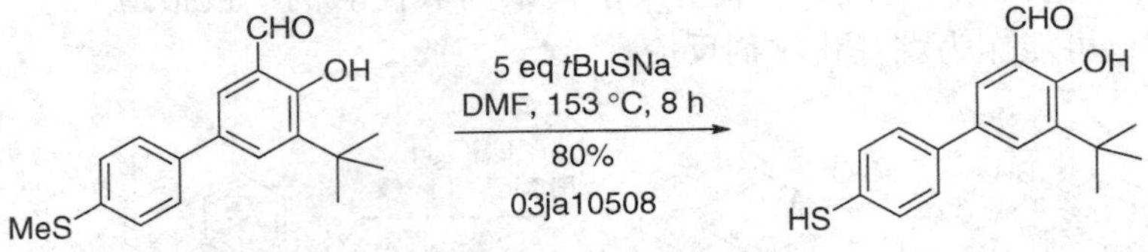

图式 8.33 芳基醚作为烷基化试剂[143](更多实例:[144])

8.2.7 其他苯酚衍生亲电

芳基磺酸酯是适合于芳香亲核取代反应的亲电试剂,但会发生一些 S—O 键的亲核裂解反应[145,146]。例如烷基磺酸芳基酯对碱不稳定,能够转化醇化物成为磺酸酯(图式 8.34)。使用铜或者钯催化剂可以避免 S—O 键的断裂和酚的形成。

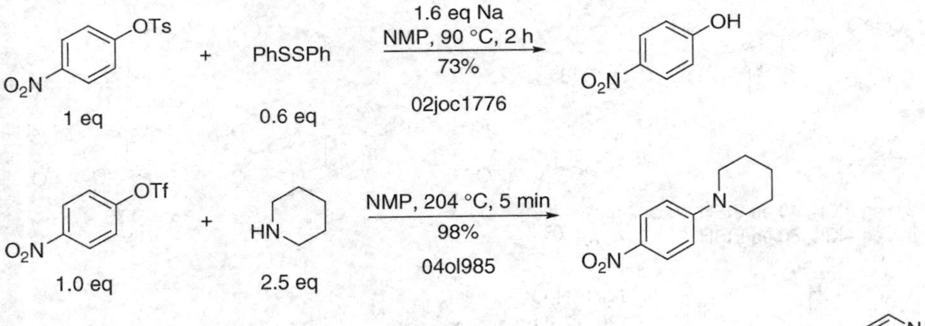

图式 8.34 芳基磺酸酯作为亲电试剂的芳香亲核取代反应[147-149]

由苯酚衍生的酯、碳酸酯和氨基甲酸酯与磺酸酯比较,更容易被亲核试剂裂解,因此很少用作芳香亲核取代反应中的亲电试剂。此外,借助过渡金属催化剂这种反应有时候也能发生(图式 8.35)。

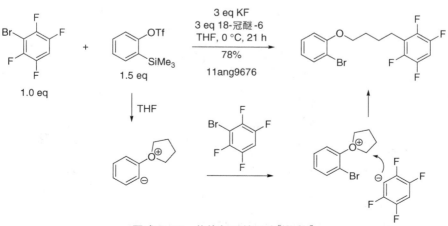

图式 8.35 在芳香亲核取代反应中羧酸盐和氨基甲酸酯作为离去基团[150,151]（更多实例：[152]）

8.2.8 芳炔

芳炔是特别活泼的一类亲电芳基化试剂。这些中间体可通过芳基卤、重氮化的氨基苯甲酸或者(2-硅基芳基)三氟甲磺酸酯的 β-消除产生。芳炔的高能量使其能够与弱亲核试剂进行芳基化反应。其两性离子中间体能够进行异常丰富但也会反复无常的化学转化（图式 8.36）。

图式 8.36 芳炔与醚的反应[153,154]

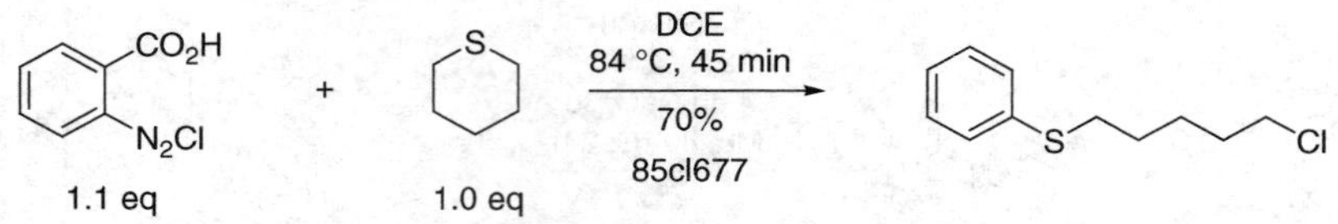

8.3 亲核试剂的问题

8.3.1 烯醇盐

跟其他两性亲核试剂一样,烯醇盐能产生异构体产物的混合物。在非极性溶剂(如醚)中,金属烯醇化合物不完全解离并且与氧结合的金属可以阻止 O-芳基化。尽管如此,烯醇碱金属化物的芳基化通常需要通过反复试验进行优化(图式 8.37)。

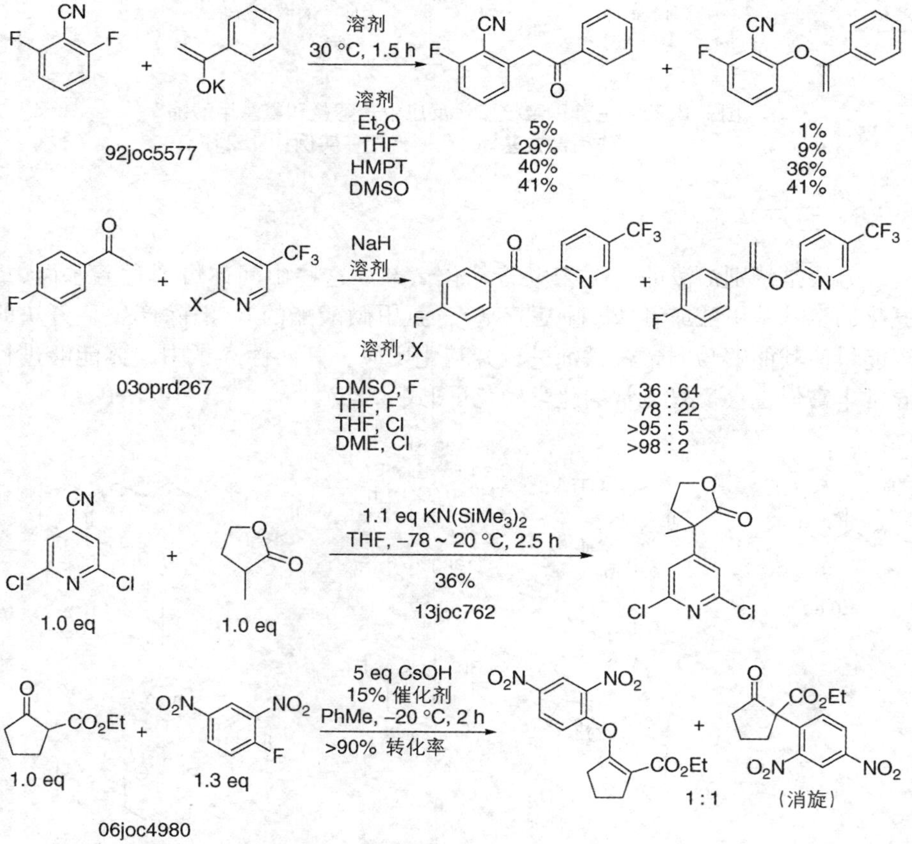

图式 8.37 酮和内酯衍生的烯醇碱进行 C—和 O—芳基化[155-158]

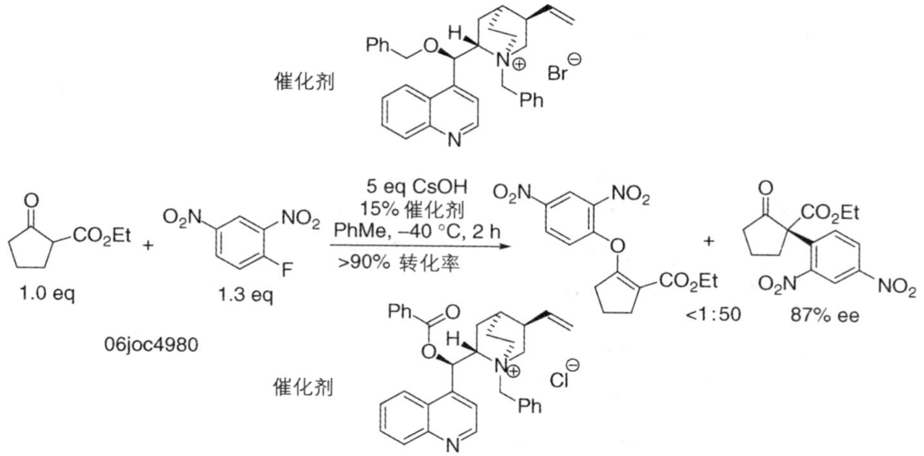

续图式 8.37

酚与酮烯醇化物相似,与亲电芳基化试剂反应时,可以是 C—也可以是 O—上的芳基化(图式 8.38)。许多参数可以影响这一转化的选择性,发现实现高收率的反应条件有时很困难。

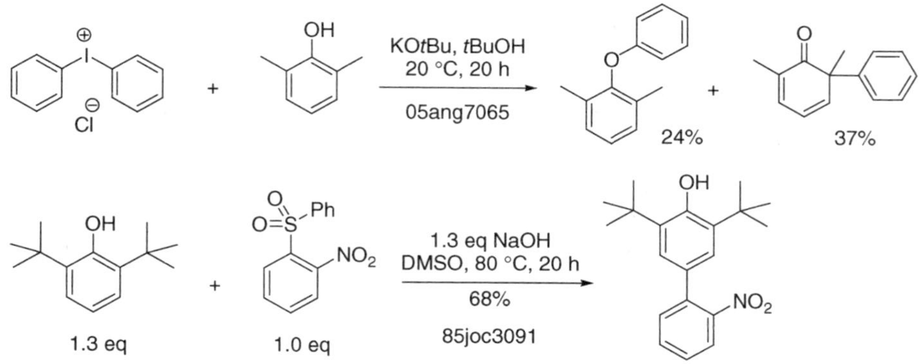

图式 8.38 酚的芳基化反应[159,160]

酮的 C—芳基化可以通过过渡金属催化来实现,O—芳基化在这样的反应中很少见(图式 8.39)。在钯催化的酮 C—芳基化反应中,钯与碳(而不是氧)键合随后还原消除主要导致 C—C 键形成[161]。

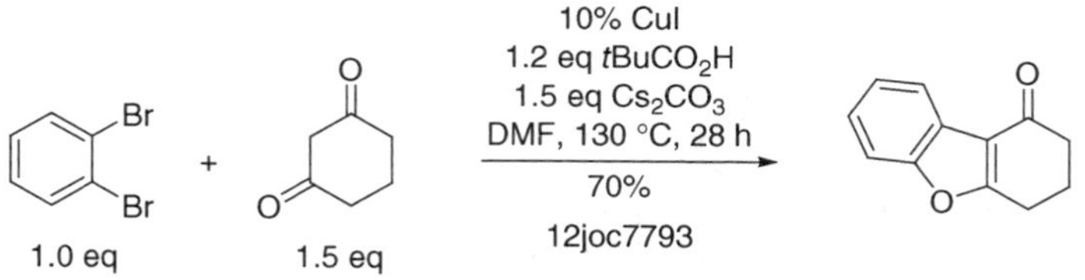

图式 8.39 1,3-二酮经 O—和 C—芳基化制备苯并呋喃[162]

8.3.2 有机镁及相关有机金属化合物

极性有机金属试剂因其化学硬度,通常在反应中更倾向于作为碱、单电子供体,或者进行卤素-金属交换,而并非作为亲核试剂。例如,室温下在四氢呋喃中用丁基锂处理芳基卤代物,丁基芳烃不是通过芳香亲核取代反应而是通过芳基锂与溴丁烷的反应得到[163,164]。此外,许多官能团也可以跟极性有机金属反应。因此,由于这些试剂可以通过许多不同的方式转化缺电子芳烃,故很少用作芳香亲核取代反应的亲核试剂(图式 8.40)。

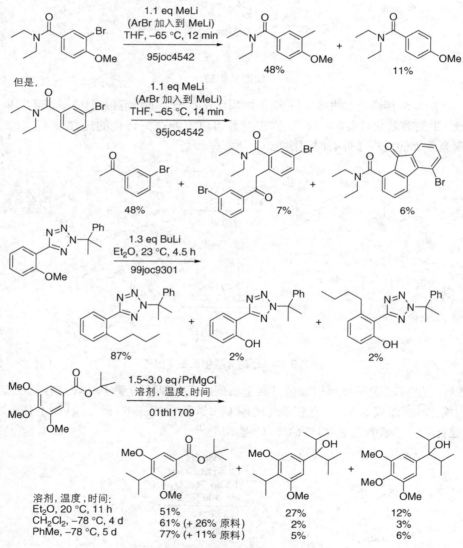

图式 8.40 极性有机金属试剂的非催化亲核取代反应[165-167](更多实例:[168])

过渡金属络合物有时可以通过将极性的"硬"有机金属化合物转化为较软的有机过渡金属络合物,或者通过断裂芳烃—离去基团的化学键,来催化芳烃离去基团的取代。已有许多过渡金属催化的极性有机金属化合物的芳基化反应被报道,多数使用铁[169]、镍[170,171]、铂、钯[172]的络合物(图式 8.41)。

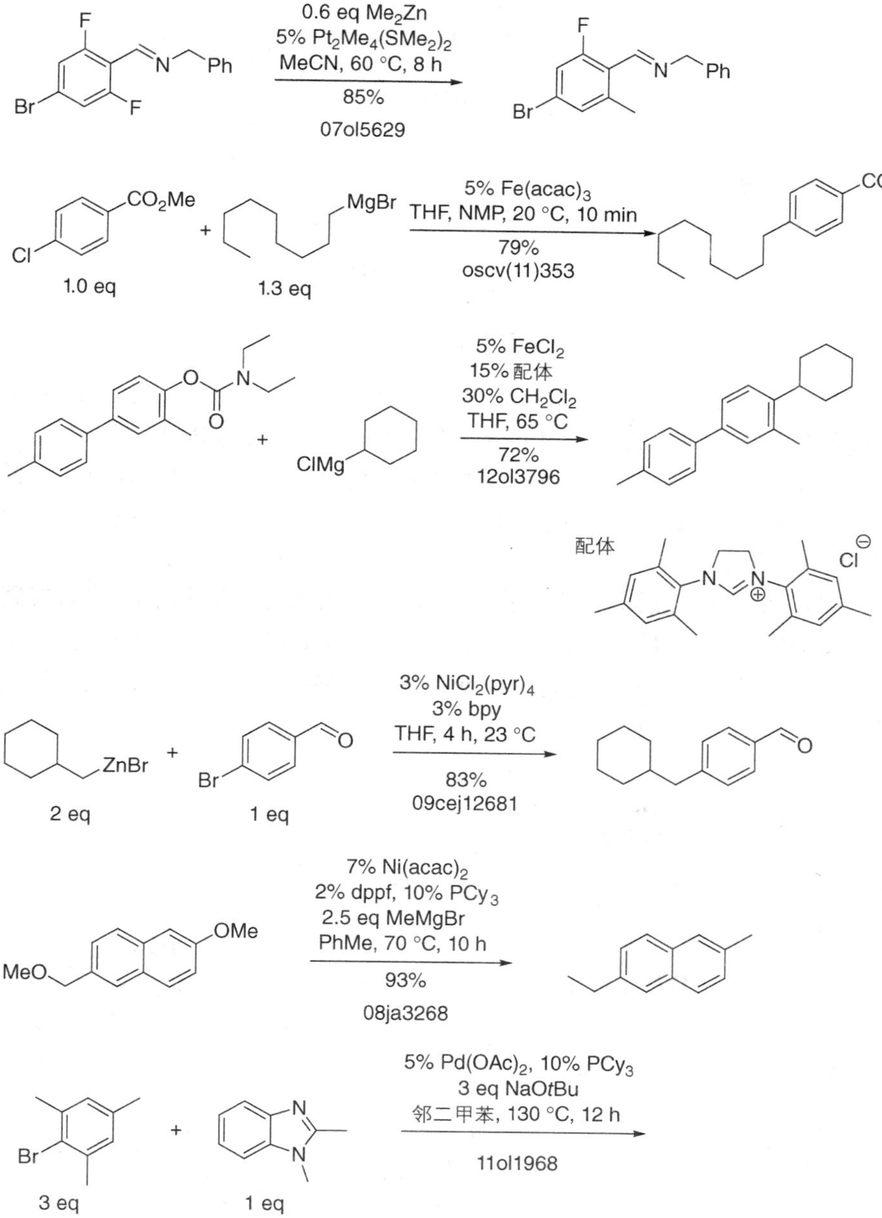

图式 8.41　过渡金属催化的有机金属化合物亲核取代反应[169,173-178]

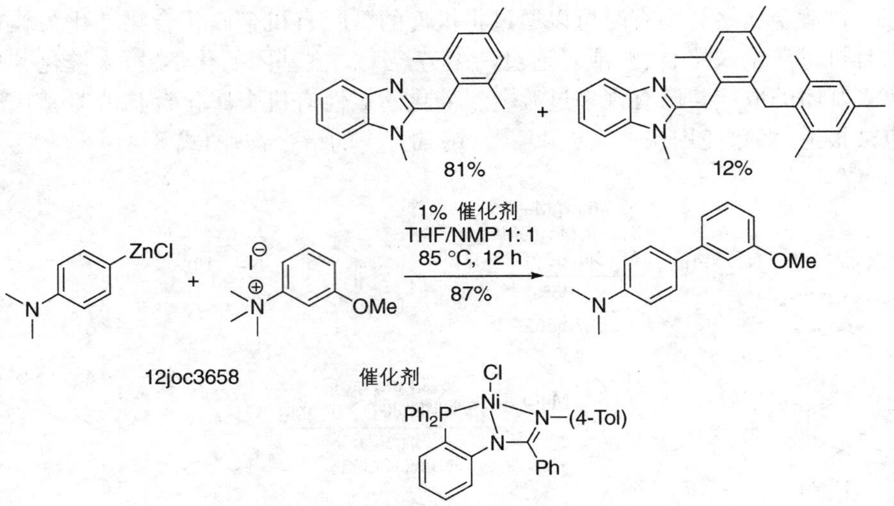

极性有机金属化合物与强缺电子芳烃的反应可被单电子转移和其他氧化还原过程复杂化。典型的副产物包括脱卤芳烃和芳基卤自偶联形成的对称联芳基。格氏试剂可以直接加成到硝基芳烃上或将其还原为亚硝基芳烃。后者对氮原子和氧原子都是亲电性的，并且可以与过量的格氏试剂形成 C—N 键或 C—O 键（图式 8.42）。

与图式 8.42 的最后一个实例密切相关的是使用过量烯基格氏试剂与硝基芳烃反应制备吲哚[鲍尔托利(Bartoli)反应[182]，图式 8.43]。据推测，该反应也经由亚硝基芳烃中间体进行。其低收率也可能由于过量的格氏试剂与亚硝基芳烃反应的区域选择性较差。

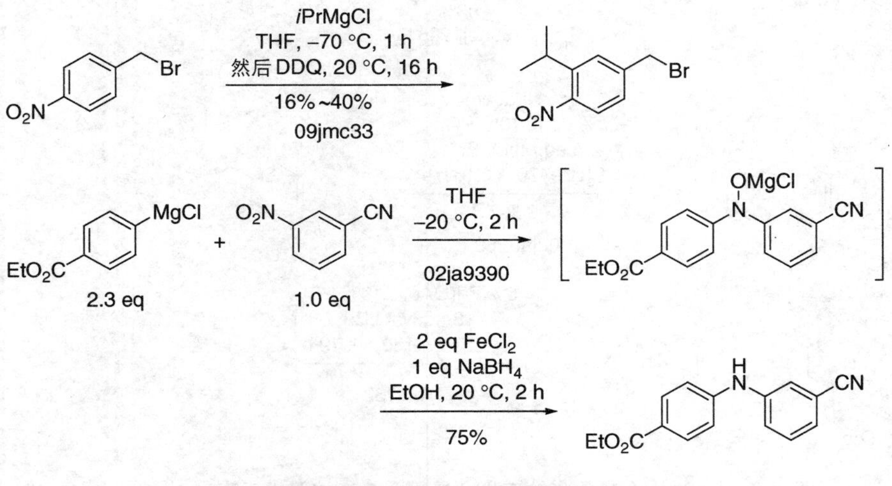

图式 8.42 硝基芳烃与格氏试剂的反应[179-181]

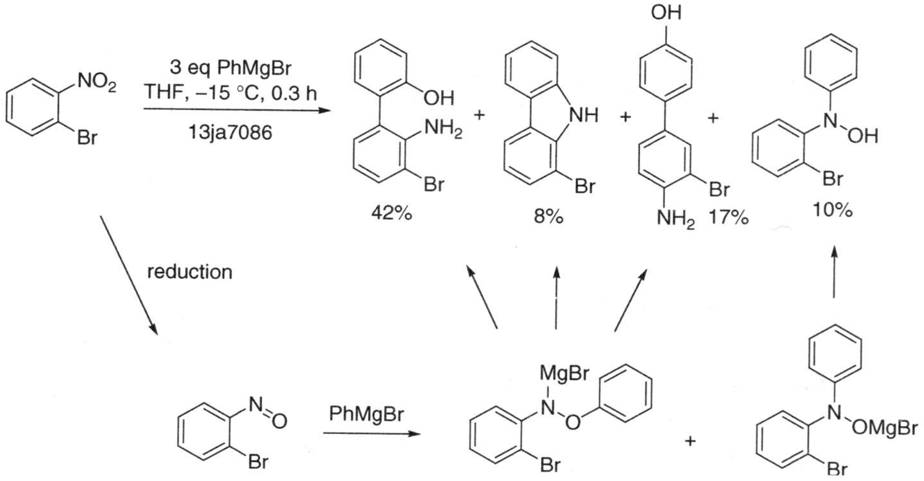

续图式 8.42

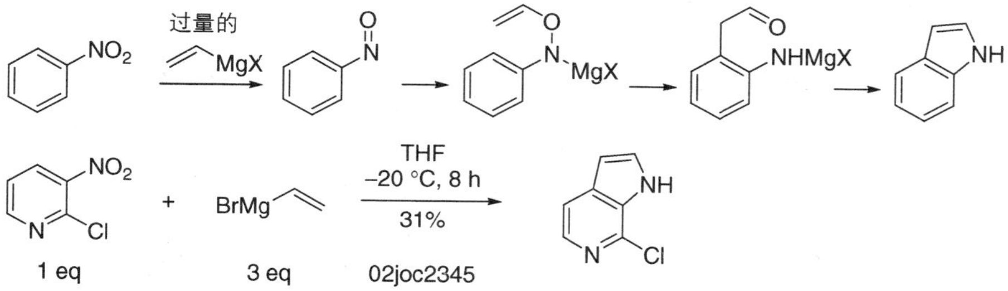

图式 8.43 从硝基芳烃制备吲哚[183]

一些芳烃有足够强的酸性足以被有机镁或相关的有机金属化合物去质子化。在氧化剂（例如硝基芳烃）的存在下，可以形成联芳基副产物。在无氧化剂存在下，使用极性有机金属试剂处理缺电子芳基卤也可以形成联芳基（图式8.44）。

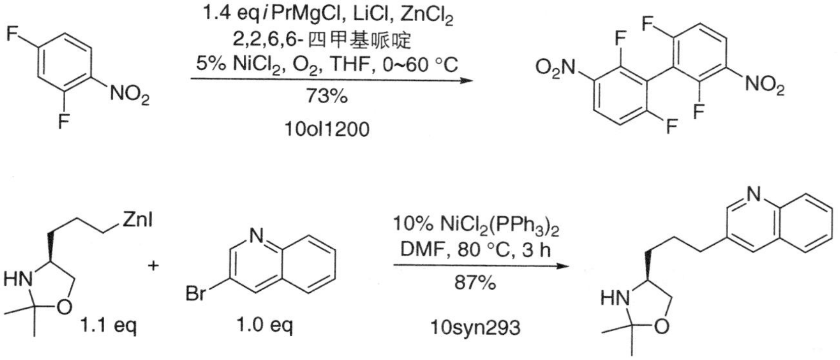

图式 8.44 由金属化芳烃的氧化二聚和根岸（Negishi）偶合形成联芳基[184,185]（更多实例[186]）

但是

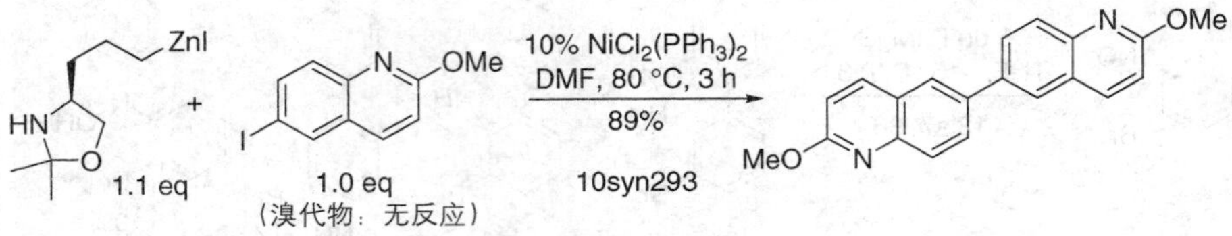

续图式 8.44

有机锂和有机镁试剂很容易引发多卤代芳烃的卤素-金属交换。产生的金属化芳烃足够稳定,在水解后得到脱卤素的芳烃,或者发生 β-消除形成芳炔。强亲电性芳炔能够被过量有机金属试剂芳基化(图式 8.45)。

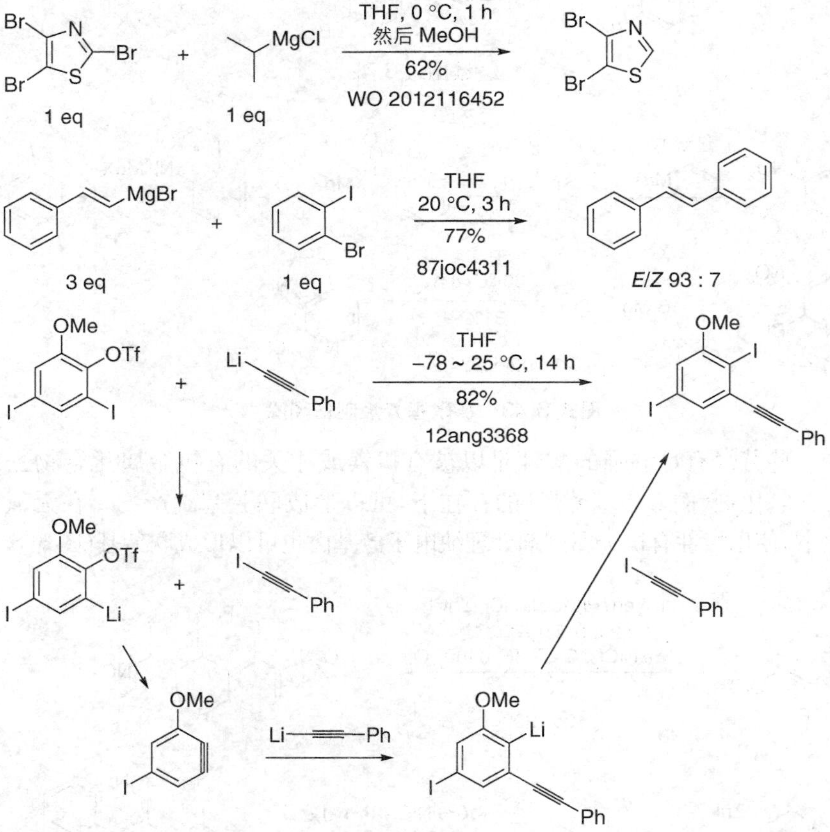

图式 8.45 芳基卤脱卤和形成芳炔[187-189]

8.3.3 氨

许多与氨成功进行芳香亲核取代反应的例子已经被报道(图式 8.46)。虽然也可以形成二芳基胺或三芳基胺，但通常这些可能的副产物能够通过使用大大过量的氨而避免。如果这个策略失败，则可使用酰胺[190]、叠氮化物（第8.3.6节）或六甲基（更昂贵）来代替氨。

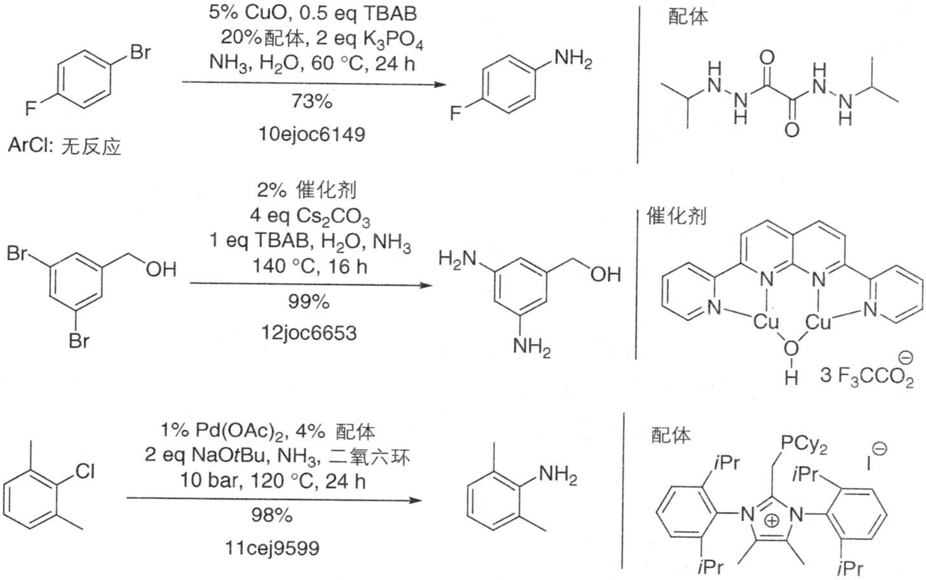

图式8.46 氨的单芳基化反应[191-193]（更多实例：[194]）

氨的亲核性和碱性较脂肪胺都更弱。因此，存在于反应体系中的其他亲核试剂（水、醇或新形成的苯胺）也可以和氨以相似的速率与亲电子试剂反应。其他副反应包括亲电试剂的还原和联芳基的形成（图式8.47）。

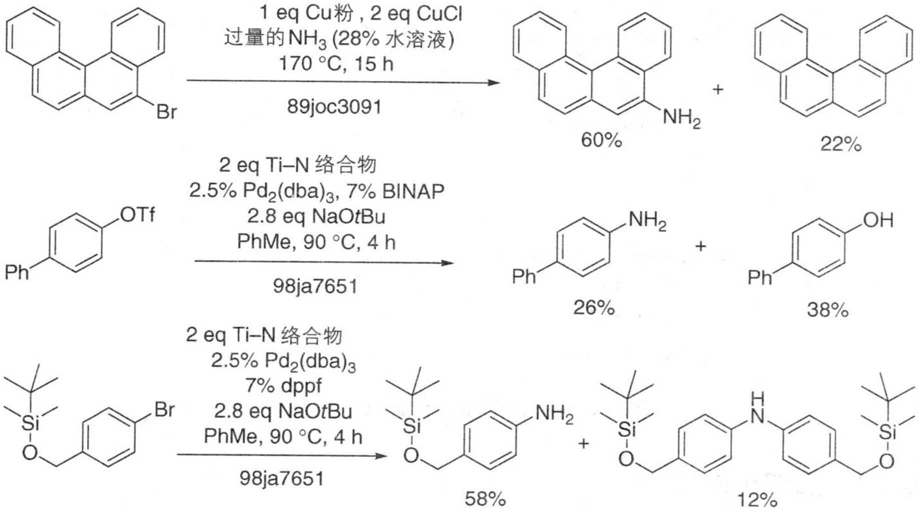

图式8.47 与氨及其合成等价物进行的芳香亲核取代反应中形成副产物[195-197]

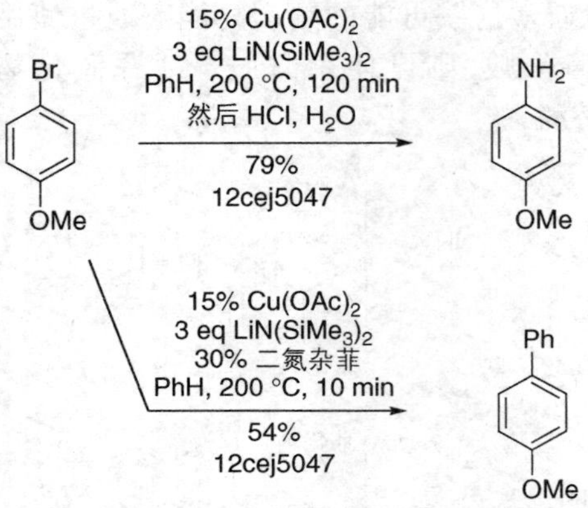

续图式 8.47

氨的另一个潜在的合成等价体是亚硝酸盐。尽管亚硝酸化合物与烷基卤剂反应会产生硝基烷烃和亚硝酸烷基酯的混合物,但亚硝酸盐的芳基化主要生成硝基芳烃(图式 8.48),它可以还原为苯胺,例如,通过氢化。

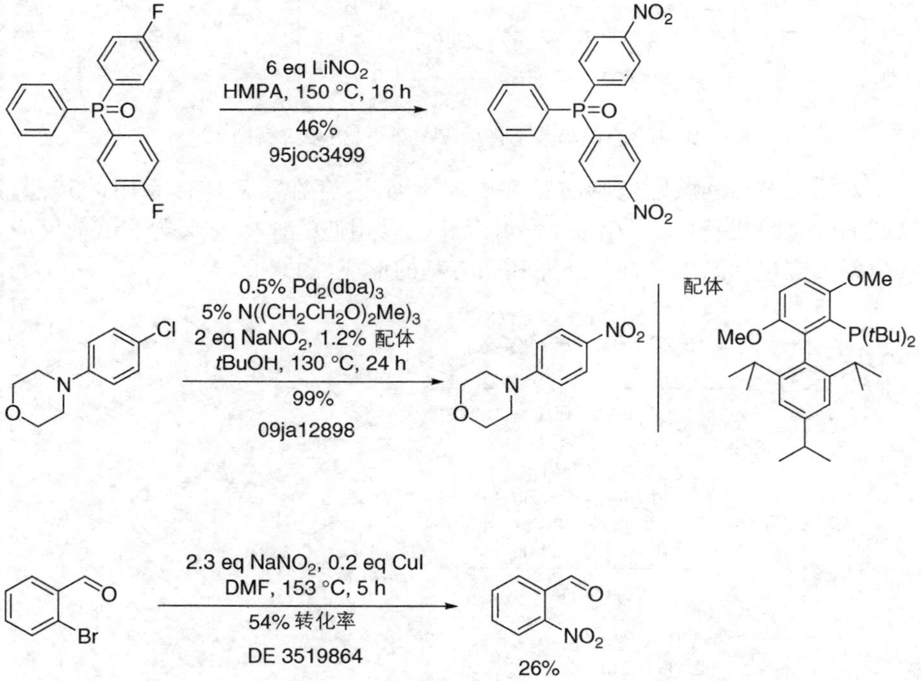

图式 8.48 亚硝酸盐的芳基化[198-200]

8.3.4 伯胺和仲胺

大多数胺与缺电子芳烃反应以预期的方式进行,因此芳香亲核取代反应是制备有取代的苯胺的一种有价值的常用策略(图式 8.49)。不过,偶尔也可能出现意外的副反应。对于催化芳基亲核取代反应,由于许多胺可以与铜等过渡金属形成稳定的无活性的络合物,故发现合适的催化剂是个难题。

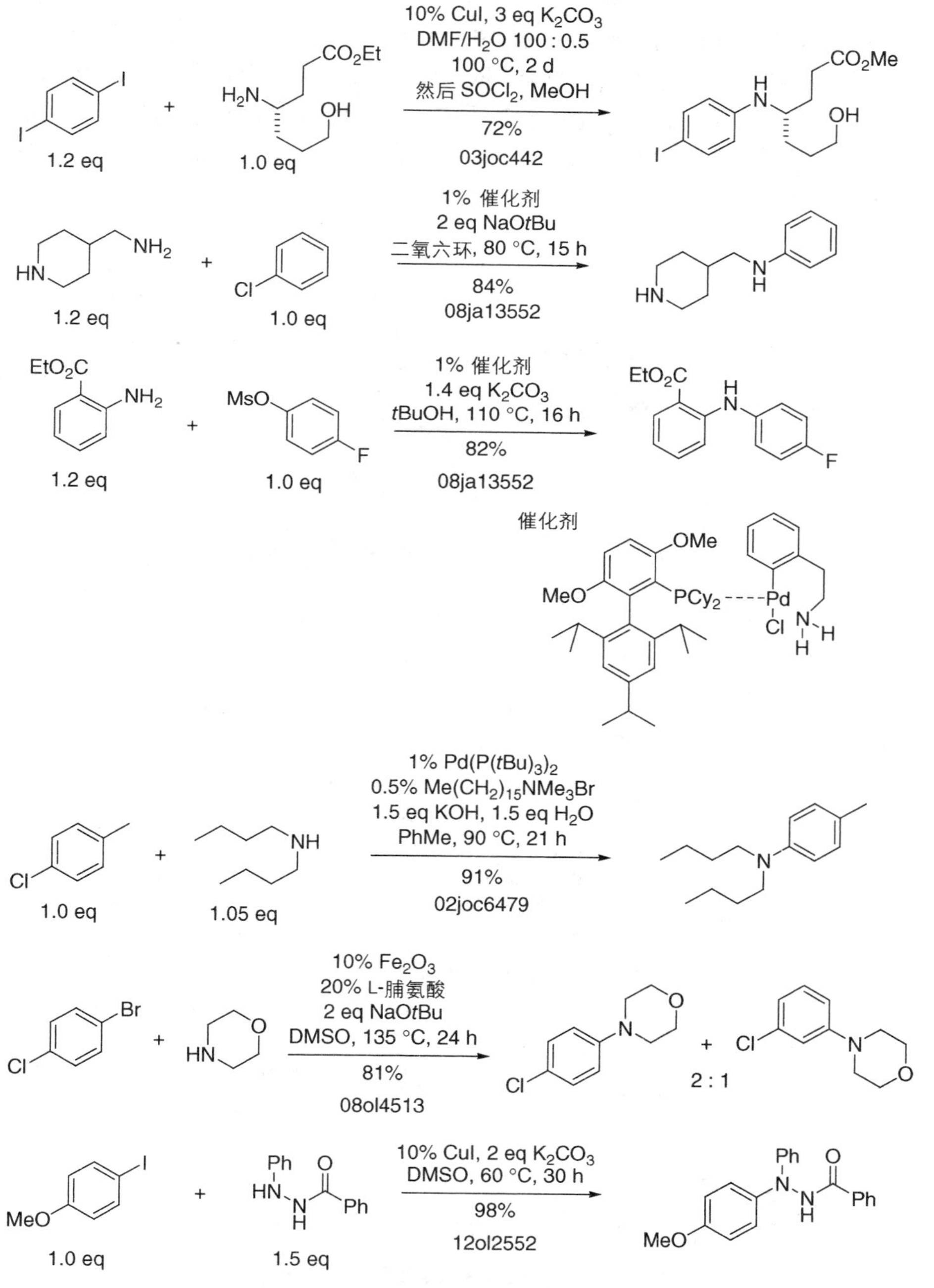

图式 8.49 胺的氮芳基化[201-205]

对于不活泼的芳基卤,可能需要过渡金属催化和强碱,有时会因此形成芳炔和区域异构产物的混合物。作为低沸点胺的替代品,其甲酰胺有时可用于芳香亲核取代反应[206,207]。

适合脂肪胺或芳香族胺的氮芳基化的碱包括氢氧化物、醇化物和碳酸盐。这表明胺作为亲核试剂的反应速度要比氧基团快得多。尽管如此,酚类偶尔能够导致副产品产生[208]。容易被脱氢的胺(例如苄胺)能够引起亲电试剂的还原反应。二芳基胺和三芳基胺也容易被氧化(例如咔唑),这类化合物的制备必须严格隔绝空气,特别是在使用强碱的情况下。强碱总能引发一些奇怪的副产物(图式 8.50),因此应尽可能避免。

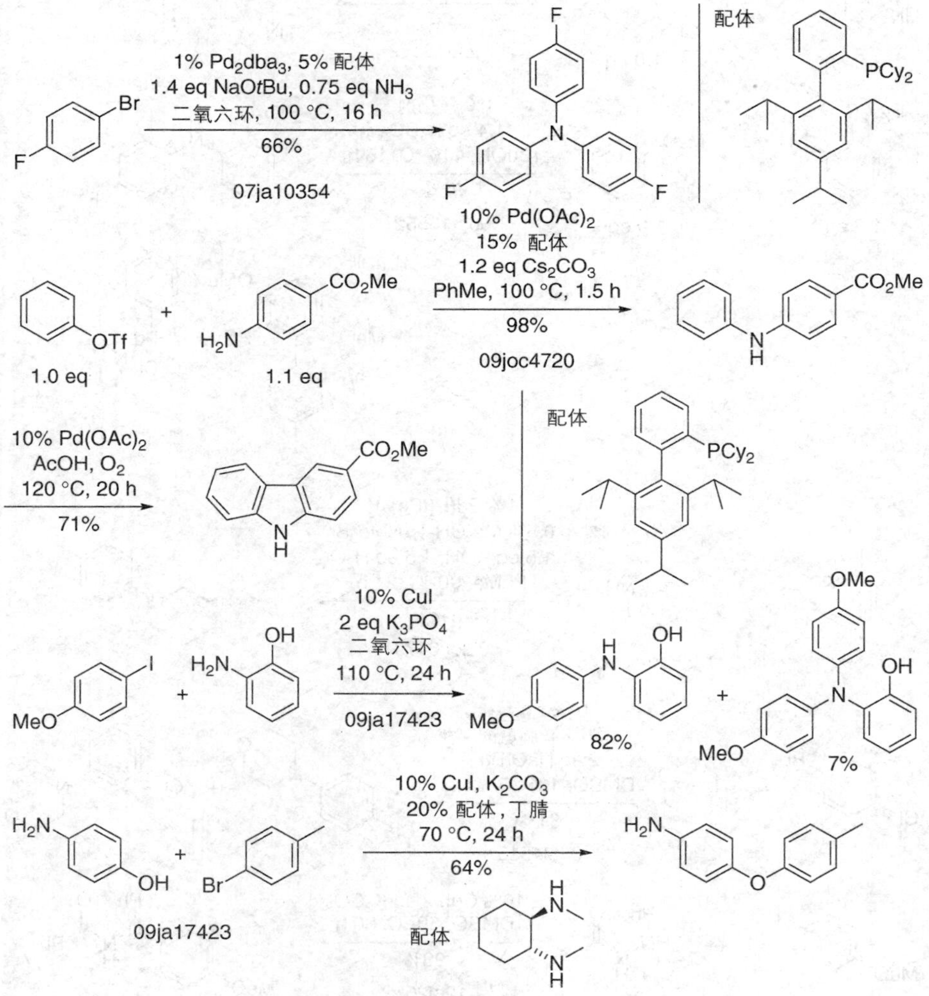

图式 8.50 烷基胺和苯胺的氮芳基化[132,209-211]

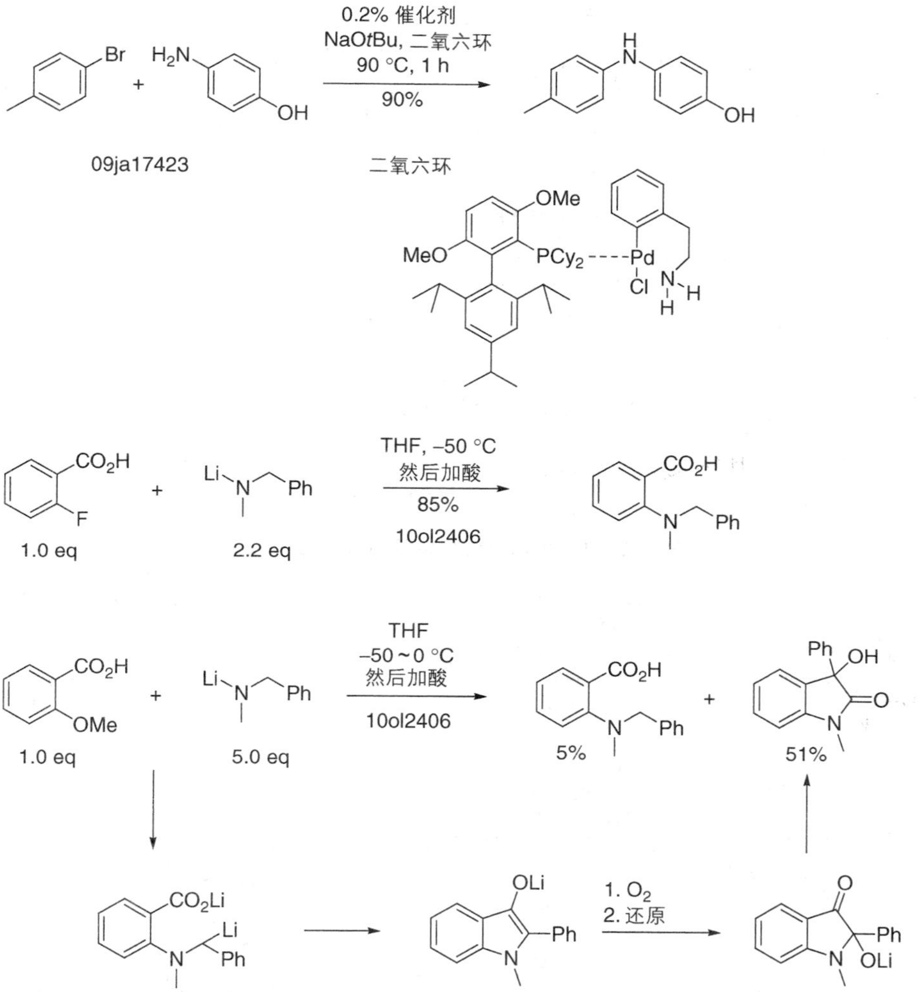

8.3.5 叔胺

由于产物能够被叔胺快速脱烷基，N-芳基季铵盐很难通过叔胺的氮芳基化制备(图式8.51)。因为叔胺的亲核性并不总是足以进行芳香亲核取代反应，要实现其氮芳基化，需要强亲电性的芳基化试剂，如苯炔。

8.3.6 叠氮化物

与金属叠氮化物的芳香亲核取代反应往往得到苯胺而非芳基叠氮化物(图式8.52)。最初形成的叠氮化物可以被弱还原剂还原，例如叔丁醇钠[215]、乙醇[216]、溴化氢或过量的金属叠氮化物[217]。然而，也有许多由芳基卤化物成

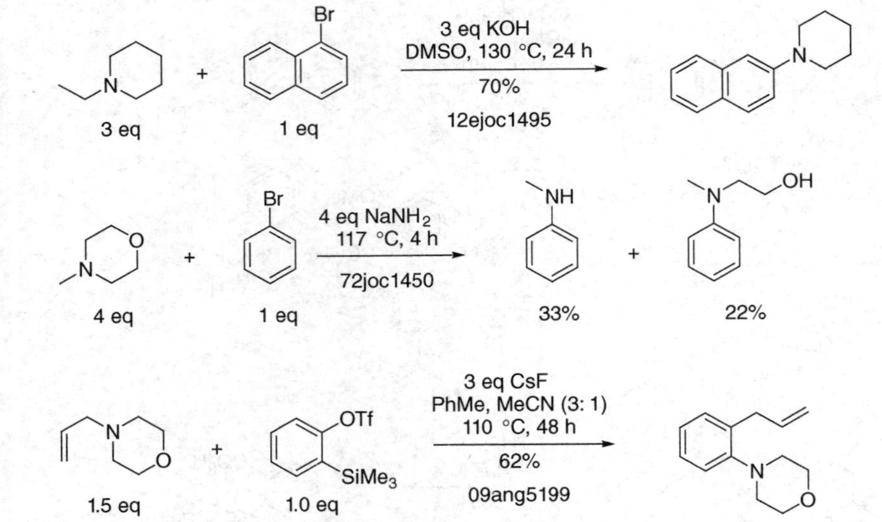

图式 8.51　用苯炔进行叔胺的芳基化[212-214]

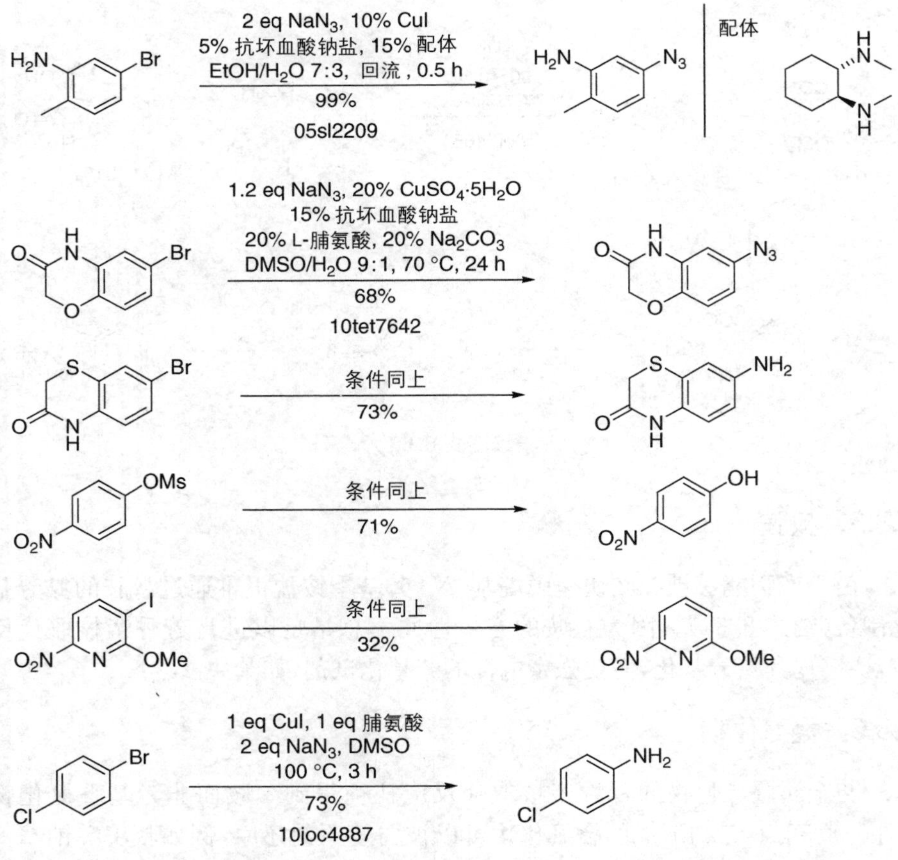

图式 8.52　与叠氮化物的亲核取代反应[216,217,219,220]（更多实例：[221]）

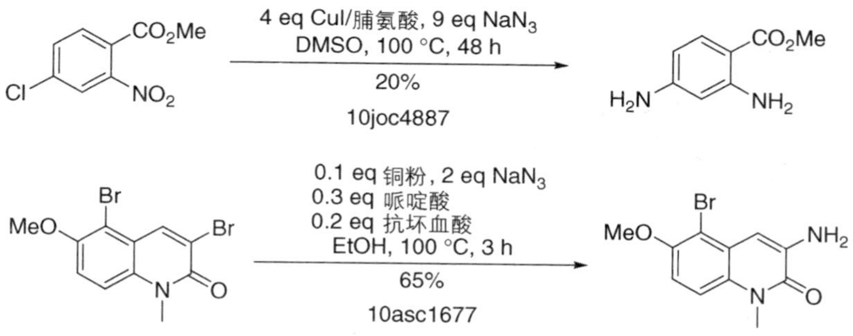

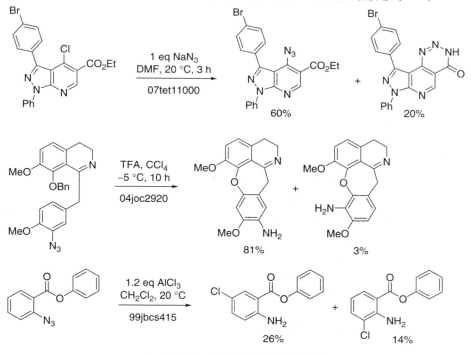

续图式 8.52

功制备芳基叠氮化物的例子被报道[218]。最好的结果是以一价铜二胺络合物作为催化剂，在抗坏血酸（从二价铜再生的一价铜）的存在下获得的。除了叠氮化物中间体被还原为苯胺，与叠氮化物进行芳香亲核取代反应过程中典型的副反应还包括硝基被还原及底物被还原为未取代的芳烃。

有机叠氮化物可进行 1,3-偶极环加成，并且作为亲核试剂能与许多官能团反应。例如，在酸性条件下，酮与叠氮化物反应得到酰胺，与 2-叠氮苯甲酸酯反应可以环化得到苯并三嗪酮（图式 8.53）。此外，芳基叠氮化物是热不稳定的，会与各种酸反应得到取代的苯胺（如 N-二芳基羟胺）。酸性反应条件通常会加速叠氮化物的分解，而大多数叠氮化物在碱的存在下是相当稳定的[222]。

图式 8.53 芳基叠氮的反应[223-225]

8.3.7 氢氧化物

氢氧化物的芳香亲核取代反应的主要副反应是形成二芳基醚和官能团的转化(图式 8.54)。从氯苯制备苯酚的副产物包括二苯基醚、及 2-苯基苯酚和 4-苯基苯酚(图式 8.16)。

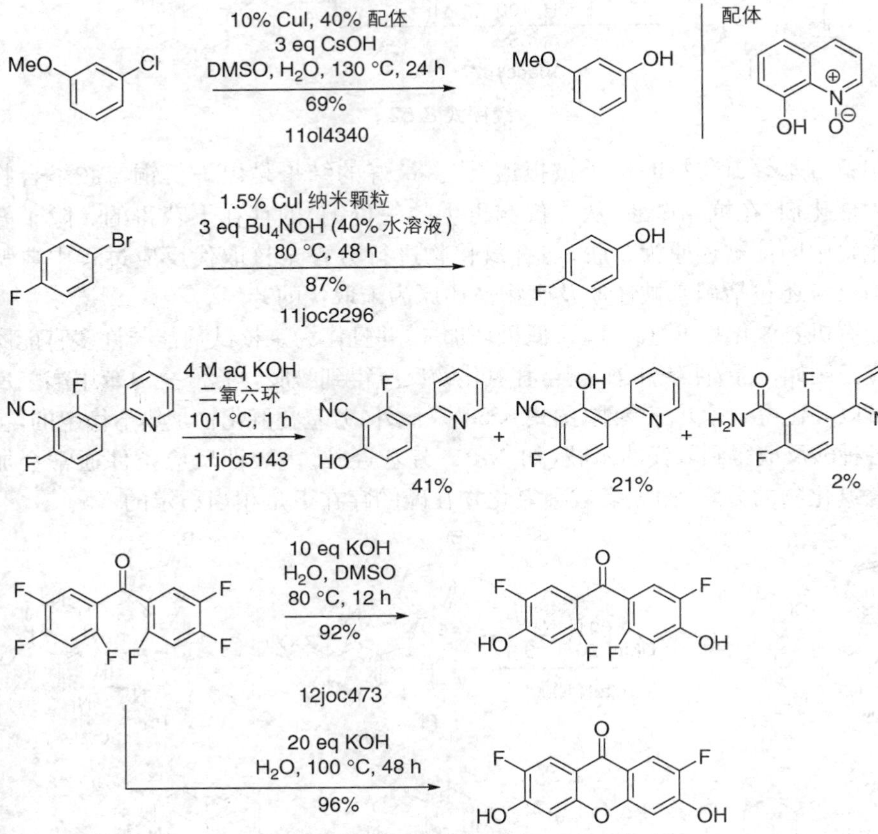

图式 8.54 卤代芳烃转化为酚[226-229]

用碳酸盐或羧酸盐处理芳基卤也能形成酚。这些亲核试剂的碱性比氢氧根弱,通常不能皂化羧酸酯(图式 8.55)。

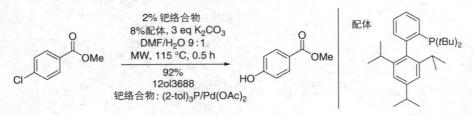

图式 8.55 碳酸盐和羧酸盐的芳基化[235-238]

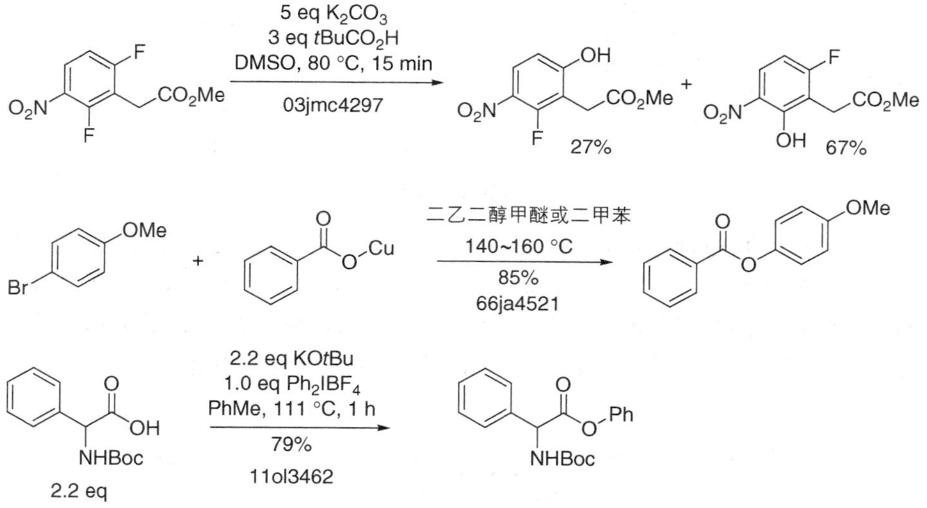

续图式 8.55

在钯和其他过渡金属络合物存在下，苯甲酸和芳基亚磺酸可以发生脱羧/脱亚磺酰化取代反应。这是过渡金属催化的苯甲酸和芳基亚磺酸氧芳基化的一个潜在副反应。

8.3.8 醇

因为醇盐的硬度，醇盐在反应中常常用作碱而非亲核试剂。无论如何，有许多成功的用醇盐进行芳香亲核取代的例子已经被报道（图式 8.56）。

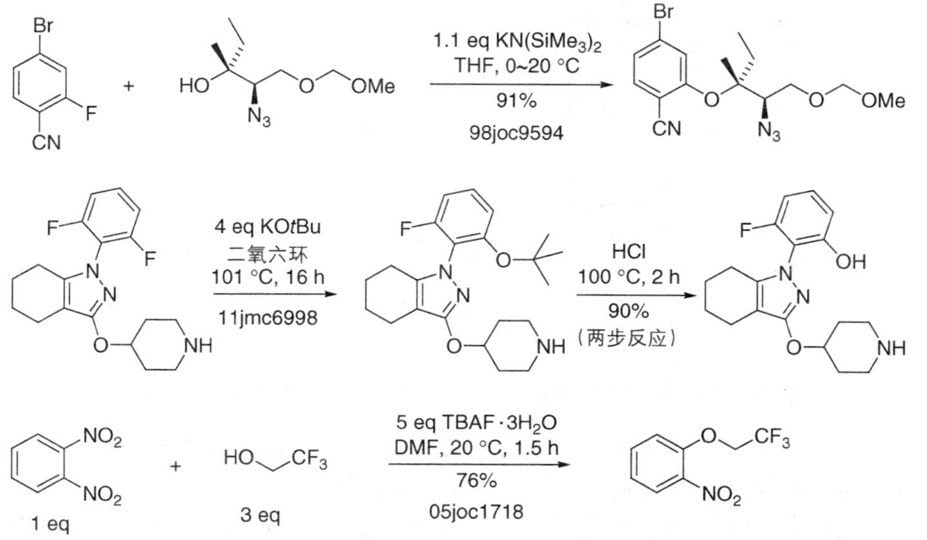

图式 8.56 醇的芳香亲核取代反应[239-242]

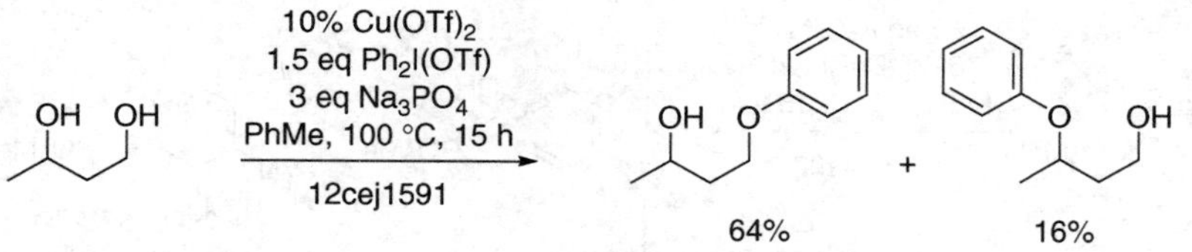

有α-氢的脂肪醇脱质子后不仅是强碱,也是强的氢供体,因而也是还原试剂(干燥的醇盐在空气中会燃烧)。因此,醇盐的O-烷基化或O-芳基化并不总是能产生高收率的醚,β-消除、还原以及羧酸或碳酸衍生物的裂解常常被视为副反应。图式8.57的最后一个例子的还原可能由碘引起。

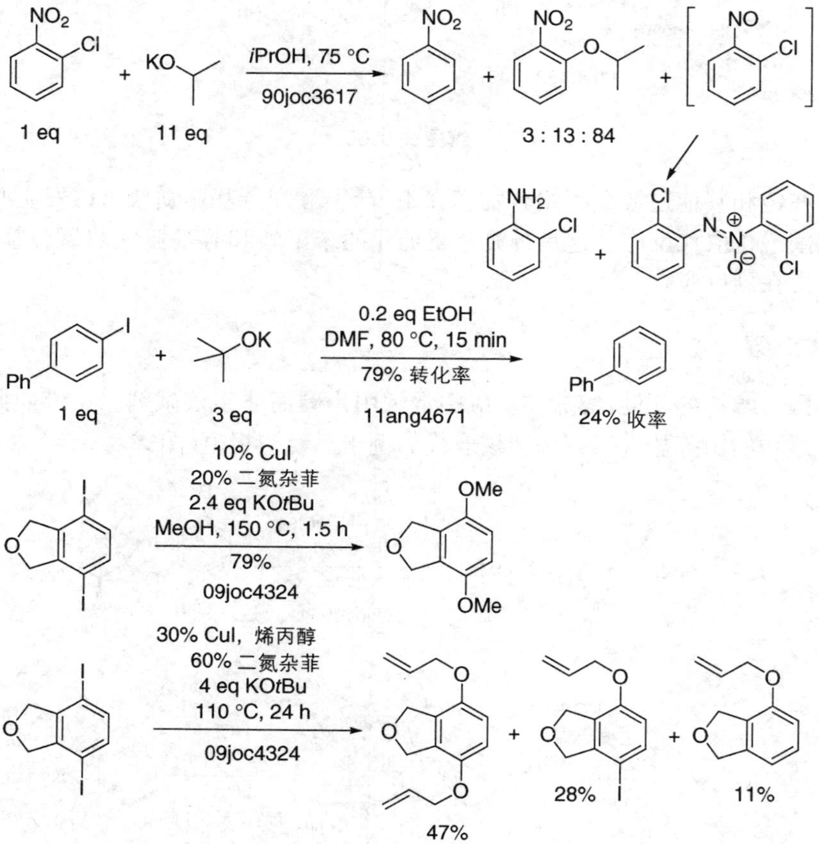

图式8.57 醇化物与芳基卤化物的反应[243-245]（更多实例：[246]）

氨基醇可以被选择性地在氮原子或氧原子上芳基化(图式8.58)。用强碱处理所得产物能引起芳基从氧原子向氮原子迁移[247]。

脂肪醇的氧芳基化也能够被钯络合物催化。然而,即使在这些催化剂的存在下,也可能出现众多的副产物(图式8.59)。

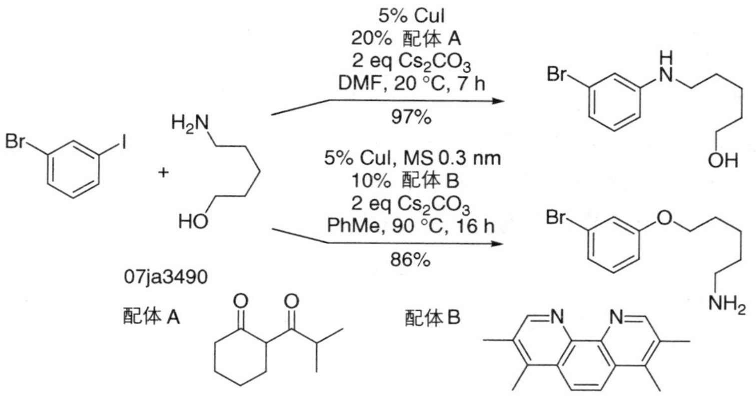

图式 8.58 氨基醇的芳基化反应[248]

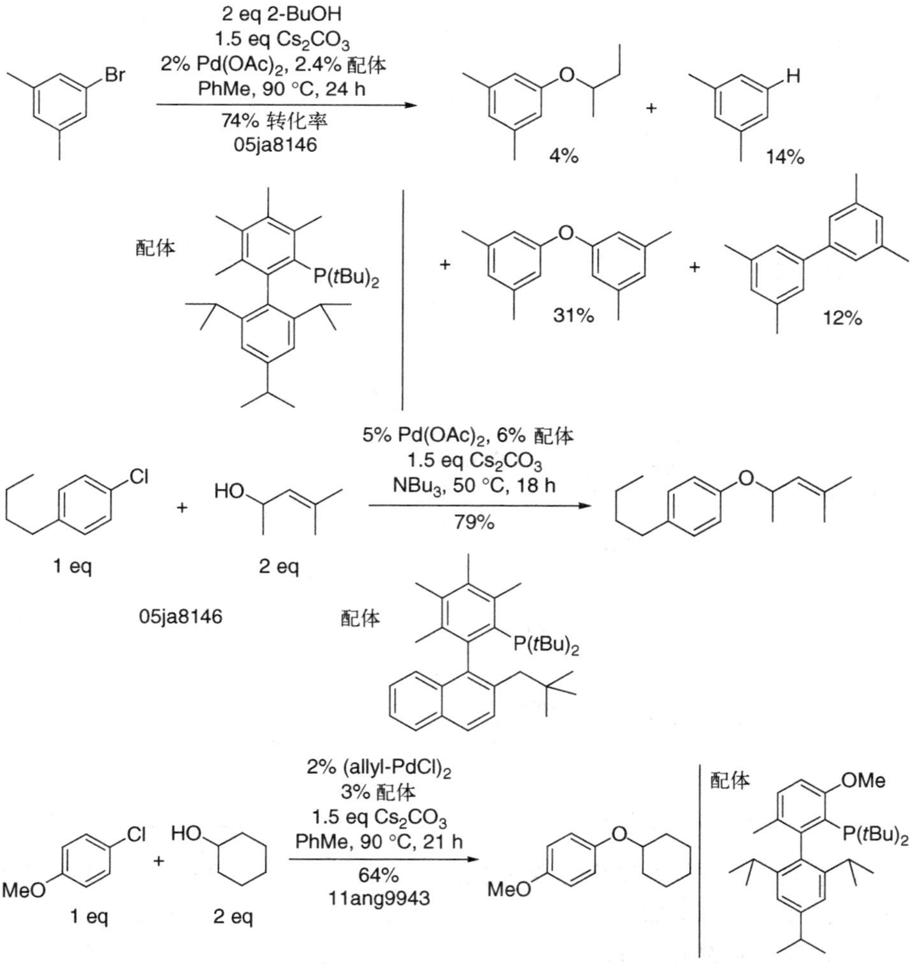

图式 8.59 钯催化下与醇的芳香亲核取代[249,250]（更多实例：[251,252]）

8.3.9 硫醇

虽然硫醇是强亲核试剂并且原则上也适合用于芳香亲核取代反应,但硫醇也是较强的还原剂且能够与许多官能团反应。尽管如此,有许多高收率的硫醇芳香亲核取代的例子已经被报道(图式 8.60)。尽管许多硫—金属键很强,但这些反应有时也能够被过渡金属络合物催化。

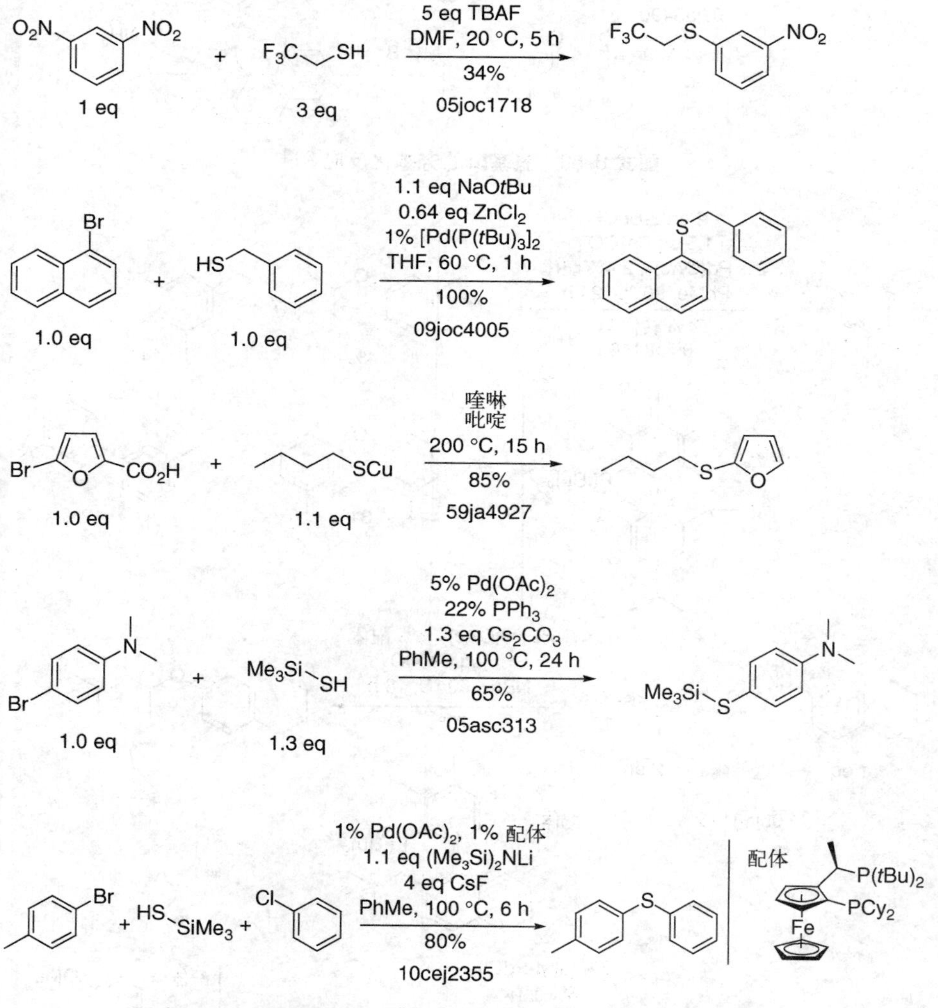

图式 8.60　硫醇的芳基化[241,253-256]（更多实例：[44,257-260]）

烷基硫醇与芳基卤化物的芳基化反应中常见的副反应是芳基卤化物的还原脱卤以及形成苯硫酚、二芳基硫醚和二烷基硫醚,后者由芳基烷基硫醚中间体被过量硫醇断裂所致(图式 8.61),它们的形成能够通过催化反应和降低反应温度阻止。

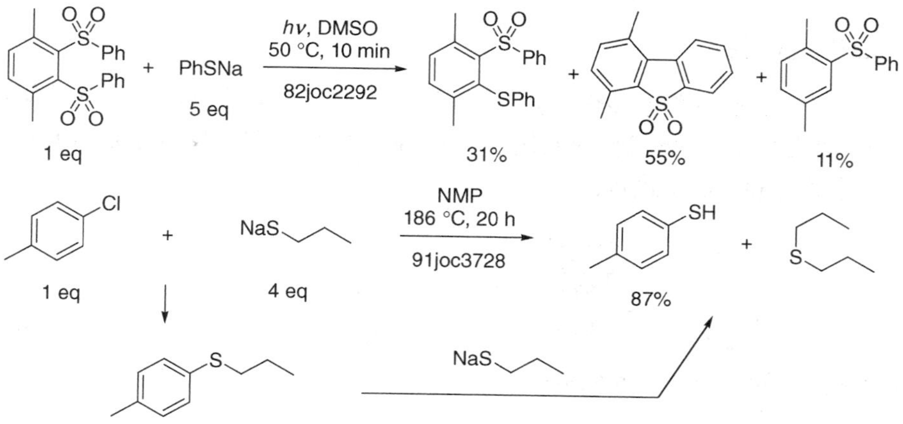

图式 8.61　硫化物与芳基化试剂的反应[19,261]

由于硫醇的高亲核性，甚至弱的亲电试剂也能够有效地与卤代芳烃竞争。例如，烷基酯或碳酸酯容易烷基化硫醇，因此它们不应该用作反应溶剂[262]。

硫氰酸盐（MSCN）是硫化氢的合成等价体。卤代芳烃通常与硫氰酸盐在硫原子上反应，得到的芳基硫氰酸酯可以重排成异硫氰酸酯。在水的存在下，硫氰酸酯中间体也可以水解为苯硫酚，后者可以被亲电试剂芳基化（图式 8.62）。

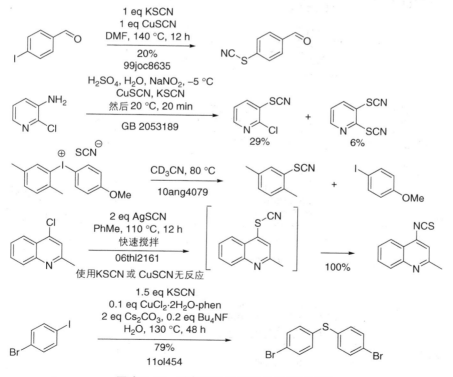

图式 8.62　硫氰酸盐的芳香化反应[263-267]

8.3.10 卤代物

由于卤代物的低亲核性,与它们进行的芳香亲核取代反应往往只能缓慢进行(图式 8.63)。碘代物的取代最容易,而氟代物的取代反应通常需要在严格无水条件和高温下进行。

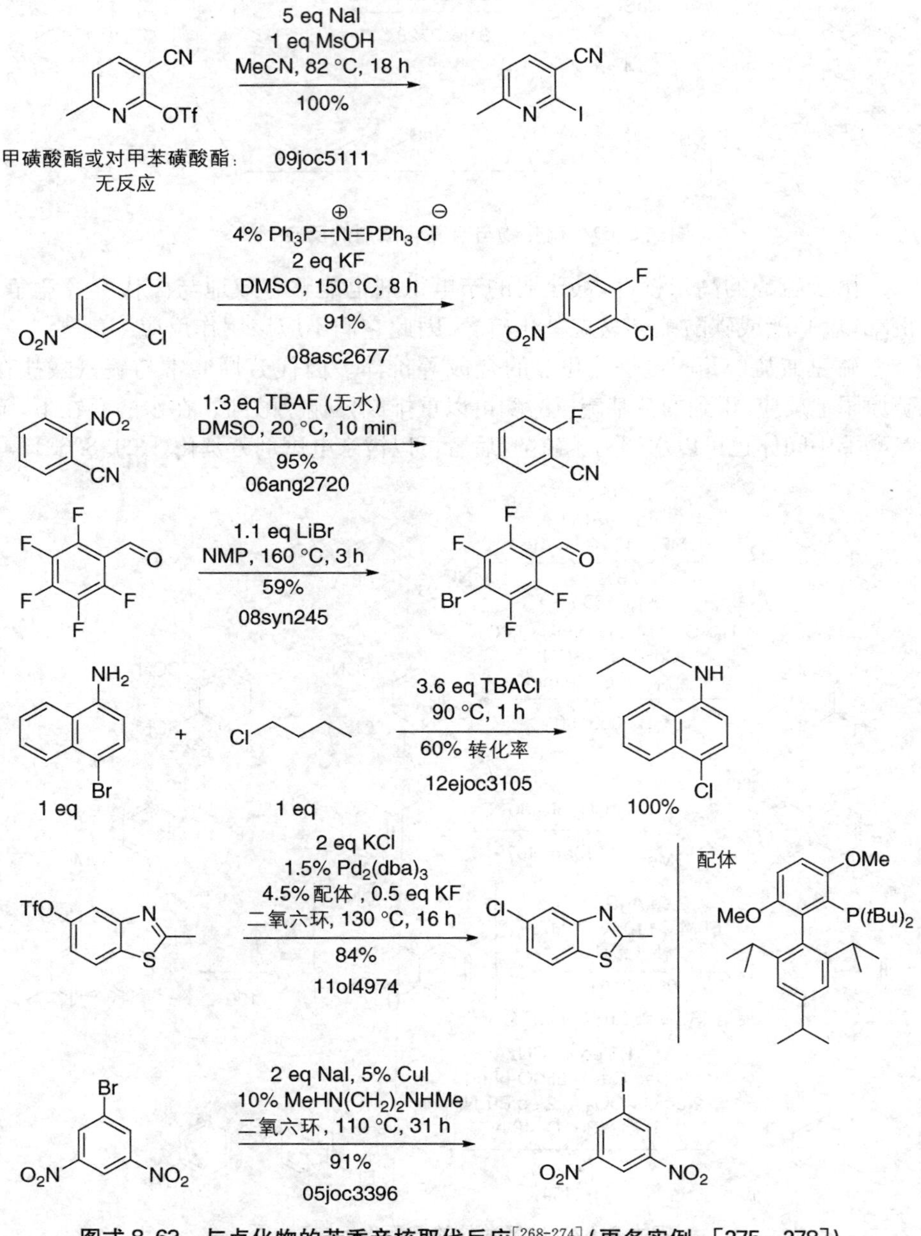

图式 8.63 与卤化物的芳香亲核取代反应[268-274] (更多实例:[275-278])

因为氟是碱性最强的卤化物,所以它容易引起许多副反应(醚的裂解、醛氧化成羧酸及与其他亲核试剂的竞争或脱羧反应)。

碘是一个很好的还原剂,并且可以还原苄醇[279]并使芳基卤脱卤。缺电子芳烃,如多硝基苯或吡啶,特别容易进行这种还原反应(图式 8.64)。

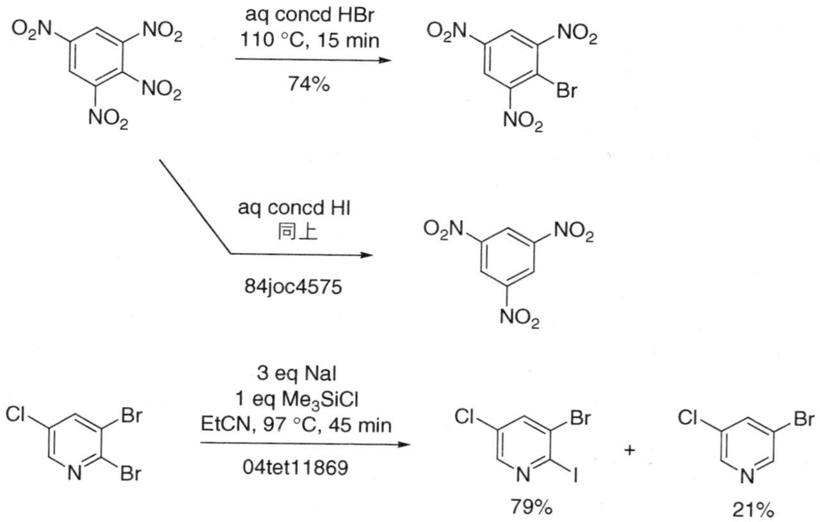

图式 8.64 碘的还原[280,281]

酚可以被 2-卤代咪唑卤化物原位还原,然后可进行卤代脱氧反应(图式 8.65)。另外,酚可以转化为三氟甲磺酸酯或其他氟代磺酸酯,而后与氟原子通过钯催化的芳香亲核取代反应转化为芳基氟化物[282]。

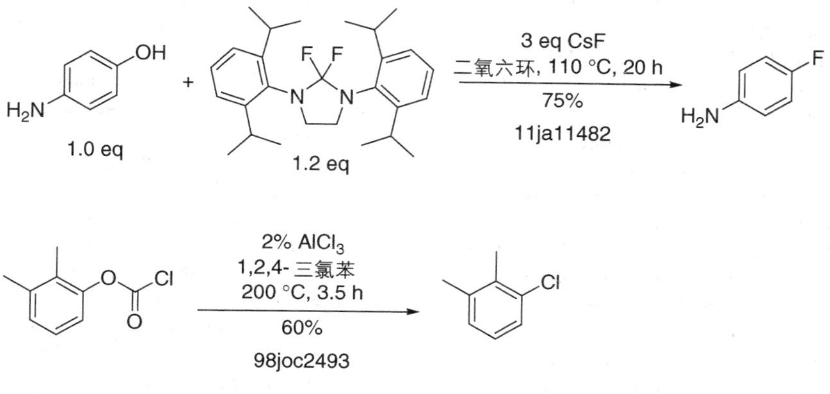

图式 8.65 酚及其衍生物被转化成为芳基卤化物[283,284]

参考文献

1. Buncel, E., Dust, J.M., and Terrier, F. (1995) Rationalizing the regioselectivity in polynitroarene anionic σ-adduct formation. Relevance to nucleophilic aromatic substitution. *Chem. Rev.*, **95**, 2261–2280.
2. Ellis, G.P. and Romney-Alexander, T.M. (1987) Cyanation of aromatic halides. *Chem. Rev.*, **87**, 779–794.
3. Traynham, J.G. (1979) Ipso substitution in free-radical aromatic substitution reactions. *Chem. Rev.*, **79**, 323–330.
4. Bunnett, J.F. and Zahler, R.E. (1951) Aromatic nucleophilic substitution reactions. *Chem. Rev.*, **51**, 273–412.
5. López Ortiz, F., Iglesias, M.J., Fernández, I., Andújar Sánchez, C.M., and Ruiz Gómez, G. (2007) Nucleophilic dearomatizing (DNAr) reactions of aromatic C,H-systems. A mature paradigm in organic synthesis. *Chem. Rev.*, **107**, 1580–1691.
6. Huffman, J.W., Wu, M.-J., and Lu, J. (1998) A very facile S_NAr reaction with elimination of methoxide. *J. Org. Chem.*, **63**, 4510–4514.
7. Murugananthan, R. and Namboothiri, I. (2010) Phosphonylpyrazoles from Bestmann–Ohira reagent and nitroalkenes: synthesis and dynamic NMR studies. *J. Org. Chem.*, **75**, 2197–2205.
8. Bunnett, J.F., Brotherton, T.K., and Williamson, S.M. (1973) N-β-Naphthylpiperidine. *Org. Synth.*, Coll. Vol. **5**, 816–818.
9. Allen, C.F.H. and Bell, A. (1955) 3-Amino-2-naphthoic acid. *Org. Synth.*, Coll. Vol. **3**, 78–82.
10. Um, I.-H., Im, L.-R., Kang, J.-S., Bursey, S.S., and Dust, J.M. (2012) Mechanistic assessment of S_NAr displacement of halides from 1-halo-2,4-dinitrobenzenes by selected primary and secondary amines: Brønsted and Mayr analyses. *J. Org. Chem.*, **77**, 9738–9746.
11. Imoto, M., Matsui, Y., Takeda, M., Tamaki, A., Taniguchi, H., Mizuno, K., and Ikeda, H. (2011) A probable hydrogen-bonded Meisenheimer complex: an unusually high S_NAr reactivity of nitroaniline derivatives with hydroxide ion in aqueous media. *J. Org. Chem.*, **76**, 6356–6361.
12. Um, I.-H., Min, S.-W., and Dust, J.M. (2007) Choice of solvent (MeCN vs H_2O) decides rate-limiting step in S_NAr aminolysis of 1-fluoro-2,4-dinitrobenzene with secondary amines: importance of Brønsted-type analysis in acetonitrile. *J. Org. Chem.*, **72**, 8797–8803.
13. Liu, J. and Robins, M.J. (2007) S_NAr displacements with 6-(fluoro, chloro, bromo, iodo, and alkylsulfonyl)purine nucleosides: synthesis, kinetics, and mechanism. *J. Am. Chem. Soc.*, **129**, 5962–5968.
14. Hintermann, L., Masuo, R., and Suzuki, K. (2008) Solvent-controlled leaving-group selectivity in aromatic nucleophilic substitution. *Org. Lett.*, **10**, 4859–4862.
15. Suwiński, J. and Świerczek, K. (2001) *Cine-* and *tele-*substitution reactions. *Tetrahedron*, **57**, 1639–1662.
16. Poirier, M., Goudreau, S., Poulin, J., Savoie, J., and Beaulieu, P.L. (2010) Metal-free coupling of azoles with 2- and 3-haloindoles providing access to novel 2- or 3-(azol-1-yl)indole derivatives. *Org. Lett.*, **12**, 2334–2337.
17. Carpenter, R.D. and Verkman, A.S. (2010) Synthesis of a sensitive and selective potassium-sensing fluoroionophore. *Org. Lett.*, **12**, 1160–1163.
18. Enguehard, C., Allouchi, H., Gueiffier, A., and Buchwald, S.L. (2003) *Ipso-* or *cine-*substitutions of 6-haloimidazo[1,2-a]pyridine derivatives with different azoles depending on the reaction conditions. *J. Org. Chem.*, **68**, 5614–5617.
19. Novi, M., Dell'Erba, C., Garbarino, G., and Sancassan, F. (1982) Normal S_NAr, telesubstitution, and electron-transfer pathways in the reactions of methyl-substituted o-bis(phenylsulfonyl)benzene derivatives with sodium arenethiolates in dimethyl sulfoxide. *J. Org. Chem.*, **47**, 2292–2298.
20. Giannopoulos, T., Ferguson, J.R., Wakefield, B.J., and Varvounisa, G.

(2000) Tele nucleophilic aromatic substitutions in 1-nitro-3- and 1,3-dinitro-5-trichloromethylbenzene, and 3-trichloromethylbenzonitrile. A new synthesis of the 1,4-benzothiazine-3(4H)-one ring system from 3-nitrobenzoic acid. *Tetrahedron*, **56**, 447–453.

21. Surowiec, M., Belekos, D., Ma̧kosza, M., and Varvounis, G. (2010) Tele nucleophilic substitutions of hydrogen in m-(trichloromethyl)nitrobenzenes with cyano and ester carbanions. *Eur. J. Org. Chem.*, **2010**, 3501–3506.

22. Agrawal, K.C., Bears, K.B., Sehgal, R.K., Brown, J.N., Rist, P.E., and Rupp, W.D. (1979) Potential radiosensitizing agents. Dinitroimidazoles. *J. Med. Chem.*, **22**, 583–586.

23. Li, Y.-X., Ji, K.-G., Wang, H.-X., Ali, S., and Liang, Y.-M. (2011) Iodine-induced regioselective C–C and C–N bonds formation of N-protected indoles. *J. Org. Chem.*, **76**, 744–747.

24. Hooper, M.W., Utsunomiya, M., and Hartwig, J.F. (2003) Scope and mechanism of palladium-catalyzed amination of five-membered heterocyclic halides. *J. Org. Chem.*, **68**, 2861–2873.

25. Torr, J.E., Large, J.M., Horton, P.N., Hursthouse, M.B., and McDonald, E. (2006) On the nucleophilic *tele*-substitution of dichloropyrazines by metallated dithianes. *Tetrahedron Lett.*, **47**, 31–34.

26. Ma̧kosza, M., Varvounis, G., Surowiec, M., and Giannopoulo, T. (2003) Tele vs. oxidative substitution of hydrogen in meta monochloromethyl, dichloromethyl, and trichloromethyl nitrobenzenes in the reaction with Grignard reagents. *Eur. J. Org. Chem.*, **2003**, 3791–3797.

27. Mach, M.H. and Bunnett, J.F. (1980) Participation of oligochlorobenzenes in the base-catalyzed halogen dance. *J. Org. Chem.*, **45**, 4660–4666.

28. Hamel, P., Girard, Y., and Atkinson, J.G. (1992) Acid-catalyzed isomerization of 3-indolyl sulfides to 2-indolyl sulfides: first synthesis of 3-unsubstituted 2-(arylthio)indoles. Evidence for a complex intermolecular process. *J. Org. Chem.*, **57**, 2694–2699.

29. Miller, J. and Parker, A.J. (1961) Dipolar aprotic solvents in bimolecular aromatic nucleophilic substitution reactions. *J. Am. Chem. Soc.*, **83**, 117–123.

30. Rueeger, H., Lueoend, R., Rogel, O., Rondeau, J.-M., Möbitz, H., Machauer, R., Jacobson, L., Staufenbiel, M., Desrayaud, S., and Neumann, U. (2012) Discovery of cyclic sulfone hydroxyethylamines as potent and selective β-site APP-cleaving enzyme 1 (BACE1) inhibitors: structure-based design and in vivo reduction of amyloid β-peptides. *J. Med. Chem.*, **55**, 3364–3386.

31. Beugelmans, R., Singh, G.P., Bois-Choussy, M., Chastanet, J., and Zhu, J. (1994) S_NAr-based macrocyclization: an application to the synthesis of vancomycin family models. *J. Org. Chem.*, **59**, 5535–5542.

32. Zhang, S., Wang, Y., Feng, X., and Bao, M. (2012) Palladium-catalyzed amination of chloromethylnaphthalene and chloromethylanthracene derivatives with various amines. *J. Am. Chem. Soc.*, **134**, 5492–5495.

33. Bodar-Houillon, F., Elissami, Y., Marsura, A., Ghermani, N.E., Espinosa, E., Bouhmaida, N., and Thalal, A. (1999) Synthesis and experimental electron density of bis(heterocyclic) azines: the case of 6,6′-bis(chloromethyl)-2,2′-bipyrazine. *Eur. J. Org. Chem.*, **1999**, 1427–1440.

34. Philipp, A. and Jirkovsky, I. (1979) Derivatives of fused 3-hydroxymethyl-pyran-4-ones as a mobile keto–allyl system. *Can. J. Chem.*, **57**, 3292–3295.

35. Bebot, M., Coudert, P., Rubat, C., Vallee-Goyet, D., Gardette, D., Mavel, S., Albuisson, E., and Couquelet, J. (1997) Synthesis and pharmacological evaluation in mice of new non-classical antinociceptive agents, 5-(4-arylpiperazin-1-yl)-4-benzyl-1,2-oxazin-6-ones. *Chem. Pharm. Bull.*, **45**, 659–667.

36. Torigoe, K., Kariya, N., Soranaka, K., and Pfleiderer, W. (2007) Side-chain transformations of 6- and 7-substituted 1,3-dimethyllumazines (= 1,3-dimethylpteridine-2,4(1H,3H)-diones). *Helv. Chim. Acta*, **90**, 1190–1205.

37. He, G. and Chen, G. (2011) A practical strategy for the structural diversification of aliphatic scaffolds through the palladium-catalyzed picolinamide-directed remote functionalization of unactivated $C(sp^3)$–H bonds. *Angew. Chem. Int. Ed.*, **50**, 5192–5196.
38. Wolfe, A.L., Duncan, K.K., Parelkar, N.K., Weir, S.J., Vielhauer, G.A., and Boger, D.L. (2012) A novel, unusually efficacious duocarmycin carbamate prodrug that releases no residual byproduct. *J. Med. Chem.*, **55**, 5878–5886.
39. Adams, J., Faitg, T., Kasparec, J., Peng, X., Ralph, J., Rheault, T.R., and Waterson, A.G. (2011) Benzene sulfonamide thiazole and oxazole compounds. WO Patent 2011/059610.
40. Littke, A., Soumeillant, M., Kaltenbach, R.F. III, Cherney, R.J., Tarby, C.M., and Kiau, S. (2007) Mild and general methods for the palladium-catalyzed cyanation of aryl and heteroaryl chlorides. *Org. Lett.*, **9**, 1711–1714.
41. Okuro, K., Furuune, M., Miura, M., and Nomura, M. (1993) Copper-catalyzed reaction of aryl iodides with active methylene compounds. *J. Org. Chem.*, **58**, 7606–7607.
42. Giri, R. and Hartwig, J.F. (2010) Cu(I)-Amido complexes in the Ullmann reaction: reactions of Cu(I)-amido complexes with iodoarenes with and without autocatalysis by CuI. *J. Am. Chem. Soc.*, **132**, 15860–15863.
43. Jones, G.O., Liu, P., Houk, K.N., and Buchwald, S.L. (2010) Computational explorations of mechanisms and ligand-directed selectivities of copper-catalyzed Ullmann-type reactions. *J. Am. Chem. Soc.*, **132**, 6205–6213.
44. Fernández-Rodríguez, M.A., Shen, Q., and Hartwig, J.F. (2006) Highly efficient and functional-group-tolerant catalysts for the palladium-catalyzed coupling of aryl chlorides with thiols. *Chem. Eur. J.*, **12**, 7782–7796.
45. Margolis, B.J., Long, K.A., Laird, D.L.T., Ruble, J.C., and Pulley, S.R. (2007) Assembly of 4-aminoquinolines via palladium catalysis: a mild and convenient alternative to S_NAr methodology. *J. Org. Chem.*, **72**, 2232–2235.
46. Sang, P., Xie, Y., Zou, J., and Zhang, Y. (2012) Copper-catalyzed sequential Ullmann *N*-arylation and aerobic oxidative C–H amination: a convenient route to indolo[1,2-*c*]quinazoline derivatives. *Org. Lett.*, **14**, 3894–3897.
47. Chu, L., Yue, X., and Qing, F.-L. (2010) Cu(II)-mediated methylthiolation of aryl C–H bonds with DMSO. *Org. Lett.*, **12**, 1644–1647.
48. Sawant, D.N., Wagh, Y.S., Tambade, P.J., Bhatte, K.D., and Bhanagea, B.M. (2011) Cyanides-free cyanation of aryl halides using formamide. *Adv. Synth. Catal.*, **353**, 781–787.
49. Li, J.-H., Xie, Y.-X., and Yin, D.-L. (2003) New role of CO_2 as a selective agent in palladium-catalyzed reductive Ullmann coupling with zinc in water. *J. Org. Chem.*, **68**, 9867–9869.
50. Monopoli, A., Caló, V., Ciminale, F., Cotugno, P., Angelici, C., Cioffi, N., and Nacci, A. (2010) Glucose as a clean and renewable reductant in the Pd-nanoparticle-catalyzed reductive homocoupling of bromo- and chloroarenes in water. *J. Org. Chem.*, **75**, 3908–3911.
51. Bergeron-Brlek, M., Giguère, D., Shiao, T.C., Saucier, C., and Roy, R. (2012) Palladium-catalyzed Ullmann-type reductive homocoupling of iodoaryl glycosides. *J. Org. Chem.*, **77**, 2971–2977.
52. Hassan, J., Sévignon, M., Gozzi, C., Schulz, E., and Lemaire, M. (2002) Aryl–aryl bond formation one century after the discovery of the Ullmann reaction. *Chem. Rev.*, **102**, 1359–1469.
53. Wipf, P. and Lynch, S.M. (2003) Synthesis of highly oxygenated dinaphthyl ethers via S_NAr reactions promoted by Barton's base. *Org. Lett.*, **5**, 1155–1158.
54. Bryan, C.S., Braunger, J.A., and Lautens, M. (2009) Efficient synthesis of benzothiophenes on the basis of an unusual palladium-catalyzed vinylic C–S coupling. *Angew. Chem. Int. Ed.*, **48**, 7064–7068.
55. Price, C.C. and Stacy, G.W. (1955) *p*-Aminophenyl disulfide. *Org. Synth.*, Coll. Vol. **3**, 86–87.
56. Bunnett, J.F. and Weiss, R.H. (1988) Radical anion arylation: diethyl

phenylphosphonate. *Org. Synth.*, Coll. Vol. **6**, 451–454.

57. Gandhi, S.S., Gibson, M.S., Kaldas, M.L., and Vines, S.M. (1979) Reactions of some aromatic nitro compounds with alkali metal amides. *J. Org. Chem.*, **44**, 4705–4707.

58. Weissermel, K. and Arpe, H.-J. (1993) *Industrial Organic Chemistry*, VCH Verlagsgesellschaft mbH, Weinheim.

59. Goryunov, L.I., Grobe, J., Le Van, D., Shteingarts, V.D., Mews, R., Lork, E., and Würthwein, E.-U. (2010) Di- and trifluorobenzenes in reactions with Me_2EM (E = P, N; M = $SiMe_3$, $SnMe_3$, Li) reagents: evidence for a concerted mechanism of aromatic nucleophilic substitution. *Eur. J. Org. Chem.*, **2010**, 1111–1123.

60. Sandin, H., Swanstein, M.-L., and Wellner, E. (2004) A fast and parallel route to cyclic isothioureas and guanidines with use of microwave-assisted chemistry. *J. Org. Chem.*, **69**, 1571–1580.

61. Diness, F. and Fairlie, D.P. (2012) Catalyst-free *N*-arylation using unactivated fluorobenzenes. *Angew. Chem. Int. Ed.*, **51**, 8012–8016.

62. Beringer, F.M., Brierley, A., Drexler, M., Gindler, E.M., and Lumpkin, C.C. (1953) Diaryliodonium salts. 11. The phenylation of organic and inorganic bases. *J. Am. Chem. Soc.*, **75**, 2708–2712.

63. Sukata, K. and Akagawa, T. (1989) Poly(ethylene glycol)s and their dimethyl ethers as catalysts for the reaction of aryl halides with diphenylamine in the presence of potassium hydroxide. *J. Org. Chem.*, **54**, 1476–1479.

64. Tripathy, S., LeBlanc, R., and Durst, T. (1999) Formation of 2-substituted iodobenzenes from iodobenzene via benzyne and ate complex intermediates. *Org. Lett.*, **1**, 1973–1975.

65. Hartman, W.W. (1941) *p*-Cresol. *Org. Synth.*, Coll. Vol. **1**, 175–176.

66. Weston, A.W. and Suter, C.M. (1955) 3,5-Dihydroxybenzoic acid. *Org. Synth.*, Coll. Vol. **3**, 288–290.

67. Agnello, L.A. and Williams, W.H. (1960) Synthetic phenol. *Ind. Eng. Chem.*, **52**, 894–900.

68. Loughhead, D.G., Flippin, L.A., and Weikert, R.J. (1999) Synthesis of mexiletine stereoisomers and related compounds via S_NAr nucleophilic substitution of a $Cr(CO)_3$-complexed aromatic fluoride. *J. Org. Chem.*, **64**, 3373–3375.

69. Brenner, E., Baldwin, R.M., and Tamagnan, G. (2005) Asymmetric synthesis of (+)-(*S,S*)-reboxetine via a new (*S*)-2-(hydroxymethyl)morpholine preparation. *Org. Lett.*, **7**, 937–939.

70. Kundu, S.K., Tan, W.S., Yan, J.-L., and Yang, J.-S. (2010) Pentiptycene building blocks derived from nucleophilic aromatic substitution of pentiptycene triflates and halides. *J. Org. Chem.*, **75**, 4640–4643.

71. Itakura, H., Mizuno, H., Hirai, K., and Tomioka, H. (2000) Generation, characterization, and kinetics of triplet di[1,2,3,4,5,6,7,8-octahydro-1,4:5,8-di(ethano)anthryl]carbene. *J. Org. Chem.*, **65**, 8797–8806.

72. Lee, S., Chataigner, I., and Piettre, S.R. (2011) Facile dearomatization of nitrobenzene derivatives and other nitroarenes with *N*-benzyl azomethine ylide. *Angew. Chem. Int. Ed.*, **50**, 472–476.

73. Jiang, Y., Qin, Y., Xie, S., Zhang, X., Dong, J., and Ma, D. (2009) A general and efficient approach to aryl thiols: CuI-catalyzed coupling of aryl iodides with sulfur and subsequent reduction. *Org. Lett.*, **11**, 5250–5253.

74. Gilman, H. and Gainer, G.C. (1949) Some 6-quinolyl sulfides and sulfones. *J. Am. Chem. Soc.*, **71**, 1747–1751.

75. Beier, P., Pastyříková, T., Vida, N., and Iakobson, G. (2011) S_NAr reactions of nitro(pentafluorosulfanyl)benzenes to generate SF_5 aryl ethers and sulfides. *Org. Lett.*, **13**, 1466–1469.

76. Harris, N.V., Smith, C., and Bowden, K. (1990) Antifolate and antibacterial activities of 5-substituted 2,4-diaminoquinazolines. *J. Med. Chem.*, **33**, 434–444.

77. Russell, A. and Tebbens, W.G. (1955) 2,6-Dimethoxybenzonitrile. *Org. Synth.*, Coll. Vol. **3**, 293–294.

78. Mauthner, F. (1929) Untersuchungen über die γ-Resorcylsäure. *J. Prakt. Chem.*, **117**, 259–265.

79. Beck, J.R. and Yahner, J.A. (1978) Nitro displacement by methanethiol anion. Synthesis of bis-, tris-, pentakis-, and hexakis(methylthio)benzenes. *J. Org. Chem.*, **43**, 2048–2052.

80. Cogolli, P., Testaferri, L., Tingoli, M., and Tiecco, M. (1979) Alkyl thioether activation of the nitro displacement by alkanethiol anions. A useful process for the synthesis of poly[(alkylthio)benzenes]. *J. Org. Chem.*, **44**, 2636–2642.

81. Montanari, S., Paradisi, C., and Scorrano, G. (1991) Influence of ion pairing, steric effects, and other specific interactions on the reactivity of thioanions with chloronitrobenzenes. Nucleophilic aromatic substitution vs reduction. *J. Org. Chem.*, **56**, 4274–4279.

82. Beier, P., Pastyříková, T., and Iakobson, G. (2011) Preparation of SF$_5$ aromatics by vicarious nucleophilic substitution reactions of nitro(pentafluorosulfanyl)benzenes with carbanions. *J. Org. Chem.*, **76**, 4781–4786.

83. Mąkosza, M., Chromiński, M., and Sulikowski, D. (2011) Synthesis of nitroaryl derivatives of glycine via oxidative nucleophilic substitution of hydrogen in nitroarenes. *Arkivoc*, 82–91.

84. Mąkosza, M. (2010) Nucleophilic substitution of hydrogen in electron-deficient arenes, a general process of great practical value. *Chem. Soc. Rev.*, **39**, 2855–2868.

85. Mąkosza, M. and Wojciechowski, K. (2004) Nucleophilic substitution of hydrogen in heterocyclic chemistry. *Chem. Rev.*, **104**, 2631–2666.

86. Verbeeck, S., Herrebout, W.A., Gulevskaya, A.V., van der Veken, B.J., and Maes, B.U.W. (2010) ONSH: optimization of oxidative alkylamination reactions through study of the reaction mechanism. *J. Org. Chem.*, **75**, 5126–5133.

87. Gulevskaya, A.V., Verbeeck, S., Burov, O.N., Meyers, C., Korbukova, I.N., Herrebout, W., and Maes, B.U.W. (2009) Synthesis of (alkylamino)nitroarenes by oxidative alkylamination of nitroarenes. *Eur. J. Org. Chem.*, **2009**, 564–574.

88. Dantas de Araujo, A., Christensen, C., Buchardt, J., Kent, S.B.H., and Alewood, P.F. (2011) Synthesis of tripeptide mimetics based on dihydroquinolinone and benzoxazinone scaffolds. *Chem. Eur. J.*, **17**, 13983–13986.

89. Lawrence, N.J., Liddle, J., Bushell, S.M., and Jackson, D.A. (2002) A three-component coupling process based on vicarious nucleophilic substitution (VNS$_{AR}$)-alkylation reactions: an approach to indoprofen and derivatives. *J. Org. Chem.*, **67**, 457–464.

90. Florio, S., Lorusso, P., Luisi, R., Granito, C., Ronzini, L., and Troisi, L. (2004) Vicarious nucleophilic substitution of (chloroalkyl)heterocycles with nitroarenes. *Eur. J. Org. Chem.*, **2004**, 2118–2124.

91. Kawakami, T. and Suzuki, H. (1999) Masked acylation of *m*-dinitrobenzene and derivatives with nitroalkanes under basic conditions: nitromethylation and α-(hydroxyimino)alkylation. *Tetrahedron Lett.*, **40**, 1157–1160.

92. Lemek, T., Groszek, G., and Cmoch, P. (2008) Vicarious nucleophilic substitutions of hydrogen in 1,1,1-trifluoro-*N*-[oxido(phenyl)(trifluoromethyl)-λ^4-sulfanylidene]methanesulfonamide. *Eur. J. Org. Chem.*, **2008**, 4206–4209.

93. Gomez, R., Jolly, S.J., Williams, T., Vacca, J.P., Torrent, M., McGaughey, G., Lai, M.-T., Felock, P., Munshi, V., DiStefano, D., Flynn, J., Miller, M., Yan, Y., Reid, J., Sanchez, R., Liang, Y., Paton, B., Wan, B.-L., and Anthony, N. (2011) Design and synthesis of conformationally constrained inhibitors of non-nucleoside reverse transcriptase. *J. Med. Chem.*, **54**, 7920–7933.

94. Pews, R.G., Lysenko, Z., and Vosejpka, P.C. (1997) A safe cost-efficient synthesis of 4,6-diaminoresorcinol. *J. Org. Chem.*, **62**, 8255–8256.
95. Makosza, M. and Sienkiewicz, K. (1998) Hydroxylation of nitroarenes with alkyl hydroperoxide anions via vicarious nucleophilic substitution of hydrogen. *J. Org. Chem.*, **63**, 4199–4208.
96. Kawakami, T., Uehata, K., and Suzuki, H. (2000) NaH-mediated one-pot cyclocondensation of 6-nitroquinoline with aromatic hydrazones to form [1.2.4]triazino[6,5-*f*]quinolines and/or pyrazolo[3,4-*f*]quinolines. *Org. Lett.*, **2**, 413–415.
97. Katritzky, A.R. and Laurenzo, K.S. (1988) Alkylaminonitrobenzenes by vicarious nucleophilic amination with 4-(alkylamino)-1,2,4-triazoles. *J. Org. Chem.*, **53**, 3978–3982.
98. Bunnett, J.F., Rauhut, M.H., Knutson, D., and Bussell, G.E. (1954) Studies on the conditions, scope and mechanism of the von Richter reaction. *J. Am. Chem. Soc.*, **76**, 5755–5761.
99. Arvela, R.K. and Leadbeater, N.E. (2003) Rapid, easy cyanation of aryl bromides and chlorides using nickel salts in conjunction with microwave promotion. *J. Org. Chem.*, **68**, 9122–9125.
100. Sundermeier, M., Zapf, A., and Beller, M. (2003) A convenient procedure for the palladium-catalyzed cyanation of aryl halides. *Angew. Chem. Int. Ed.*, **42**, 1661–1664.
101. Weissman, S.A., Zewge, D., and Chen, C. (2005) Ligand-free palladium-catalyzed cyanation of aryl halides. *J. Org. Chem.*, **70**, 1508–1510.
102. Sundermeier, M., Zapf, A., Mutyala, S., Baumann, W., Sans, J., Weiss, S., and Beller, M. (2003) Progress in the palladium-catalyzed cyanation of aryl chlorides. *Chem. Eur. J.*, **9**, 1828–1836.
103. Ren, Y., Liu, Z., He, S., Zhao, S., Wang, J., Niu, R., and Yin, W. (2009) Development of an open-air and robust method for large-scale palladium-catalyzed cyanation of aryl halides: the use of *i*-PrOH to prevent catalyst poisoning by oxygen. *Org. Process Res. Dev.*, **13**, 764–768.
104. Ushkov, A.V. and Grushin, V.V. (2011) Rational catalysis design on the basis of mechanistic understanding: highly efficient Pd-catalyzed cyanation of aryl bromides with NaCN in recyclable solvents. *J. Am. Chem. Soc.*, **133**, 10999–11005.
105. Schareina, T., Zapf, A., Mägerlein, W., Müller, N., and Beller, M. (2007) A state-of-the-art cyanation of aryl bromides: a novel and versatile copper catalyst system inspired by nature. *Chem. Eur. J.*, **13**, 6249–6254.
106. Guthrie, R.D., Hartmann, C., Neill, R., and Nutter, D.E. (1987) Carbanions: electron transfer vs. proton capture. 8. Use of sterically protected aromatic nitro compounds as base-resistant, one-electron oxidants. *J. Org. Chem.*, **52**, 736–740.
107. Müller, E. (1957) *Methoden der Organischen Chemie (Houben-Weyl)*, Band **11/1**, Stickstoffverbindungen II, Georg Thieme Verlag, Stuttgart, p. 462.
108. Lauer, W.M., Sprung, M.M., and Langkammerer, C.M. (1936) The Piria reaction. III. Mechanism studies. *J. Am. Chem. Soc.*, **58**, 225–228.
109. Wróbel, Z. (2003) Double thioalkylation/arylation of nitroarenes with the reduction of nitro- to amino group. *Tetrahedron*, **59**, 101–110.
110. Halama, A., Kaválek, J., Macháček, V., and Weidlich, T. (1999) Aromatic nucleophilic substitution of hydrogen: mechanism of reaction of 6-nitroquinoline with cyanide ions, with and without participation of methyl cyanoacetate. *J. Chem. Soc., Perkin Trans. 1*, 1839–1845.
111. Tomioka, Y., Mochiike, A., Himeno, J., and Yamazaki, M. (1981) Studies on aromatic nitro compounds. I. Reaction of 6-nitroquinoline with active methylene compounds in the presence of bases. *Chem. Pharm. Bull.*, **29**, 1286–1291.
112. Shang, R., Ji, D.-S., Chu, L., Fu, Y., and Liu, L. (2011) Synthesis of α-aryl nitriles through palladium-catalyzed decarboxylative coupling of cyanoacetate salts with aryl halides and

113. Song, B., Rudolphi, F., Himmler, T., and Gooßen, L.J. (2011) Practical synthesis of 2-arylacetic acid esters via palladium-catalyzed dealkoxycarbonylative coupling of malonates with aryl halides. *Adv. Synth. Catal.*, **353**, 1565–1574.

114. Yang, D., Yang, H., and Fu, H. (2011) Copper-catalyzed aerobic oxidative synthesis of aromatic carboxylic acids. *Chem. Commun.*, **47**, 2348–2350.

115. Jiang, Z., Huang, Q., Chen, S., Long, L., and Zhoua, X. (2012) Copper-catalyzed cyanation of aryl iodides with malononitrile: an unusual cyano group transfer process from $C(sp^3)$ to $C(sp^2)$. *Adv. Synth. Catal.*, **354**, 589–592.

116. He, L., Wang, J.-Q., Gong, Y., Liu, Y.-M., Cao, Y., He, H.-Y., and Fan, K.-N. (2011) Titania-supported iridium subnanoclusters as an efficient heterogeneous catalyst for direct synthesis of quinolines from nitroarenes and aliphatic alcohols. *Angew. Chem. Int. Ed.*, **50**, 10216–10220.

117. Dell'Erba, C., Gabellini, A., Novi, M., Petrillo, G., Tavani, C., Cosimelli, B., and Spinelli, D. (2001) Ring opening of 2-substituted 4-nitrothiophenes with pyrrolidine. Access to new functionalized nitro-unsaturated building blocks. *Tetrahedron*, **57**, 8159–8165.

118. Petrillo, G., Novi, M., Garbarino, G., and Filiberti, M. (1989) The reaction between arenediazonium tetrafluoroborates and alkaline thiocarboxylates in DMSO: a convenient access to aryl thiolesters and other aromatic sulfur derivatives. *Tetrahedron*, **45**, 7411–7420.

119. DeTar, D.F. and Kosuge, T. (1958) Mechanisms of diazonium salt reactions. VI. The reaction of diazonium salts with alcohols under acidic conditions; evidence for hydride transfer. *J. Am. Chem. Soc.*, **80**, 6072–6077.

120. Lipilin, D.L., Smirnov, O.Y., Churakov, A.M., Strelenko, Y.A., Tyurin, A.Y., Ioffe, S.L., and Tartakovsky, V.A. (2004) 2-Alkyl-1,2,3,4-benzotetrazinium tetrafluoroborates: their reaction with nucleophiles. *Eur. J. Org. Chem.*, **2004**, 4794–4801.

121. Broxton, T.J. and McLeish, M.J. (1983) Reactions of aryl diazonium salts and alkyl arylazo ethers. 9. Studies of the carbanionic and free radical mechanisms of dediazoniation of substituted 2-chlorobenzenediazonium salts. *J. Org. Chem.*, **48**, 191–195.

122. Dollings, P.J., Dietrich, A.J., and Wrobel, J.E. (2000) Naphtho[2,3-b]heteroar-4-yl derivatives. US Patent 6121271.

123. Swain, C.G., Sheats, J.E., and Harbison, K.G. (1975) Evidence for phenyl cation as an intermediate in reactions of benzenediazonium salts in solution. *J. Am. Chem. Soc.*, **97**, 783–790.

124. Robinson, M.K., Kochurina, V.S., and Hanna, J.M. (2007) Palladium-catalyzed homocoupling of arenediazonium salts: an operationally simple synthesis of symmetrical biaryls. *Tetrahedron Lett.*, **48**, 7687–7690.

125. Trimmer, R.W., Stover, L.R., and Skjold, A.C. (1985) Solid-state aromatic S_N2 reactions: displacement of the nitro moiety in arenediazonium salts. *J. Org. Chem.*, **50**, 3612–3614.

126. Oku, A. and Matsui, A. (1979) Diazotization of nitroanthranilic acids. Effect of carboxyl group on the nucleophilic ipso substitution of the nitro group by chloride ion. *J. Org. Chem.*, **44**, 3342–3344.

127. Lauer, W.M. and Langkammerer, C.M. (1934) The action of sodium bisulfite on resorcinol. *J. Am. Chem. Soc.*, **56**, 1628–1629.

128. Charoonniyomporn, P., Thongpanchang, T., Witayakran, S., Thebtaranonth, Y., Phillips, K.E.S., and Katz, T.J. (2004) An efficient one-pot synthesis of bisalkylthioarenes. *Tetrahedron Lett.*, **45**, 457–459.

129. Node, M., Kawabata, T., Ohta, K., Fujimoto, M., Fujita, E., and Fuji, K. (1984) Hard acid and soft nucleophile systems. 8. Reductive dehalogenation of o- and p-halophenols and their derivatives. *J. Org. Chem.*, **49**, 3641–3643.

130. Delprato, I. and Bertoldi, M. (1997) Process for preparation of 4-mercapto-1-naphthol compounds. EP Patent 0763526.
131. Mino, T., Yamada, H., Komatsu, S., Kasai, M., Sakamoto, M., and Fujita, T. (2011) Atropisomerism at C−N bonds of acyclic amines: synthesis and application to palladium-catalyzed asymmetric allylic alkylations. *Eur. J. Org. Chem.*, **2011**, 4540–4542.
132. Belaud-Rotureau, M., Le, T.T., Phan, T.H.T., Nguyen, T.H., Aissaoui, R., Gohier, F., Derdour, A., Nourry, A., Castanet, A., Nguyen, K.P.P., and Mortier, J. (2010) Synthesis of *N*-aryl and *N*-alkyl anthranilic acids via S_NAr reaction of unprotected 2-fluoro- and 2-methoxybenzoic acids by lithioamides. *Org. Lett.*, **12**, 2406–2409.
133. Mulzer, M. and Coates, G.W. (2011) A catalytic route to ampakines and their derivatives. *Org. Lett.*, **13**, 1426–1428.
134. Wempe, M.F., Jutabha, P., Quade, B., Iwen, T.J., Frick, M.M., Ross, I.R., Rice, P.J., Anzai, N., and Endou, H. (2011) Developing potent human uric acid transporter 1 (hURAT1) inhibitors. *J. Med. Chem.*, **54**, 2701–2713.
135. Wu, D., Pisula, W., Haberecht, M.C., Feng, X., and Müllen, K. (2009) Oxygen- and sulfur-containing positively charged polycyclic aromatic hydrocarbons. *Org. Lett.*, **11**, 5686–5689.
136. Guilarte, V., Castroviejo, M.P., García-García, P., Fernández-Rodríguez, M.A., and Sanz, R. (2011) Approaches to the synthesis of 2,3-dihaloanilines. Useful precursors of 4-functionalized-1*H*-indoles. *J. Org. Chem.*, **76**, 3416–3437.
137. Imakura, Y., Okimoto, K., Konishi, T., Hisazumi, M., Yamazaki, J., and Kobayashi, S. (1992) Regioselective cleavage reaction of the aromatic methylenedioxy ring. V. Cleavage with sodium alkoxides–alcohols, potassium *tert*-butoxide–alcohols, dimsyl anion–methyl alcohol, metallic sodium–alcohols, and sodium cyanide in dipolar aprotic solvents. *Chem. Pharm. Bull.*, **40**, 1691–1696.
138. Caubère, C., Caubère, P., Renard, P., Bizot-Espiart, J.-G., Ianelli, S., Nardelli, M., and Jamart-Grégoire, B. (1994) Functionalisation of the alkoxy group of alkyl aryl ethers. Demethylation, alkylthiolation and reduction of 5-methoxyindoles. *Tetrahedron*, **50**, 13433–13448.
139. Trivedi, A.R., Siddiqui, A.B., and Shah, V.H. (2008) Design, synthesis, characterization and antitubercular activity of some 2-heterocycle-substituted phenothiazines. *Arkivoc*, **2**, 210–217.
140. Barrett, O.V., Downer-Riley, N.K., and Jackson, Y.A. (2012) Thermally induced cyclization of electron-rich *N*-arylthiobenzamides to benzothiazoles. *Synthesis*, **44**, 2579–2586.
141. Pirkle, W.H. and Finn, J.M. (1983) Useful routes to 9-anthryl ethers and sulfides. *J. Org. Chem.*, **48**, 2779–2780.
142. Eberhart, A.J., Imbriglio, J.E., and Procter, D.J. (2011) Nucleophilic *ortho* allylation of aryl and heteroaryl sulfoxides. *Org. Lett.*, **13**, 5882–5885.
143. Gianneschi, N.C., Bertin, P.A., Nguyen, S.T., and Mirkin, C.A. (2003) A supramolecular approach to an allosteric catalyst. *J. Am. Chem. Soc.*, **125**, 10508–10509.
144. Pinehart, A., Dallaire, C., Van Bierbeek, A., and Gingras, M. (1999) Efficient formation of aromatic thiols from thiomethylated precursors. *Tetrahedron Lett.*, **40**, 5479–5482.
145. Sridhar, M., Kumar, B.A., and Narender, R. (1998) Expedient and simple method for regeneration of alcohols, from toluenesulfonates using Mg–MeOH. *Tetrahedron Lett.*, **39**, 2847–2850.
146. Sabitha, G., Abraham, S., Reddy, B.V.S., and Yadav, J.S. (1999) Microwave assisted selective cleavage of sulfonates and sulfonamides in dry media. *Synlett*, 1745–1746.
147. Chakraborti, A.K., Nayak, M.K., and Sharma, L. (2002) Diphenyl disulfide and sodium in NMP as an efficient protocol for in situ generation of thiophenolate anion: selective deprotection of aryl alkyl ethers and alkyl/aryl esters under nonhydrolytic conditions. *J. Org. Chem.*, **67**, 1776–1780.

148. Xu, G. and Wang, Y.-G. (2004) Microwave-assisted amination from aryl triflates without base and catalyst. *Org. Lett.*, **6**, 985–987.

149. Sach, N.W., Richter, D.T., Cripps, S., Tran-Dubé, M., Zhu, H., Huang, B., Cui, J., and Sutton, S.C. (2012) Synthesis of aryl ethers via a sulfonyl transfer reaction. *Org. Lett.*, **14**, 3886–3889.

150. Shimasaki, T., Tobisu, M., and Chatani, N. (2010) Nickel-catalyzed amination of aryl pivalates by the cleavage of aryl C–O bonds. *Angew. Chem. Int. Ed.*, **49**, 2929–2932.

151. Hie, L., Ramgren, S.D., Mesganaw, T., and Garg, N.K. (2012) Nickel-catalyzed amination of aryl sulfamates and carbamates using an air-stable precatalyst. *Org. Lett.*, **14**, 4182–4185.

152. Muto, K., Yamaguchi, J., and Itami, K. (2012) Nickel-catalyzed C–H/C–O coupling of azoles with phenol derivatives. *J. Am. Chem. Soc.*, **134**, 169–172.

153. Yoshida, H., Asatsu, Y., Mimura, Y., Ito, Y., Ohshita, J., and Takaki, K. (2011) Three-component coupling of arynes and organic bromides. *Angew. Chem. Int. Ed.*, **50**, 9676–9679.

154. Nakayama, J., Hoshino, K., and Hoshino, M. (1985) Carbon–sulfur bond cleavage by benzyne generated from 2-carboxybenzenediazonium chloride. *Chem. Lett.*, 677–678.

155. Guedira, N.-E. and Beugelmans, R. (1992) Ambident behavior of ketone enolate anions in S_NAr substitutions on fluorobenzonitrile substrates. *J. Org. Chem.*, **57**, 5577–5585.

156. Renga, J.M., McLaren, K.L., and Ricks, M.J. (2003) Process optimization and synthesis of 3-(4-fluorophenyl)-4,5-dihydro-N-[4-(trifluoromethyl)phenyl]-4-[5-(trifluoromethyl)-2-pyridyl]-1H-pyrazole-1-carboxamide. *Org. Process Res. Dev.*, **7**, 267–271.

157. Thompson, A.D. and Huestis, M.P. (2013) Cyanide anion as a leaving group in nucleophilic aromatic substitution: synthesis of quaternary centers at azine heterocycles. *J. Org. Chem.*, **78**, 762–769.

158. Kobbelgaard, S., Bella, M., and Jørgensen, K.A. (2006) Improved asymmetric S_NAr reaction of β-dicarbonyl compounds catalyzed by quaternary ammonium salts derived from cinchona alkaloids. *J. Org. Chem.*, **71**, 4980–4987.

159. Ozanne-Beaudenon, A. and Quideau, S. (2005) Regioselective hypervalent-iodine(III)-mediated dearomatizing phenylation of phenols through direct ligand coupling. *Angew. Chem. Int. Ed.*, **44**, 7065–7069.

160. Stahly, G.P. (1985) Synthesis of unsymmetrical biphenyls by reaction of nitroarenes with phenols. *J. Org. Chem.*, **50**, 3091–3094.

161. Cao, C., Wang, L., Cai, Z., Zhang, L., Guo, J., Pang, G., and Shi, Y. (2011) Palladium-catalyzed α-ketone arylation under mild conditions. *Eur. J. Org. Chem.*, **2011**, 1570–1574.

162. Aljaar, N., Malakar, C.C., Conrad, J., Strobel, S., Schleid, T., and Beifuss, U. (2012) Cu-catalyzed reaction of 1,2-dihalobenzenes with 1,3-cyclohexanediones for the synthesis of 3,4-dihydrodibenzo[b,d]furan-1(2H)-ones. *J. Org. Chem.*, **77**, 7793–7803.

163. Merrill, R.E. and Negishi, E. (1974) Tetrahydrofuran-promoted aryl-alkyl coupling involving organolithium reagents. *J. Org. Chem.*, **39**, 3452–3453.

164. Bailey, W.F., Luderer, M.R., and Jordan, K.P. (2006) Effect of solvent on the lithium–bromine exchange of aryl bromides: reactions of n-butyllithium and tert-butyllithium with 1-bromo-4-tert-butylbenzene at 0°C. *J. Org. Chem.*, **71**, 2825–2828.

165. Patterson, J.W. (1995) The synthesis of mycophenolic acid from 2,4-dihydroxybenzoic acid. *J. Org. Chem.*, **60**, 4542–4548.

166. Norman, D.P.G., Bunnell, A.E., Stabler, S.R., and Flippin, L.A. (1999) Nucleophilic aromatic substitution reactions of novel 5-(2-methoxyphenyl)tetrazole derivatives with organolithium reagents. *J. Org. Chem.*, **64**, 9301–9306.

167. Kojima, T., Ohishi, T., Yamamoto, I., Matsuoka, T., and Kotsukia, H. (2001)

A new practical method for regioselective nucleophilic aromatic alkylation of *ortho*- or *para*-methoxy-substituted aromatic esters with Grignard reagents. *Tetrahedron Lett.*, **42**, 1709–1712.

168. Shindo, M., Koga, K., and Tomioka, K. (1992) A catalytic method for asymmetric nucleophilic aromatic substitution giving binaphthyls. *J. Am. Chem. Soc.*, **114**, 8732–8733.

169. Silberstein, A.L., Ramgren, S.D., and Garg, N.K. (2012) Iron-catalyzed alkylations of aryl sulfamates and carbamates. *Org. Lett.*, **14**, 3796–3799.

170. Kumada, M., Tamao, K., and Sumitani, K. (1988) Phosphine-nickel complex catalyzed cross-coupling of Grignard reagents with aryl and alkenyl halides: 1,2-dibutylbenzene. *Org. Synth.*, Coll. Vol. **6**, 407–411.

171. Kim, C.-B., Jo, H., Ahn, B.-K., Kim, C.K., and Park, K. (2009) Nickel *N*-heterocyclic carbene catalyst for cross-coupling of neopentyl arenesulfonates with methyl and primary alkyl Grignard reagents. *J. Org. Chem.*, **74**, 9566–9569.

172. Cooper, T., Novak, A., Humphreys, L.D., Walker, M.D., and Woodward, S. (2006) User-friendly methylation of aryl and vinyl halides and pseudohalides with DABAL-Me$_3$. *Adv. Synth. Catal.*, **348**, 686–690.

173. Wang, T., Alfonso, B.J., and Love, J.A. (2007) Platinum(II)-catalyzed cross-coupling of polyfluoroaryl imines. *Org. Lett.*, **9**, 5629–5631.

174. Fürstner, A., Leitner, A., and Seidel, G. (2009) 4-nonylbenzoic acid. *Org. Synth.*, Coll. Vol. **11**, 353–358.

175. Phapale, V.B., Guisán-Ceinos, M., Buñuel, E., and Cárdenas, D.J. (2009) Nickel-catalyzed cross-coupling of alkyl zinc halides for the formation of C(sp^2)–C(sp^3) bonds: scope and mechanism. *Chem. Eur. J.*, **15**, 12681–12688.

176. Guan, B.-T., Xiang, S.-K., Wang, B.-Q., Sun, Z.-P., Wang, Y., Zhao, K.-Q., and Shi, Z.-J. (2008) Direct benzylic alkylation via Ni-catalyzed selective benzylic sp^3 C–O activation. *J. Am. Chem. Soc.*, **130**, 3268–3269.

177. Song, G., Su, Y., Gong, X., Han, K., and Li, X. (2011) Pd(0)-Catalyzed diarylation of sp^3 C–H bond in (2-azaaryl)methanes. *Org. Lett.*, **13**, 1968–1971.

178. Zhang, X.-Q. and Wang, Z.-X. (2012) Cross-coupling of aryltrimethylammonium iodides with arylzinc reagents catalyzed by amido pincer nickel complexes. *J. Org. Chem.*, **77**, 3658–3663.

179. Gillespie, R.J., Bamford, S.J., Botting, R., Comer, M., Denny, S., Gaur, S., Griffin, M., Jordan, A.M., Knight, A.R., Lerpiniere, J., Leonardi, S., Lightowler, S., McAteer, S., Merrett, A., Misra, A., Padfield, A., Reece, M., Saadi, M., Selwood, D.L., Stratton, G.C., Surry, D., Todd, R., Tong, X., Ruston, V., Upton, R., and Weiss, S.M. (2009) Antagonists of the human A$_{2A}$ adenosine receptor. 4. Design, synthesis, and preclinical evaluation of 7-aryltriazolo[4,5-*d*]pyrimidines. *J. Med. Chem.*, **52**, 33–47.

180. Sapountzis, I. and Knochel, P. (2002) A new general preparation of polyfunctional diarylamines by the addition of functionalized arylmagnesium compounds to nitroarenes. *J. Am. Chem. Soc.*, **124**, 9390–9391.

181. Gao, H., Ess, D.H., Yousufuddin, M., and Kürti, L. (2013) Transition-metal-free direct arylation: synthesis of halogenated 2-amino-2′-hydroxy-1,1′-biaryls and mechanism by DFT calculations. *J. Am. Chem. Soc.*, **135**, 7086–7089.

182. Dalpozzo, R. and Bartoli, G. (2005) Bartoli indole synthesis. *Curr. Org. Chem.*, **9**, 163–178.

183. Zhang, Z., Yang, Z., Meanwell, N.A., Kadow, J.F., and Wang, T. (2002) A general method for the preparation of 4-and 6-azaindoles. *J. Org. Chem.*, **67**, 2345–2347.

184. Truong, T., Alvarado, J., Tran, L.D., and Daugulis, O. (2010) Nickel, manganese, cobalt, and iron-catalyzed deprotonative arene dimerization. *Org. Lett.*, **12**, 1200–1203.

185. Suhartono, M., Schneider, A.E., Dürner, G., and Göbel, M.W. (2010) Synthetic aromatic amino acids from a Negishi cross-coupling reaction. *Synthesis*, 293–303.

186. Do, H.-Q. and Daugulis, O. (2009) An aromatic Glaser–Hay reaction. *J. Am. Chem. Soc.*, **131**, 17052–17053.
187. Wu, F., Lu, E., and Barden, C. (2012) Antimicrobial/adjuvant compounds and methods. WO Patent 2012116452.
188. Du, C.-J.F. and Hart, H. (1987) Aryne reactions of polyhalobenzenes with alkenyl and alkynyl Grignard reagents. *J. Org. Chem.*, **52**, 4311–4314.
189. Hamura, T., Chuda, Y., Nakatsuji, Y., and Suzuki, K. (2012) Catalytic generation of arynes and trapping by nucleophilic addition and iodination. *Angew. Chem. Int. Ed.*, **51**, 3368–3372.
190. Romero, M., Harrak, Y., Basset, J., Orúe, J.A., and Pujol, M.D. (2009) Direct synthesis of primary arylamines via C–N cross-coupling of aryl bromides and triflates with amides. *Tetrahedron*, **65**, 1951–1956.
191. Meng, F., Zhu, X., Li, Y., Xie, J., Wang, B., Yao, J., and Wan, Y. (2010) Efficient copper-catalyzed direct amination of aryl halides using aqueous ammonia in water. *Eur. J. Org. Chem.*, **2010**, 6149–6152.
192. Liao, B.-S. and Liu, S.-T. (2012) Diamination of phenylene dihalides catalyzed by a dicopper complex. *J. Org. Chem.*, **77**, 6653–6656.
193. Dumrath, A., Lübbe, C., Neumann, H., Jackstell, R., and Beller, M. (2011) Recyclable catalysts for palladium-catalyzed aminations of aryl halides. *Chem. Eur. J.*, **17**, 9599–9604.
194. Huang, M., Lin, X., Zhu, X., Peng, W., Xie, J., and Wan, Y. (2011) A highly versatile catalytic system for N-arylation of amines with aryl chlorides in water. *Eur. J. Org. Chem.*, **2011**, 4523–4527.
195. Mirsadeghi, S., Prasad, G.K.B., Whittaker, N., and Thakker, D.R. (1989) Synthesis of the K-region monofluoro- and difluorobenzo[c]phenanthrenes. *J. Org. Chem.*, **54**, 3091–3096.
196. Hori, K. and Mori, M. (1998) Synthesis of nonsubstituted anilines from molecular nitrogen via transmetalation of arylpalladium complex with titanium–nitrogen fixation complexes. *J. Am. Chem. Soc.*, **120**, 7651–7652.
197. Pieber, B., Cantillo, D., and Kappe, C.O. (2012) Direct arylation of benzene with aryl bromides using high-temperature/high-pressure process windows: expanding the scope of C–H activation chemistry. *Chem. Eur. J.*, **18**, 5047–5055.
198. Whitaker, C.M., Kott, K.L., and McMahon, R.J. (1995) Synthesis and solid-state structure of substituted arylphosphine oxides. *J. Org. Chem.*, **60**, 3499–3508.
199. Fors, B.P. and Buchwald, S.L. (2009) Pd-catalyzed conversion of aryl chlorides, triflates, and nonaflates to nitroaromatics. *J. Am. Chem. Soc.*, **131**, 12898–12899.
200. Holm, B. (1985) Verfahren zur Herstellung von *o*- und *p*-Nitrobenzaldehyd. DE Patent 3519864.
201. Ma, D., Xia, C., Jiang, J., Zhang, J., and Tang, W. (2003) Aromatic nucleophilic substitution or CuI-catalyzed coupling route to martinellic acid. *J. Org. Chem.*, **68**, 442–451.
202. Fors, B.P., Watson, D.A., Biscoe, M.R., and Buchwald, S.L. (2008) A highly active catalyst for Pd-catalyzed amination reactions: cross-coupling reactions using aryl mesylates and the highly selective monoarylation of primary amines using aryl chlorides. *J. Am. Chem. Soc.*, **130**, 13552–13554.
203. Kuwano, R., Utsunomiya, M., and Hartwig, J.F. (2002) Aqueous hydroxide as a base for palladium-catalyzed amination of aryl chlorides and bromides. *J. Org. Chem.*, **67**, 6479–6486.
204. Guo, D., Huang, H., Xu, J., Jiang, H., and Liu, H. (2008) Efficient iron-catalyzed N-arylation of aryl halides with amines. *Org. Lett.*, **10**, 4513–4516.
205. Xiong, X., Jiang, Y., and Ma, D. (2012) Assembly of N,N-disubstituted hydrazines and 1-aryl-1H-indazoles via copper-catalyzed coupling reactions. *Org. Lett.*, **14**, 2552–2555.
206. Petersen, T.P., Larsen, A.F., Ritzén, A., and Ulven, T. (2013) Continuous flow nucleophilic aromatic substitution with dimethylamine generated in situ by decomposition of DMF. *J. Org. Chem.*, **78**, 4190–4195.

207. Chen, W.-X. and Shao, L.-X. (2012) N-Heterocyclic carbene-palladium(II)-1-methylimidazole complex catalyzed amination between aryl chlorides and amides. *J. Org. Chem.*, **77**, 9236–9239.
208. Withbroe, G.J., Singer, R.A., and Sieser, J.E. (2008) Streamlined synthesis of the bippyphos family of ligands and cross-coupling applications. *Org. Process Res. Dev.*, **12**, 480–489.
209. Surry, D.S. and Buchwald, S.L. (2007) Selective palladium-catalyzed arylation of ammonia: synthesis of anilines as well as symmetrical and unsymmetrical di- and triarylamines. *J. Am. Chem. Soc.*, **129**, 10354–10355.
210. Watanabe, T., Oishi, S., Fujii, N., and Ohno, H. (2009) Palladium-catalyzed direct synthesis of carbazoles via one-pot N-arylation and oxidative biaryl coupling: synthesis and mechanistic study. *J. Org. Chem.*, **74**, 4720–4726.
211. Maiti, D. and Buchwald, S.L. (2009) Orthogonal Cu- and Pd-based catalyst systems for the O- and N-arylation of aminophenols. *J. Am. Chem. Soc.*, **131**, 17423–17429.
212. Fang, Y., Zheng, Y., and Wang, Z. (2012) Direct base-assisted C–N bond formation between aryl halides and aliphatic tertiary amines under transition-metal-free conditions. *Eur. J. Org. Chem.*, **2012**, 1495–1498.
213. Kametani, T., Kigasawa, K., Hiiragi, M., and Aoyama, T. (1972) Studies on the syntheses of heterocyclic compounds. CDLX. Benzyne reaction. XIII. Benzyne reaction of halogenobenzenes with N-alkylmorpholines. *J. Org. Chem.*, **37**, 1450–1453.
214. Cant, A.A., Bertrand, G.H.V., Henderson, J.L., Roberts, L., and Greaney, M.F. (2009) The benzyne aza-Claisen reaction. *Angew. Chem. Int. Ed.*, **48**, 5199–5202.
215. Burnley, J., Carbone, G., and Moses, J.E. (2013) Catalytic reduction of ortho- and para-azidonitrobenzenes via tert-butoxide ion mediated electron transfer. *Synlett*, 652–656.
216. Messaoudi, S., Brion, J.-D., and Alami, M. (2010) An expeditious copper-catalyzed access to 3-aminoquinolinones, 3-aminocoumarins and anilines using sodium azide. *Adv. Synth. Catal.*, **352**, 1677–1687.
217. Markiewicz, J.T., Wiest, O., and Helquist, P. (2010) Synthesis of primary aryl amines through a copper-assisted aromatic substitution reaction with sodium azide. *J. Org. Chem.*, **75**, 4887–4890.
218. D'Anna, F., Marullo, S., and Noto, R. (2008) Ionic liquids/[bmim][N_3] mixtures: promising media for the synthesis of aryl azides by S_NAr. *J. Org. Chem.*, **73**, 6224–6228.
219. Andersen, J., Madsen, U., Björkling, F., and Liang, X. (2005) Rapid synthesis of aryl azides from aryl halides under mild conditions. *Synlett*, 2209–2213.
220. Goriya, Y. and Ramana, C.V. (2010) The [Cu]-catalyzed S_NAr reactions: direct amination of electron deficient aryl halides with sodium azide and the synthesis of arylthioethers under Cu(II)-ascorbate redox system. *Tetrahedron*, **66**, 7642–7650.
221. Zhu, W. and Ma, D. (2004) Synthesis of aryl azides and vinyl azides via proline-promoted CuI-catalyzed coupling reactions. *Chem. Commun.*, 888–889.
222. Boyer, J.H. and Canter, F.C. (1954) Alkyl and aryl azides. *Chem. Rev.*, **54**, 1–57.
223. Kendre, D.B., Toche, R.B., and Jachak, M.N. (2007) Synthesis of novel dipyrazolo[3,4-b:3,4-d]pyridines and study of their fluorescence behavior. *Tetrahedron*, **63**, 11000–11004.
224. Rodrigues, J.A.R., Abramovitch, R.A., de Sousa, J.D.F., and Leiva, G.C. (2004) Diastereoselective synthesis of cularine alkaloids via enium ions and an easy entry to isoquinolines by aza-Wittig electrocyclic ring closure. *J. Org. Chem.*, **69**, 2920–2928.
225. de Carvalho, M., Sorrilha, A.E.P.M., and Rodrigues, J.A.R. (1999) Reaction of aromatic azides with strong acids: formation of fused nitrogen heterocycles and arylamines. *J. Braz. Chem. Soc.*, **10**, 415–420.
226. Yang, K., Li, Z., Wang, Z., Yao, Z., and Jiang, S. (2011) Highly efficient synthesis of phenols by copper-catalyzed

hydroxylation of aryl iodides, bromides, and chlorides. *Org. Lett.*, **13**, 4340–4343.

227. Xu, H.-J., Liang, Y.-F., Cai, Z.-Y., Qi, H.-X., Yang, C.-Y., and Feng, Y.-S. (2011) CuI-nanoparticles-catalyzed selective synthesis of phenols, anilines, and thiophenols from aryl halides in aqueous solution. *J. Org. Chem.*, **76**, 2296–2300.

228. Kozhevnikov, V.N., Dahms, K., and Bryce, M.R. (2011) Nucleophilic substitution of fluorine atoms in 2,6-difluoro-3-(pyridin-2-yl)benzonitrile leading to soluble blue-emitting cyclometalated Ir(III) complexes. *J. Org. Chem.*, **76**, 5143–5148.

229. Woydziak, Z.R., Fu, L., and Peterson, B.R. (2012) Synthesis of fluorinated benzophenones, xanthones, acridones, and thioxanthones by iterative nucleophilic aromatic substitution. *J. Org. Chem.*, **77**, 473–481.

230. Zhou, C., Liu, Q., Li, Y., Zhang, R., Fu, X., and Duan, C. (2012) Palladium-catalyzed desulfitative arylation by C–O bond cleavage of aryl triflates with sodium arylsulfinates. *J. Org. Chem.*, **77**, 10468–10472.

231. Becht, J.-M. and Le Drian, C. (2011) Formation of carbon–sulfur and carbon–selenium bonds by palladium-catalyzed decarboxylative cross-couplings of hindered 2,6-dialkoxybenzoic acids. *J. Org. Chem.*, **76**, 6327–6330.

232. Bhadra, S., Dzik, W.I., and Goossen, L.J. (2012) Decarboxylative etherification of aromatic carboxylic acids. *J. Am. Chem. Soc.*, **134**, 9938–9941.

233. Duan, Z., Ranjit, S., Zhang, P., and Liu, X. (2009) Synthesis of aryl sulfides by decarboxylative C–S cross-couplings. *Chem. Eur. J.*, **15**, 3666–3669.

234. Makhlynets, O.V., Das, P., Taktak, S., Flook, M., Mas-Ballesté, R., Rybak-Akimova, E.V., and Que, L. (2009) Iron-promoted *ortho-* and/or *ipso-*hydroxylation of benzoic acids with H_2O_2. *Chem. Eur. J.*, **15**, 13171–13180.

235. Yu, C.-W., Chen, G.S., Huang, C.-W., and Chern, J.-W. (2012) Efficient microwave-assisted Pd-catalyzed hydroxylation of aryl chlorides in the presence of carbonate. *Org. Lett.*, **14**, 3688–3691.

236. Parlow, J.J., Stevens, A.M., Stegeman, R.A., Stallings, W.C., Kurumbail, R.G., and South, M.S. (2003) Synthesis and crystal structures of substituted benzenes and benzoquinones as tissue factor VIIa inhibitors. *J. Med. Chem.*, **46**, 4297–4312.

237. Cohen, T. and Lewin, A.H. (1966) The production of organocopper intermediates from radicals in the reactions of aromatic halides and diazonium ions with cuprous benzoate. New synthetic methods for aryl benzoates. *J. Am. Chem. Soc.*, **88**, 4521–4522.

238. Petersen, T.B., Khan, R., and Olofsson, B. (2011) Metal-free synthesis of aryl esters from carboxylic acids and diaryliodonium salts. *Org. Lett.*, **13**, 3462–3465.

239. Woiwode, T.F., Rose, C., and Wandless, T.J. (1998) A simple and efficient method for the preparation of hindered alkyl–aryl ethers. *J. Org. Chem.*, **63**, 9594–9596.

240. Osborne, R., Clarke, N., Glossop, P., Kenyon, A., Liu, H., Patel, S., Summerhill, S., and Jones, L.H. (2011) Efficient conversion of a nonselective norepinephrin reuptake inhibitor into a dual muscarinic antagonist-$β_2$-agonist for the treatment of chronic obstructive pulmonary disease. *J. Med. Chem.*, **54**, 6998–7002.

241. Tejero, I., Huertas, I., González-Lafont, A., Lluch, J.M., and Marquet, J. (2005) A fast radical chain mechanism in the polyfluoroalkoxylation of aromatics through NO_2 group displacement. Mechanistic and theoretical studies. *J. Org. Chem.*, **70**, 1718–1727.

242. Kuriyama, M., Hamaguchi, N., and Onomura, O. (2012) Copper(II)-catalyzed monoarylation of vicinal diols with diaryliodonium salts. *Chem. Eur. J.*, **18**, 1591–1594.

243. Arca, V., Paradisi, C., and Scorrano, G. (1990) Competition between radical and nonradical reactions of halonitrobenzenes in alkaline alcoholic solutions. *J. Org. Chem.*, **55**, 3617–3621.

244. Shirakawa, E., Zhang, X., and Hayashi, T. (2011) Mizoroki–Heck-type reaction mediated by potassium *tert*-butoxide. *Angew. Chem. Int. Ed.*, **50**, 4671–4674.

245. Yamamoto, Y., Takuma, R., Hotta, T., and Yamashita, K. (2009) Synthesis of 2,5-dihydrofuran-fused quinones from ether-tethered diiododiyne. *J. Org. Chem.*, **74**, 4324–4328.

246. Paradisi, C., Quintily, U., and Scorrano, G. (1983) Anion activation in the synthesis of ethers from oxygen anions and 1-chloro-4-nitrobenzene. *J. Org. Chem.*, **48**, 3022–3026.

247. Sekiguchi, S. and Okada, K. (1975) Aromatic nucleophilic substitution. V. Confirmation of the spiro Janovsky complex in base-catalyzed rearrangement of *N*-acetyl-β-aminoethyl-2,4-dinitrophenyl ether with simultaneous migration of acetyl group. *J. Org. Chem.*, **40**, 2782–2786.

248. Shafir, A., Lichtor, P.A., and Buchwald, S.L. (2007) *N*- versus *O*-arylation of aminoalcohols: orthogonal selectivity in copper-based catalysts. *J. Am. Chem. Soc.*, **129**, 3490–3491.

249. Vorogushin, A.V., Huang, X., and Buchwald, S.L. (2005) Use of tunable ligands allows for intermolecular Pd-catalyzed C–O bond formation. *J. Am. Chem. Soc.*, **127**, 8146–8149.

250. Wu, X., Fors, B.P., and Buchwald, S.L. (2011) A single phosphine ligand allows palladium-catalyzed intermolecular C–O bond formation with secondary and primary alcohols. *Angew. Chem. Int. Ed.*, **50**, 9943–9947.

251. Kataoka, N., Shelby, Q., Stambuli, J.P., and Hartwig, J.F. (2002) Air stable, sterically hindered ferrocenyl dialkylphosphines for palladium-catalyzed C–C, C–N, and C–O bond-forming cross-couplings. *J. Org. Chem.*, **67**, 5553–5566.

252. Torraca, K.E., Huang, X., Parrish, C.A., and Buchwald, S.L. (2001) An efficient intermolecular palladium-catalyzed synthesis of aryl ethers. *J. Am. Chem. Soc.*, **123**, 10770–10771.

253. Eichman, C.C. and Stambuli, J.P. (2009) Zinc-mediated palladium-catalyzed formation of carbon–sulfur bonds. *J. Org. Chem.*, **74**, 4005–4008.

254. Adams, R. and Ferretti, A. (1959) Thioethers from halogen compounds and cuprous mercaptides. II. *J. Am. Chem. Soc.*, **81**, 4927–4931.

255. Kreis, M. and Bräse, S. (2005) A general and efficient method for the synthesis of silyl-protected arenethiols from aryl halides or triflates. *Adv. Synth. Catal.*, **47**, 313–319.

256. Fernández Rodríguez, M.A. and Hartwig, J.F. (2010) One-pot synthesis of unsymmetrical diaryl thioethers by palladium-catalyzed coupling of two aryl bromides and a thiol surrogate. *Chem. Eur. J.*, **16**, 2355–2359.

257. Fernández Rodríguez, M.A. and Hartwig, J.F. (2009) A general, efficient, and functional-group-tolerant catalyst system for the palladium-catalyzed thioetherification of aryl bromides and iodides. *J. Org. Chem.*, **74**, 1663–1672.

258. Sayah, M. and Organ, M.G. (2011) Carbon–sulfur bond formation of challenging substrates at low temperature by using Pd–PEPPSI-IPent. *Chem. Eur. J.*, **17**, 11719–11722.

259. Rout, L., Sen, T.K., and Punniyamurthy, T. (2007) Efficient CuO-nanoparticle-catalyzed C–S cross-coupling of thiols with iodobenzene. *Angew. Chem. Int. Ed.*, **46**, 5583–5586.

260. Baldovino-Pantaleón, O., Hernández-Ortega, S., and Morales-Morales, D. (2006) Alkyl- and arylthiolation of aryl halides catalyzed by fluorinated bis-imino-nickel NNN pincer complexes [NiCl$_2${C$_5$H$_3$N-2,6-(CHNAr$_f$)$_2$}]. *Adv. Synth. Catal.*, **348**, 236–242.

261. Shaw, J.E. (1991) Preparation of thiophenols from unactivated aryl chlorides and sodium alkanethiolates in *N*-methyl-2-pyrrolidone. *J. Org. Chem.*, **56**, 3728–3729.

262. Maiolo, F., Testaferri, L., Tiecco, M., and Tingoli, M. (1981) Fragmentation of aryl alkyl sulfides. A simple, one-pot synthesis of polymercaptobenzenes from polychlorobenzenes. *J. Org. Chem.*, **46**, 3070–3073.

263. Gryko, D.T., Clausen, C., and Lindsey, J.S. (1999) Thiol-derivatized porphyrins

264. Edinberry, M.N., Gymer, G.E., and Jevons, S. (1981) Antifungal pyridine-2-thiones. GB Patent 2053189.
265. Wang, B., Graskemper, J.W., Qin, L., and DiMagno, S.G. (2010) Regiospecific reductive elimination from diaryliodonium salts. *Angew. Chem. Int. Ed.*, **49**, 4079–4083.
266. Zhong, B., Al-Awar, R.S., Shih, C., Grimes, J.H., Vieth, M., and Hamdouchi, C. (2006) Novel route to the synthesis of 4-quinolyl isothiocyanates. *Tetrahedron Lett.*, **47**, 2161–2164.
267. Ke, F., Qu, Y., Jiang, Z., Li, Z., Wu, D., and Zhou, X. (2011) An efficient copper-catalyzed carbon–sulfur bond formation protocol in water. *Org. Lett.*, **13**, 454–457.
268. Maloney, K.M., Nwakpuda, E., Kuethe, J.T., and Yin, J. (2009) One-pot iodination of hydroxypyridines. *J. Org. Chem.*, **74**, 5111–5114.
269. Lacour, M.-A., Zablocka, M., Duhayon, C., Majoral, J.-P., and Taillefer, M. (2008) Efficient phosphorus catalysts for the halogen-exchange (halex) reaction. *Adv. Synth. Catal.*, **350**, 2677–2682.
270. Sun, H. and DiMagno, S.G. (2006) Room-temperature nucleophilic aromatic fluorination: experimental and theoretical studies. *Angew. Chem. Int. Ed.*, **45**, 2720–2725.
271. Franzke, A. and Pfaltz, A. (2008) Synthesis of functionalized borate building blocks for the anionic derivatization of neutral compounds. *Synthesis*, 245–252.
272. Monopoli, A., Cotugno, P., Cortese, M., Calvano, C.D., Ciminale, F., and Nacci, A. (2012) Selective N-alkylation of arylamines with alkyl chloride in ionic liquids: scope and applications. *Eur. J. Org. Chem.*, **2012**, 3105–3111.
273. Pan, J., Wang, X., Zhang, Y., and Buchwald, S.L. (2011) An improved palladium-catalyzed conversion of aryl and vinyl triflates to bromides and chlorides. *Org. Lett.*, **13**, 4974–4976.
274. for attachment to electroactive surfaces. *J. Org. Chem.*, **64**, 8635–8647.
274. Jian, H. and Tour, J.M. (2005) Preparative fluorous mixture synthesis of diazonium-functionalized oligo(phenylene vinylene)s. *J. Org. Chem.*, **70**, 3396–3424.
275. Maggini, M., Passudetti, M., Gonzales-Trueba, G., Prato, M., Quintily, U., and Scorrano, G. (1991) A general procedure for the fluorodenitration of aromatic substrates. *J. Org. Chem.*, **56**, 6406–6411.
276. Maimone, T.J., Milner, P.J., Kinzel, T., Zhang, Y., Takase, M.K., and Buchwald, S.L. (2011) Evidence for in situ catalyst modification during the Pd-catalyzed conversion of aryl triflates to aryl fluorides. *J. Am. Chem. Soc.*, **133**, 18106–18109.
277. Kuduk, S.D., DiPardo, R.M., and Bock, M.G. (2005) Tetrabutylammonium salt induced denitration of nitropyridines: synthesis of fluoro-, hydroxy-, and methoxypyridines. *Org. Lett.*, **7**, 577–579.
278. Janmanchi, K.M. and Dolbier, W.R. (2008) Highly reactive and regenerable fluorinating agent for oxidative fluorination of aromatics. *Org. Process Res. Dev.*, **12**, 349–354.
279. Milne, J.E., Storz, T., Colyer, J.T., Thiel, O.R., Seran, M.D., Larsen, R.D., and Murry, J.A. (2011) Iodide-catalyzed reductions: development of a synthesis of phenylacetic acids. *J. Org. Chem.*, **76**, 9519–9524.
280. Nielsen, A.T., Chafin, A.P., and Christian, S.L. (1984) Nitrocarbons. 4. Reaction of polynitrobenzenes with hydrogen halides. Formation of polynitrohalobenzenes. *J. Org. Chem.*, **49**, 4575–4580.
281. Cottet, F. and Schlosser, M. (2004) Logistic flexibility in the preparation of isomeric halopyridinecarboxylic acids. *Tetrahedron*, **60**, 11869–11874.
282. Wannberg, J., Wallinder, C., Ünlüsoy, M., Sköld, C., and Larhed, M. (2013) One-pot, two-step, microwave-assisted palladium-catalyzed conversion of aryl alcohols to aryl fluorides via aryl nonaflates. *J. Org. Chem.*, **78**, 4184–4189.

283. Tang, P., Wang, W., and Ritter, T. (2011) Deoxyfluorination of phenols. *J. Am. Chem. Soc.*, **133**, 11482–11484.
284. Lui, N., Marhold, A., and Rock, M.H. (1998) Liquid-phase decarboxylation of aromatic haloformates: a new access to chloro- and fluoroaromatics. *J. Org. Chem.*, **63**, 2493–2496.

后记：化学研究的质量[①]

人们在浏览化学杂志时会有这样的印象，许多化学家更关心怎样提高其文章的数量而不是解决重要问题（在这方面，我也不是完全无可指责的）。主要原因是为其提供资金的机构设置了错误的激励模式，从而促使无关紧要的工作快速发表，通过灌输科学就是炫耀性地发表和拔高与自己相关项目的观念使得一代又一代的年轻科学家们误入歧途[1]。尽管优秀的化学研究在大多数大学进行着，一些学术研究团体并不能足够客观地评估他们研究项目的质量。那么，什么是"好"的研究项目？

大多数读者可能生活在繁荣的社会里，并且相信能够负担得起无关紧要的研究。一个人对工作的满意度与其对工作重要性的认知成正比，化学家也不例外。无论如何，当人类当今面临着严重威胁着人类的中期生存的更紧迫的问题（海洋的破坏、细菌耐药性的增加、大气层不可逆的毁坏、普遍缺乏抗病毒药物等）时，没有人承受得起不相关的研究。

我将"好的研究"定义为"重要的研究"，"重要"意指那些或多或少能够提高生产力从而提高每个人的生活和繁荣质量的事情。因此，"差的研究"就是"无关紧要的研究"，不能以任何方式提升我们的生活质量和繁荣程度，纯属浪费金钱。

虽然许多重要的研究项目在商业上也是有价值的，但并非必须如此。事实上，有一些高度重要的研究，如对于臭氧空洞或其他人为环境变化的研究，并无实际商业意义，但对我们未来繁荣确是至关重要的，因此必须大力推动并且由政府提供经费资助。

化学家处在一个能通过提高生产率和降低价格提升社会繁荣的上好位置。工业化学研究基本上是处理诸如"怎样能让一个产品生产得更便宜？"和"怎么能在不增加成本的情况下提高产品的质量？"这样的问题。对于依赖于专门材料（医药、农药、化妆品等）的公司，一个更重要的问题是"哪个化合物或混合物有我们寻找的性质"。因此，最重要的化学研究是发展更高效的（即更

[①] 原著后记为"Economics，Politics，and the Quality of Chemical Research"，感兴趣的读者可在 http://onlinelibrary.wiley.com/doi/10.1002/9783527687800.epil/pdf 网址中查阅。

便宜、更清洁、更安全、更简短、更简单、更高收率、更有选择性的)合成方法(而不是代价昂贵的另外一种替代方法),以及发展能够制备具有潜在价值属性的新型化合物合成方法的研究。机理研究如果产生的知识有助于改进合成方法或新型化合物合成的发展,它也是必要的。

一些无关要紧的化学研究项目为使其免受外界批评常常被伪装成"基础研究",它们包括:

- 通过新的、昂贵的方法和试剂制备廉价化合物(而这些化合物本身很容易通过廉价的途径获得);
- 一些无关痛痒的天然产物或非天然化合物的合成(真正的目的是发展新的有价值的合成方法除外);
- 无关紧要的反应机理、动力学、立体选择性或区域选择性的研究。

一些被学术型化学家常常忽略的有机合成的关键问题包括:
- 所需催化剂的价格(对一种合成方法而言,它是最重要的评价标准);
- 贵金属催化剂的回收利用;
- 生成废弃物数量和类型(理想的:无废弃物;可接受的:水、二氧化碳、氮气、可燃有机物;勉强接受的:硫酸、磷酸、钠、钾、镁、钙、铝、硅盐;不可接受的:化学计量过渡金属盐);
- 反应的热安全性和可放大性。

关于最后一点,近年来普遍使用微波加热和密闭小瓶反应(来实现温度高于溶剂沸点的反应)。开发出的封闭体系的新化学反应很难放大,因此,不如在更现实反应条件下进行的新化学反应有价值。

所以,化学研究的一个极重要的领域是对已有的重要有机反应进行改进,使其仅需较少的昂贵催化剂、试剂和溶剂,生成不太危险的废弃物,以及降低热失控事件的风险。目前一些十分重要的领域包括:

- 能在温和条件下实现用氢气还原酯、酰胺、脲、羧酸、醛和酮的催化剂(氢气是最便宜、最环境友好的还原剂);
- 能在温和条件下实现脂肪族碳—杂原子键氢解的催化剂;
- 用氢作为还原剂还原形成 C—C 键的催化剂;
- 用空气选择性氧化有机化合物的催化剂(例如氧化烯烃、醇、烷基卤化物或者醚到醛、酮或羧酸;羟基化;醚化;卤化;胺化;酰胺化;脱氢;醚到缩醛或酯,甲基到醇、醛或羧酸);
- 仅仅有氢气、水、氮气、一氧化碳或二氧化碳等副产物的 C—C 键形成反应;
- 瞬态转化醇/烯醇成为强亲电体的催化剂(例如使醇成为烷基化试剂,酚成为芳基化试剂,用于将酮类转化成乙烯卤或联烯等);
- S_N2 反应的催化剂(例如,在温和条件下通过亲核取代反应将不活泼的氯化物、醇、醚、酯、硫醚、胺等转化为其他化合物);

· 能更好地预测反应结果(区域选择性、立体选择性、可能的副产物、溶剂效应)的软件。

研究可以规划,其结果和重要性却不能。许多能有力推动社会繁荣的科学发现的影响是无法预见的。突破性的、高影响力的发现通常源自无人问津或者与既有观点冲突的项目。但这些项目常常不受同行青睐,也很难获得资金支持。因此,不用急切地逼迫学术界和工业界的化学家们拿出可以快速发表或有利可图的成果,而应该使他们有机会从事富有远见的、长期的、高风险和高回报的项目,这一点至关重要。而一旦能得到公共资金资助的项目也将应该雄心勃勃地以有价值的相关重要目标为基础。

参考文献

1. Hampe, M. (2013) Science on the market: what does competition do to research? *Angew. Chem. Int. Ed.*, **52**, 6550–6551.

索 引

a

acetals, 缩醛
    as electrophiles, 作为亲电试剂　13, 111
    oxidation, 氧化　151
acetanilides, 乙酰苯胺
    acylation, 酰化　91, 94, 99
    alkylation, 烷基化　184
    arylation, 芳基化　182, 184
    conversion to indolines, 转化成二氢吲哚　184
    cyclization to benzoxazoles, 环合成苯并噁唑　142, 194
    halogenation, 卤代　129, 131, 142
    hydroxylation, 羟基化　182
    nitration, 硝化　172, 175
    olefination, 烯基化　49
acetic formic anhydride, 乙酸甲酸酐　106, 111
acetone, 丙酮
    arylation, 芳基化　27
    preparation, 制备　30
acetophenones, 乙酰苯
    arylation, 芳基化　32
    ortho alkylation, 邻位烷基化　8
    ortho olefination, 邻位烯基化　48
    reaction with XeF_2, 与氟化氙的反应　132
acetoxylation, 乙酰氧化　172, 181
acetylenes, see alkynes, 乙炔, 见炔烃

acetyl nitrate, 乙酸硝酸酐　164, 170
acid catalysis, 酸催化
    ether cleavage, 醚键断裂　89
    isomerizations, 异构化　89, 99, 210
    $S_N Ar$, 芳香亲核取代反应　211
acrylates, 丙烯酸酯
    alkylation with, 烷基化试剂　10, 23
    oxidative arylation, 氧化芳基化　48-53
acryloyl chloride, 丙烯酰氯　87, 100
active hydrogen compounds, arylation, 含活泼氢化合物, 芳基化　209, 214, 221, 226, 234, 235
acylation of, 酰化反应
    acylarenes, 酰基芳烃　89, 93
    alkenes, 烯烃　88-90
    alkynes, 炔烃　88-90
    anilines, 苯胺　90-92
    arenes, 芳烃　85-111
    arylketones, 芳香酮　89, 93
    enamines, 烯胺　100
    indoles, 吲哚　97, 108
    nitroarenes, 芳香硝基化合物　92
    phenols, 苯酚　90-92
    phthalic anhydride, 邻苯二甲酸酐　100
    pyridines, 吡啶　94
    thiophenols, 硫酚　90-92
acyl fluorides, 酰氟　86
acyl halides, 酰卤
    decarbonylation, 脱羰基化　95
    isomerization, 异构化　99
acyl peroxides, as precursors to radicals,

247

酰基过氧化物,作为自由基前体　67
alcohols,醇
　　arylation,芳基化　217,218,250
　　as electrophiles,作为亲电试剂　94
　　oxidation,氧化　94,227
　　reaction with diazonium salts,与重氮盐反应　228
　　reduction,还原　259
aldehydes,醛
　　as electrophiles,作为亲电试剂　13,94
　　conversion to acyl radicals,转化成酰基自由基　110
　　halogenation,卤代　151
　　oxidation,氧化　94,151
alkali fusion of aryl sulfonates,芳香磺酸盐的碱熔　218
alkanesulfonyl halides,arylation,烷基磺酰卤,芳基化　192
alkenes,烯烃
　　acylation,酰化　88-90
　　alkylation with,烷基化试剂　4,8-10,14,16-23
　　arylation,芳基化　45-54
　　halogenation,卤代　148
　　olefination with,烯基化试剂　22,45-52
alkylation of,烷基化
　　arenes,芳烃　1-34
　　aryl halides,芳基卤　214,221,226,233,234,236,237,240,248
alkyl chloroformates,烷基氯甲酸酯　105
alkyl esters, as electrophiles,烷基酯,作为亲电试剂　98,101-103
alkyl fluorides, as electrophiles,烷基氟,作为亲电试剂　6
alkyl groups,烷基
　　amidation,酰胺化　183
　　arylation,芳基化　11,213
　　dehydrogenation,脱氢　32,170
　　halogenation,卤代　126,134,138,146,148,151

hydroxylation,羟基化　164
isomerization,异构化　89,99
nitration,硝化　164
oxidation,氧化　164,166,169,170
substitution,取代　125,139,163,164,170
N-alkylpyridinium salts, as electrophiles, N-烷基吡啶盐,作为亲电试剂　1
alkynes,炔
　　acylation,酰化　90
　　arylation,芳基化　54-56
　　oxidative dimerization (Eglinton reaction),氧化二聚(艾格林顿反应)　57
　　reaction with BrCN,与溴化氰的反应　107
allenes,联烯
　　arylation,芳基化　52
　　formation,形成　55
allylic electrophiles,烯丙基亲电试剂　17,22,52,53
aluminum anilides,酰苯胺铝　18
amidation,酰胺化
　　of alkanes,烷烃的　183
　　of arenes,芳烃的　177-184
amides,酰胺
　　N-arylation, N-芳基化　173,180,245
　　N-benzylation, N-苄基化　183
　　halogenation,卤代　151,152
　　hydrolysis,水解　127
　　nitration,硝化　169
　　oxidation,氧化　152
amination of arenes,芳烃氨基化　175-177
amines,胺
　　alkylation,烷基化　4,14,17,18,258
　　arylation,芳基化　176,242-248,254
　　dealkylation,去烷基化　92,100,141,142,148,244,248
　　dehydrogenation,脱氢　100,141
　　displacement,取代　9,238

索 引

as electrophiles,作为亲电试剂 1
nitration,硝化 165,171
nitrosation,亚硝化 142,166
oxidation,氧化 100,148
reaction with acyl halides,与酰卤反应 100
reaction with diazonium salts,与重氮盐反应 227
reaction with halogens,与卤素反应 148

aminoalcohols,氨基醇
arylation,芳基化 217
as electrophiles,作为亲电试剂 33

ammonia, arylation of,氨,芳基化 241-243

anhydrides, mixed carboxylic,酸酐,混合酸 110

anilines,苯胺
acylation,酰化 90
alkylation,烷基化 13-18,258
arylation,芳基化 176,218,245,246
conversion to benzimidazoles,转化成苯并咪唑 181
conversion to phenols,转化成苯酚 207
dealkylation,脱烷基化 92,142,148
formation,形成 214,241-243
halogenation,卤代 138-143
nitration,硝化 164,166,171
nitrosation,亚硝化 142,166,174
oxidation,氧化 138,140-142,148
sulfenylation,亚磺化 199
sulfonation,磺化 191
trifluoromethylation,三氟甲基化 16
tritylation,三苯甲基化 16

anthracenes, halogenation,蒽,卤代 126

anti Markovnikov addition,反马氏加成 20

arenesulfonic acids,芳香磺酸
formation,形成 191,217,225,229
S_NAr,芳香亲核取代反应 206,218

aromatic nucleophilic substitutions,芳香亲核取代 205-259
acid-/base-catalysis,酸/碱-催化 211
mechanisms,机理 205
regioselectivity,区域选择性 205
transition-metal catalysis,过渡金属催化 211

arylations with aryl halides,与芳香卤代物的芳基化 61-69
via cationic intermediates,通过阳离子中间体 61-63
via radicals,通过自由基 63-65
by transition-metal catalysis,通过过渡金属催化 67-69
via transition-metal chelates,通过过渡金属螯合物 65-67

arylations with diazonium salts,与重氮盐的芳基化 69-73

arylations with unsubstituted arenes,与取代芳烃的芳基化 78

aryl cations,芳基阳离子 62,63

aryl esters,芳基酯
Fries rearrangement,弗里斯重排 91,106
S_NAr,亲核芳香取代反应 232

arylethers,芳基醚
dealkylation,去烷基化 88,89,103,206
S_NAr,亲核芳香取代反应 207,229,230,236,237

aryl halides,芳香卤代物
arylation,芳基化 61-69
dehalogenation,脱卤 71,145,229,236,241,243,249,254,255,259
formation,形成 121-152,253,259
homocoupling,偶联 71
hydrolysis,水解 248,250
isomerization,异构化 210

aryl thioethers,芳香硫醚
cleavage,断裂 150,231,257
formation,形成 196-201,216,225,227,229,230,252,253

249

halogenation,卤代 149-151
    isomerization,异构化 99,210
arynes,芳炔 195,218,232,233,241
azide ion, arylation of,重氮根,芳基化 247-250
azides,叠氮化合物
    amidations with,酰胺化试剂 181
    decomposition of,分解 145
    formation,形成 182,247-250
    reduction,还原 145,249,250
aziridines,氮杂环丙烷 24
azoles,唑类
    acylation,酰化 93-98
    alkylation,烷基化 19
    halogenation,卤代 144-146
    nitration,硝化 167,173

b

Bamberger cleavage,班伯格断裂 94,96
Bartoli reaction,巴尔托里反应 238
benzaldehydes,苯甲醛
    as acylating reagents,作为酰化试剂 94
    acylation,酰化 86
    as alkylating reagents,作为烷基化试剂 10,13
    arylation,芳基化 70
    formation,形成 106-110,149
    halogenation,卤代 122,129,139,151
    nitration,硝化 162,163
    olefination,烯基化 49
S_NAr,亲核芳香取代反应 216
    sulfonylation,磺酰化 191
benzamides, hydrolysis,苯甲酰胺,水解 127
benzimidazoles,苯并咪唑
    C-alkylation,C-烷基化 10,19
    N-arylation,N-芳基化 217
    formation,形成 179,181
benzofurans,苯并呋喃
    from 2-arylphenols,从 2-芳基苯酚

139
    from 1,2-dihalobenzenes,从 1,2-二卤苯 235
    dithiocarboxylation,二硫代羧基化 104
benzoic acid derivatives,苯甲酸衍生物
    acylation with,酰化试剂 86,91-96,101
    alkylation,烷基化 7,8,11
    amination,胺化 175
    arylation,芳基化 69
    decarboxylation,脱羧反应 74,136,164
    formation,形成 101-108
    halogenation,卤代 127,134-137
    nitration,硝化 163
    olefination,烯基化 49,52,55
    preparation,制备 101-108
benzonitriles,苯腈
    amination,胺化 175
    arylation,芳基化 64
    formation,形成 107,108,178,219,224,226,230
    halogenation,卤代 137
    hydrolysis,水解 137,251
    nitration,硝化 169
S_NAr of cyano group,氰基的亲核芳香取代 209
benzophenone,二苯甲酮
    as alkylating reagent,作为烷化试剂 10
    alkylation,烷基化 7
    nitration,硝化 165
benzoquinone,苯醌
    as oxidant,作为氧化试剂 46,52,53,75,76
    preparation,制备 141
benzoquinones,苯醌类
    alkylation,烷基化 11
    formation,形成 138,141,152,173
    nitration,硝化 174
benzothiazoles,苯并噻唑
    arylation,芳基化 70

formation, 形成 230
benzotriazoles, formation, 苯并三唑, 形成 179
benzoxazoles, 苯并噁唑
  acylation, 酰化 98
  aminomethylation, 胺甲基化 4
  from anilides, 从酰苯胺 142
  arylation, 芳基化 98
benzylic, 苄基
  acylation, 酰化 97
  alkylation, 烷基化 211
  arylation, 芳基化 16
  cyanation, 氰基化 211
  halogenation, 卤代 126,134,138,141,146,151
  nitration, 硝化 164,169
  substitution vs S_NAr, 取代及芳香亲核取代 212,213
biaryls, formation, 联芳基, 形成 61-79
binaphthol, 联萘酚 78
bis(chloromethyl)ether, as electrophile, 二氯甲基醚, 作为亲电试剂 7
boronic acid esters, 硼酸酯
  arylation with, 芳基化试剂 9
  olefination with, 烯基化试剂 48
boronic acids, 硼酸
  arylation with, 芳基化试剂 75
  halogenation, 卤代 137,142
  nitration, 硝化 163
boron trihalides, 三卤化硼 26,90,91
bromination of, 溴代
  arenes with BrCN, 芳烃与溴化氰 107
  benzaldehyde, 苯甲醛 150
  benzyl bromides, 苄溴 148
  hydroquinones, 氢醌 139
  indanes, 吲满 126
  phenols, 酚 6,139
  2-bromomalonates, as brominating reagent, 2-溴丙二酸酯, 作为溴代试剂 6

N-bromosuccinimide, N-溴代琥珀酰亚胺 123
tert-butyl chloride, as electrophile, 叔丁基氯, 作为亲电试剂 3,6
tert-butyl groups, 叔丁基
  introduction with pivaloyl chloride, 通过特戊酰氯引入 99
  ipso substitution, 原位取代 89
  tert-butyl iodide, 叔丁基碘 11
butyrolactone, as electrophile, 丁内酯, 作为亲电试剂 101,102

c

caffeine, arylation, 咖啡因, 芳基化 75
Cannizzaro reaction, 康尼扎罗反应 9,216,217
carbamates, 氨基甲酸酯
  as electrophiles, 作为亲电试剂 105,108,111,232,237
  as nucleophiles, 作为亲核试剂 214
carbamoyl chlorides, 氨基甲酰氯 105
carbazoles, 咔唑
  bromination, 溴代 130
  carboxylation, 羧基化 103
  dimerization, 二聚 130,179
  formation, 形成 179,184,239,246
  iodination 碘代 140
carbenes, 卡宾
  aromatic C—H insertion, 芳香 C—H 插入反应 28
  dimerization, 二聚 93,96
  reaction with phenolates, 与苯酚酯的反应 107-109
carbocations, 碳正离子
  hydride abstraction by, 氢摄取 6,89
  rearrangement, 重排 6,7,22,102
carbodiimides, cyclization to benzimidazoles, 碳二亚胺, 环化成苯并咪唑 179
carbonate ion, arylation, 碳酸根离子, 芳基化 252

carbonic acid derivatives, as electrophiles, 碳酸衍生物,作为亲电试剂　101－107,232,259
carbon dioxide, as electrophile,二氧化碳,作为亲电试剂　104
carbon disulfide, as electrophile,二硫化碳,作为亲电试剂　104
carbon monoxide,一氧化碳
　acylations with,酰化试剂　95,98
　carboxylation with,羧基化试剂　105
　extrusion from acyl cations,从酰基阳离子中释放　95,99
　formylations with,甲酰化试剂　106
carboxylation,羧基化　101－105
carboxylic acids,羧酸
　conversion to radicals,转化成自由基　73,74
　decarboxylation,脱羧反应　73,74,136,163,256
　as electrophiles,作为亲电试剂　86,99
　formation,形成　101－105,150,165,167,169
　as leaving group,作为离去基团　232
　as nucleophiles,作为亲核试剂　252
　O-arylation,O-芳基化　252
carboxylic anhydrides,羧酸酐　110,111
carboxylic esters,羧酸酯
　as electrophiles,作为亲电试剂　89,92,98－103,232,236,250
　formation,形成　104,106,108,149,252
catalysis,催化剂
　Friedel-Crafts acylation,付里德尔-克拉夫酰化反应　85
　Friedel-Crafts alkylation,付里德尔-克拉夫烷基化反应　2－5
　halogenation,卤代　128,129
　nitration,硝化　167
　poisons,毒物　50
　S_NAr,芳香亲核取代反应　211
catechols, fluorination,儿茶酚类,氟代　132
chelate formation,形成螯合物　46,48,65,68,69
chloramines,氯胺类　175,178
chlorination,氯代　121－152
chloroacetic acid,氯乙酸　25
chloroacetic esters,氯乙酸酯　25－27
chloroacetone, as electrophile,氯丙酮,作为亲电试剂　25－27
chloroacetonitrile,氯乙腈　26
chlorobenzene, conversion to phenol,氯苯,转化成酚　217
chloroformates,氯甲酸酯　105,108,259
chloromethylation,氯甲基化　7
N-chlorosuccinimide, N-氯琥珀酰亚胺　123,134,137,146,150
chromanes, formation,色满类,形成　14
chromium carbonyl complexes,六羰基铬络合物　218
cine substitution,移位取代反应　207－209
cinnamic acid anhydride, olefination with,肉桂酸酐,烯基化试剂　48
cinnamic acid esters, from acrylates,肉桂酸酯,来自丙烯酸酯　48,49
cumene (isopropylbenzene),枯烯(异丙基苯)
　conversion to phenol,转化成酚　30
　as hydride donor,作为氢供体　6
cyanation,氰基化
　electrophilic,亲电的　104,107,108
　nucleophilic,亲核的　214,219,224,226,230
cyanide ion,氰离子
　as leaving group,作为离去基团　209,234
　from DMF,来自二甲基甲酰胺　79
　in S_NAr,在芳香亲核取代反应中　214,219,224,226,230
cyanoacetic esters,腈乙酸酯
　arylation,芳基化　214

reaction with nitroarenes,与硝基芳烃反应 226
cyanogen halides,卤化腈 104,107,108
cyanohydrins, as electrophiles,腈醇,作为亲电试剂 32
cyanuric chloride, arylation,氰脲酰氯(2,4,6-三氯-1,3,5-三嗪),芳基化 61,62
cyclohexanes, arylation,环己烷,芳基化 11,213
cyclohexenones,环己酮 12,15
cyclopropanoyl chlorides, isomerization,环丙酰氯 99

d

DDT,双对氯苯基三氯乙烷 1
dealkylation of,去烷基化
    amines,胺类 92,100,142
    arenes,芳烃 89,125,139,163,170
    arylethers,芳香醚类 88,89,206,207,230,236,250
    benzylic alcohols,苄醇 125,139
    pyridines,吡啶类 169
dearomatization,去芳构化 15,126,130-132,138,139,169,205
decarbonylation, of acyl halides,脱羰基,酰卤 95,99
decarboxylation,脱羧反应 73,74,136,163,256
decyanation,脱氰基 209
dediazoniation,脱重氮化反应 226-228
deformylation,脱甲酰化反应 53,163
dehalogenation,脱卤反应
    aryl halides,芳香卤代物 71,229,236,241,243,249,254,255,259
    imidazoles,咪唑类 144,145
    thiazoles,噻唑类 145,241
dehydration, of propanol,脱水,丙醇的 30
dehydrogenation,脱氢

of alkyl groups,烷基的 32,170
of amines,胺类的 100,141
of β-ketoesters,β-酮酸酯 32
during nitrations,在硝化过程 170
with thionyl chloride,与亚硫酰氯 198
dehydroxylation, phenols,脱羟基,酚类 20,229,259
diacylperoxides, as precursors to radicals,双酰基过氧化物,作为自由基前体 11,65,67
1,1-diarylalkanes,1,1-二芳基烷烃 17,28,30,31
diarylamines,二芳基胺类
    arylation,芳基化 179,218,246
    conversion to carbazoles,转化成咔唑 179
    formation,形成 239
    oxidation,氧化 179
diazoalkanes,重氮烷类 1
diazocarbonyl compounds,重氮羰基化合物 26,28,171
diazomethane,重氮甲烷 11
diazonium salts,重氮盐
    arylation of (Gomberg-Bachmann),芳基化(刚伯格-巴赫曼反应) 226
    conversion to indazoles,转化成吲唑 69
    dediazoniation of,去重氮化 69,72-74
    hydrolysis of,水解 74
    as precursors to arynes,作为芳香炔的前体 233
    reaction with alcohols,与醇的反应 227,228
    S_NAr reactions of,芳香亲核取代反应 228
1,3-dicarbonyl compounds, arylation,1,3-二羰基化合物,芳基化 30,209,234,235
dichlorocarbene,二氯卡宾 107,108
dichloromethane, as electrophile,二氯甲烷,作为亲电试剂 4

Diels - Alder reaction, of furans,狄尔斯-阿尔德反应,呋喃的 54
1,1 - dihaloalkenes, as electrophiles,1,1-二卤烯烃,作为亲电试剂 88
diketene, arylation,双烯酮,芳基化 103
diketones, arylation,双烯酮类,芳基化 30,235
dimerization of,二聚
    arenes,芳烃 32,62,66,122,128,130,178,179,240
    aryl halides,芳香卤代物 71,240,255
    boronic acids,硼酸类 75
    diazonium salts,重氮盐 228
    fluorenes 芴类 225
    furans,呋喃类 53
    imidazoles,咪唑类 76,96
    indoles,吲哚类 97,173,194
    oxazoles,噁唑类 74
    oxiranes,环氧乙烷类 24
    phenols,酚类 78
    pyridines,吡啶类 76,96
dimethylformamide (DMF),二甲基甲酰胺
    arylation,芳基化 216
    conversion to cyanide ion,转化成氰离子 79
    formylations with,甲酰化试剂 108,130
    as solvent for Friedel - Crafts acylations,作为付里德尔-克拉夫酰化反应溶剂 92
    as source of dimethylamine,作为二甲胺对等体 243
dimethylsulfoxide (DMSO),二甲基亚砜
    as solvent,作为溶剂 197
    thiomethylation with,硫甲基化试剂 200
diols, arylation,二醇,芳基化 253
diquat,1,1′-亚乙基-2,2′-联吡啶二溴盐,敌草快(杀草剂) 76,78
disulfides,二硫化合物
    arylation,芳基化 199 - 201
    formation,形成 149
dithiocarboxylic acids, preparation,二硫代羧酸,制备 103
Dow phenol synthesis,陶氏苯酚合成 217,218
Duff reaction,达夫反应 107,108
durene (1,2,4,5 - tetramethylbenzene),杜烯(1,2,4,5-四甲基苯)
    acylation,酰化 93
    alkylation,烷基化 22
    nitration,硝化 164

e

Eglinton reaction,艾格林顿反应 57
enolates, arylation,烯醇酯,芳基化 209,214,226
epichlorohydrin,环氧氯丙烷 24
epoxides, arylation,环氧化合物,芳基化 23,24
esters,酯
    as electrophiles,作为亲电试剂 98,100 -103,232,236
    formation,形成 104,106,108,149,252
    reaction with azides,与叠氮化合物反应 250
ethanolamines, as electrophiles,乙醇胺类,作为亲电试剂 33
ethers,醚
    alkylation with,烷基化试剂 17
    cleavage,断裂 89,149,233
    halogenation,卤代 149
    nucleophilic displacement,亲核取代 207,230,236,237
ethylation, of benzene,乙基化,苯的 3
ethylene glycol, as electrophile,乙二醇,作为亲电试剂 33

f

ferrocene, acylation,二茂铁,酰化 87

fluorene, alkylation,芴,烷基化 32
fluorenones,芴酮
    as electrophiles,作为亲电试剂 16,28
    formation,形成 74,77,105
    nitration,硝化 162
fluoride ion, arylation,氟离子、芳基化 129-132,258
fluorination,氟代 129-132
    of acetanilides,乙酰苯胺 131
    of arylketones,芳香酮 129,132
    of benzamides,苯甲酰胺 127
    of benzylic alcohols,苄醇类 125
    of catechols,儿茶酚类 132
    of indoles,吲哚类 131
    of nitroarenes,硝基芳烃 258
    of phenols,酚类 259
    of pyridines,吡啶类 124
formaldehyde, reaction with,甲醛,反应 aniline,苯胺 17,18
terephthalic acid,对苯二甲酸 7
formamides, arylation,甲酰胺,芳基化 216
formic acetic anhydride,甲酸乙酸酐 106,111
formic acid, carboxylation with,甲酸,羧基化试剂 105
formylation,甲酰化 106,130
formyl fluoride,甲酰氟 106
Friedel-Crafts,付里德尔-克拉夫
    acylation,酰化 85-111
    alkylation,烷基化 1-34
    catalysts,催化剂 85
    solvents,溶剂 87,92
Fries rearrangement,弗里斯重排 91,106
furans,呋喃
    acylation,酰化 104
    alkylation,烷基化 3
    Diels-Alder reaction,狄尔斯-阿尔德反应 54
    dimerization,二聚 53
    halogenation,卤代 123
    nitration,硝化 162
    olefination,烯基化 53
    sulfonylation,磺酰化 123

g

Gattermann-Koch reaction,盖特曼-科赫反应 106
glucose, as reducing reagent,葡萄糖,作为还原剂 71,216
glycidyl ethers, as electrophiles,缩水甘油醚 23,24
glycolic acid, alkylation of indole,乙醇酸,吲哚的烷基化 33
Gomberg-Bachmann reaction,冈伯格-巴赫曼反应 69
Grignard reagents, S_NAr with,格林纳德试剂,芳香亲核取代试剂 235-237

h

Hale-Britton process,黑尔-布里顿过程 217,218
halide ions, arylation of,卤离子,芳基化 253,258
Haller-Bauer reaction,哈勒-巴沃尔反应 9,217
N-haloamides,N-卤代酰胺 142,181
N-haloamines,N-卤代胺 175,178
α-haloesters, as electrophiles,α-卤代酯,作为亲电试剂 6,24,25,27,101
haloform reaction,卤仿反应 104
halogenation,卤代 121-152
    regioselectivity,区域选择性 125
halogen dance,卤素跳舞(指在芳环上的卤素异构化) 210
N-haloimides,N-卤代酰亚胺
    as halogenating reagents,作为卤代试剂 121-123
    isomerization,异构化 123
α-haloketones, as electrophiles,α-卤代

酮,作为亲电试剂 25-27
α-halosulfones, vicarious S_NAr, α-卤代砜,芳香亲核取代反应替代试剂 219
hard and soft organometallics,硬和软的有机金属试剂
 reaction with lactones,与内酯的反应 102,103
 S_NAr,芳香亲核取代反应 235-241
Heck reaction,赫克反应 21,22,46,50,53
hexanitrobenzene,六硝基苯 161
Hofmann rearrangement,霍夫曼重排 152
homodimerization, see dimerization,同二聚,见二聚
Hunsdieker reaction,汉斯狄克反应 136
hydrazines,肼
 arylation,芳基化 230,245
 conversion to radicals,转化成自由基 73
 vicarious nucleophilic substitution,替代性亲核取代 222
hydrazones, reaction with nitroarenes,腙,与硝基芳烃的反应 222
hydride ions, as leaving group in S_NAr,氢离子,作为在亲核取代反应中的离去基团 205,219
hydrogen fluoride, as solvent,氟化氢,作为溶剂 21,85
hydrolysis, of aryl halides,水解,芳香卤代物的 248
hydroperoxides,氢过氧化物 30,222
hydroquinones, halogenation,氢醌,卤代 139
hydroxamic acids, olefination,羟肟酸,烯基化 50,52
hydroxide ion, arylation of,氢氧离子,芳基化 248
hydroxyalkylation,羟烷基化 7,10,24,30
hydroxylamine, as reducing reagent,羟胺,作为还原剂 216
hydroxylamines, isomerization,羟胺,异构化 224
hydroxylation, aromatic,羟基化,芳香族的 182
hydroxylation, as side reaction of,羟基化,作为副反应
 carboxylation,羧基化 105
 halogenation,卤代 124,131,144,146
 nitration,硝化 171
hydroxymethylation,羟甲基化 7

i

imidazoles,咪唑类
 acylation,酰化 95
 alkylation,烷基化 10,19
 arylation,芳基化 75,76
 cleavage,断裂 96
 dehalogenation,脱卤反应 145
 dimerization,二聚 76,96
 halogenation,卤代 143-145,209
 nitration,硝化 209
imides,酰亚胺
 N-arylation,N-芳基化 183
 N-halo,N-卤代 121-123
indanes,茚满
 bromination,溴代 126
 formation,形成 22,90,100,102
indazoles, from arenediazonium salts,吲唑,从芳香重氮盐 69
indenes, formation,茚,形成 55,138
indoles,吲哚
 acylation,酰化 97,98,148
 alkylation,烷基化 19,24,31,33,206
 alkynylation,炔基化 47
 amidation,酰胺化 181
 amination,胺化 124,178
 arylation,芳基化 70,208,217
 azidation,叠氮化反应 182
 conversion to quinolines,转化成喹啉 109

deformylation,去甲酰化 53
dimerization,二聚 97,173,194
fluorination,氟代 131
formation,形成 50,55,240
formylation,甲酰化 108
halogenation,卤代 124,131,143-147,150,193
nitration,硝化 173
olefination,烯基化 53,54
S$_N$Ar,芳香亲核取代反应 208,230
sulfenylation,亚磺酰化 200
sulfonylation,磺酰化 193,194
trimerization,三聚 146
interhalogens, halogenation with,卤间化合物(卤素互化物),卤代试剂 122
iodide ion, as reducing agent,碘离子,作为还原剂 259
iodonium salts,碘盐
    arylation with,芳基化 75,180,182,235,253,257
    formation, during halogenations,形成,在卤代过程 125
    iodination with,碘代试剂 181
ionic liquids, as solvents for acylations,离子液体,作为酰化溶剂 87
ipso substitution of,原位取代
    acyl groups,酰基 132,163,165
    alkoxy groups,烷氧基 9,183,229,230
    amino groups,氨基 9
    benzyl groups,苄基 170
    boron,硼 75,137,163,174
    carboxyl groups,羧基 74,136,163
    diazonium groups,重氮基 69-73,226
    ethyl groups,乙基 89
    halides,卤代物 53,61-69,164,205-259
    hydroxyalkyl groups,羟烷基 125,139,170
    hydroxyl groups,羟基 20,132,229
    methyl groups,甲基 164

    nitro groups,硝基 219
    silanes,硅基 163
    sulfonyl groups,磺酰基 195
    tert-butyl groups,叔丁基 89,163
isocyanates, as electrophiles,异氰酸酯 108
isomerization of,异构化
    alkylarenes,烷基芳烃 89
    cyclopropanoyl halides,环丙酰卤 99
    polyhaloarenes,多卤芳烃 210
    thioethers,硫醚 99,210
isopropylbenzene(cumene),异丙基苯(枯烯)
    conversion to phenol,转化成酚 30
    as hydride donor,作为氢供体 6
isothiocyanates, formation by S$_N$Ar,异硫氰酸酯,通过芳香亲核取代反应形成 257

k

ketones,酮
    cleavage,断裂 132,165
    as electrophiles,作为亲电试剂 10,25-32
    fluorination,氟代 132
    as nucleophiles,作为亲核试剂 211,233
Kolbe-Schmitt reaction,科尔贝-施密特反应 103

l

lactams,内酰胺
    N-arylation,N-芳基化 180
    halogenation,卤代 146,151
    nitration,硝化 169
    from pyrroles,从吡咯 144,178
lactones,内酯
    arylation,芳基化 234
    as electrophiles,作为亲电试剂 89,101-103
    formation,形成 7,170

lithiation vs alkylation of haloarenes, 芳香卤的锂化和烷基化 235-237

m

malonic esters, 丙二酸酯
 arylation, 芳基化 209
 2-bromo, as brominating reagent, 2-溴代, 作为溴代试剂 6
malononitrile, as source of cyanide, 丙二腈, 作为氰化物对等体 226
Markovnikov addition, 马尔科夫尼科夫加成 20
Meerwein arylation, 梅尔外因芳基化 227
Meisenheimer salts, 迈森海默盐 206
mercaptans, 硫醇
 alkylation, 烷基化 231, 257
 arylation, 芳基化 252
 halogenation, 卤代 149-151
mesitylene, 1,3,5-三甲苯
 alkylation, 烷基化 6, 25
 chlorination, 氯代 178
 olefination, 烯基化 47
mesylates, 甲磺酸酯
 cleavage, 断裂 231
 as electrophiles, 作为亲电试剂 25
meta-selective reactions, 间位选择性反应 4, 51, 140, 174
methoxy groups, 甲氧基
 dealkylation, 去烷基化 89, 230, 236
 displacement, 取代 9, 48, 230, 236, 247
 as nucleophile in S_NAr, 作为芳香亲核反应中的亲核反应试剂 220, 254
methylation, 甲基化 19, 20
Michael addition, 迈克尔加成 21, 23, 29
Mitsunobu reaction, 光延反应 11
mixed carboxylic anhydrides, 混合酸酐 110
mixed carboxylic carbonic anhydrides, 混合羧酸碳酸酐 111

n

naphthalenes, 萘
 nitration, 硝化 167
 oxidative degradation, 氧化降解 167
 S_NAr at, 芳香亲核取代反应 206, 214, 216, 222, 223, 225, 229, 230, 237, 248, 256, 258
nitramines, 四硝基甲苯胺 172
nitration of arenes, 芳烃的硝化 161-175
 catalysis, 催化 167
 electron-deficient arenes, 缺电子芳烃 167
 mechanisms, 机理 161
 regioselectivity, 区域选择性 164
 nitrile oxides, 氧化腈 27, 74
nitriles, 腈
 displacement of cyano group, 氰基取代 209
 formation by S_NAr, 通过芳香亲核反应形成 212, 214, 219, 220, 223, 224, 226
nitrite ion, arylation, 氰离子, 芳基化 244
nitroalkanes, 硝基烷烃
 as electrophiles, 作为亲电试剂 26-29
 as nucleophiles, 作为亲核试剂 221
nitroalkenes, arylation, 硝基烯烃, 芳基化 21, 29
nitroarenes, 硝基芳烃
 amination, 胺化 176, 222
 cyanation, 氰基化 220, 223, 224, 226, 230
 as dipolarophiles, 作为亲偶极试剂 219, 222
 formation, 形成 161-175, 244
 halogenation, 卤代 133, 137
 hydroxylation, 羟基化 222
 reaction with alcohols, 与醇反应 227, 253, 254
 reaction with C—H acidic compounds, 与酸性C—H化合物反应 226

reaction with halide ions,与卤离子反应 253,258

reaction with hydrazones,与腙反应 222

reaction with Grignard reagents,与格林那德试剂反应 239,240

reaction with sulfide ion,与硫离子反应 220

reaction with thiols,与硫醇反应 256

reduction,还原 220,224,225,227,259

S_NAr reactions of,芳香亲核取代反应 219

vicarious nucleophilic substitution,替代亲核取代 107,219,221

nitrobenzene,硝基苯

  acylation,酰化 92

  arylation,芳基化 69,77

  chloromethylation,氯甲基化 7

  conversion to 3-phenylindole,转化成3-苯基吲哚 55

  as solvent,作为溶剂 87,219

nitromethane, as solvent,硝基甲烷,作为溶剂 87,194

nitrosation,亚硝化 142,166

nitrosoarenes, formation,亚硝基苯、形成 142,166,239,240

norbornene, reaction with aniline,降冰片烯,与苯胺反应 17

nucleophilic substitution,亲核取代

  aromatic,芳香性 205-259

O

olefination with alkynes,与炔烃的烯基化 54-57,90

olefination with leaving-group-substituted olefins,与离去基团取代的烯烃的烯基化 45-48

olefination with unsubstituted olefins,与未取代烯烃的烯基化 46-54

olefins,烯烃

acylation,酰化 88-90

alkylation with,烷基化 4,8-10,14,16-23

arylation,芳基化 45-54

halogenation,卤代 148

olefination with,烯基化试剂 22,45-52

organomagnesium compounds, S_NAr with,有机镁化合物,芳香亲核取代 235-237

organometallics, hard and soft,有机金属化合物,硬的和软的

  acylation,酰化 94,97,102,103

  arylation,芳基化 235-241

  reaction with lactones,与内酯反应 102,103

ortho alkylation of acetophenones,苯乙酮的邻位烷基化 8

ortho-directed halogenation,邻位导向的卤代 127-129

ortho-directing groups,邻位导向基团 46,68

orthoesters, formation,原酸酯,形成 91,109

oxalyl chloride,草酰氯 92,95,104,106

oxazoles,噁唑

  acylation,酰化 98

  from anilides,从苯胺 142

  arylation,芳基化 74,98

  dimerization,二聚 74,77

oxidation of,氧化

  acetals,缩醛 151

  alcohols,醇 94,139,227

  aldehydes,醛 94,150,151

  alkyl groups,烷基 164,166,169,170

  amines,胺 100,148

  anilines,苯胺 18,138,141,148

  diarylamines,二芳基胺 179

  diarylmethanes,二芳基甲烷 18

  naphthalenes,萘 167

  phenols,酚 173

  sulfoxides,亚砜 150

thioethers,硫醚 149-151
thiols,硫醇 149-151
thiophenes,噻吩 162
oxiranes, arylation,环氧乙烷,芳基化 23,24
ozone, for nitration,臭氧,用于硝化 161

p

paraquat,百草枯 76
pararosaniline,副品红 18
peroxides,过氧化合物
    as methylating reagents,作为甲基化试剂 20
    reaction with nitroarenes,与硝基芳烃反应 222
Pfitzner-Moffatt oxidation,普菲茨纳-莫法特反应 195
phenacyl halides,苯酰卤 25
phenol, synthesis,酚,合成 28,30,217
phenols,酚
    acylation,酰化 90
    alkylation,烷基化 9,11,20,26
    arylation,芳基化 15,66,139,235,246
    carboxylation,羧基化 103,105
    conversion to aryl halides,转化成芳基卤 259
    deoxygenation,脱氧 20,132,229
    dimerization,二聚 78
    as electrophiles 作为亲电试剂 15
    formation,形成 105,171,217,248,251
    formylation,甲酰化 106-109
    halogenation,卤代 6,132,139,147
    methylation,甲基化 20
    nitration,硝化 164,173,174
    nitrosation,亚硝化 174
    oxidation,氧化 173,174
    S_NAr reactions of,芳香亲核取代反应 206,229
    sulfenylation,亚磺酰化 196,199
phenylacetic acids, halogenation,苯乙酸,卤代 127
phenylalanine, halogenation,苯丙氨酸,卤代 127
phosgene,碳酰氯 90,96,104,106
phosphines, arylation,膦化氢,芳基化 217
phthalic acids,邻苯二甲酸酐
    acylation,酰化 100
    as electrophile,作为亲电试剂 93
    formation by oxidation of naphthalenes,通过萘氧化形成 167
Piria reaction,皮睿尔反应 224,225
pivaloyl chloride, decomposition,新戊酰氯,分解 99
polyelectrophiles,多官能团亲电试剂 22,23,100,110
polyhaloarenes, isomerization,多卤芳烃,异构化 210
polynitroarenes, preparation,多硝基苯,制备 162-173
1-propanol, as electrophile,1-丙醇,作为亲电试剂 7
propiolactones, as electrophiles,丙醇酸内酯,作为亲电试剂 103
Pummerer reaction,普莫尔反应 197
purines, S_NAr,嘌呤,芳香亲核取代反应 206,213
pyridine,吡啶
    arylation,酰化 66,71,75,76
    dimerization,二聚 76,96
    halogenation,卤代 124,133-136
    nitration,硝化 168
    olefination,烯基化 56
pyridines,吡啶类
    acylation,酰化 94,110
    alkylation,烷基化 11,12,234
    amination,胺化 175,176
    arylation,芳基化 65,66,68,70,71,76
    cyanation,氰基化 214
    dealkylation,去烷基化 169
    dimerization,二聚 76,96

halogenation,卤代 124,133-136
metallation,金属化 94
nitration,硝化 168
olefination,烯基化 47,49,53
pyridine N-oxides,吡啶N-氧化物
  alkylation,烷基化 10,22
  arylation,芳基化 53,64,78
  halogenation,卤代 135
  nitration,硝化 168
  olefination,烯基化 49,53
pyrroles,吡咯类
  alkylation,烷基化 24,27,31
  amidation,酰胺化 180
  amination,胺化 135
  arylation,芳基化 66,75
  conversion to lactams,转化成酰胺 144
  conversion to pyridines,转化成吡啶 109
  dimerization,二聚 130
  formation,形成 76
  halogenation,卤代 123,130,135,143,144
  imidation,酰亚胺 123
  nitration,硝化 162
  sulfenylation,亚磺酰化 123,197
  sulfinylation,次磺酰化 197

q

quaternization of amines, by arylation,胺的季铵盐化,芳基化 244,248
quinolines,喹啉类
  formation,形成 109,227
  formylation,甲酰化 110
  from tetrahydroquinolines,从四氢喹啉 141
  S_NAr,芳香亲核取代反应 215,240
quinones,醌类
  addition of nitrite to,亚硝酸盐的加成 174
  alkylation,烷基化 11,12
  from anilines,从苯胺 141
  from benzamides,从苯甲酰胺 152
  from phenols,从酚 173,174

r

racemization of electrophiles during alkylations,在烷基化过程中亲电试剂的外消旋化 1,5
radicals, arylation,自由基,芳基化 11,12,20,27,63-67,72-74,110
rearrangement,重排
  allylic,烯丙基 21,22,52,53,55
  of carbocations,碳正离子 5,6,89,93
  of cyclopropanes,环丙烷 99
  of hydrazones,腙 179
  of polyhaloarenes,多卤代芳烃 138,210
  of sulfoxides,亚砜 197
  of thioethers,硫醚 99,210
  von Richter,冯·里克特 220,223
reduction of,还原
  aryl halides,芳基卤 71,145,229,236,241,243,249,254,255,259
  azido groups,叠氮基 145,249,250
  bromomalonates,溴代丙二酸酯 6
  nitro groups,硝基 220,225-227,239,249,254
  sulfoxides,亚砜 150,196,197
  tertiary alcohols,叔醇 31
reductive alkylation,还原烷基化 31
Reimer-Tiemann reaction,瑞莫-梯曼反应 107-109
resveratrol, iodination,白藜芦醇,碘化 147
rhodanide ion, arylation of,硫氰化离子,芳基化 257

s

saccharin, N-benzylation,糖精,N-苄基化 183
salicylic acid, preparation,水杨酸,制备 103

Sandmeyer reaction, 桑德迈尔反应 227
scavengers, for halogenations, 清除剂, 用于卤代反应 138, 147, 148
Scholl reaction, 肖勒反应 7
Schotten-Baumann procedure, 肖特-鲍曼程序 193
silanes, 硅烷
  ipso substitution, 原位取代反应 163
  nitration, 硝化 163
  reaction with benzyl esters, 与苄基酯的反应 101
  as reducing agents, 作为还原剂 31
S_NAr, 芳香亲核取代反应 205-259
  vs benzylic substitution, 相对于苄基取代 211, 212
  solvent effects, 溶剂效应 207, 234, 236
solvents for Friedel-Crafts acylation, 作为付瑞德尔-克拉夫酰化反应的溶剂 87, 92, 103
solvents for oxidations, 作为氧化反应溶剂 51
steric crowding, S_NAr, 立体拥挤, 芳香亲核取代反应 219
stilbenes, 芪类、二苯乙烯类
  formation, 形成 27, 48-50, 54, 55, 241
  halogenation, 卤代 147
styrene, 苯乙烯
  arylation, 芳基化 16
  dimerization, 二聚 90
styrenes, 苯乙烯类
  acylation, 酰化 88, 90
  formation, 形成 45-57
substitution, aromatic, 取代, 芳香性
nucleophilic, 亲核的 205-259
succinimides, N-halo, 琥珀酰亚胺, N-卤代 121-123
sulfenylation, 亚磺酰化 199-201
sulfide ion, S_NAr, 硫离子, 芳香亲核取代反应 220
sulfinates, 亚磺酸酯

  arylation with, 芳基化试剂 75
  conversion to sulfoxides, 转化成亚砜 196, 197
  formation, 形成 193
sulfinylation, 次磺酰化 195
sulfite ion, arylation, 亚硫酸盐, 芳基化 225
sulfolane, as solvent, 环丁砜, 作为溶剂 87, 92
sulfonamides, N-arylation, 磺酰胺类, N-芳基化 180, 181
sulfonates (sulfonic acid esters), 磺酸酯类
  cleavage, 断裂 231, 243
  formation, 形成 191
sulfones, 砜类
  formation, 形成 191-194
  S_NAr 芳香亲核取代反应 208, 235
sulfonic acids, 磺酸
  formation, 形成 191, 217, 225, 229
  S_NAr, 芳香亲核取代反应 206
sulfonic acid anhydrides, arylation, 磺酸酐, 芳基化 192
sulfonic acid esters, see sulfonates, 磺酸酯
sulfonium salts, formation, 锍盐, 形成 196
sulfonylation, 磺酰化 191
sulfonyl halides, 磺酰卤
  arylations with, 芳基化试剂 194
  conversion to sulfones, 转化成砜 194
  formation, 形成 191, 192
sulfoxides, 亚砜
  alkylation with, 烷基化试剂 10
  as electrophiles, 作为亲电试剂 196
  formation, 形成 196, 197
  isomerization, 异构化 197
  reduction, 还原 150
sulfur fluorides, 氟化硫 132
sulfuric acid, 硫酸
  as solvent, 作为溶剂 141
  sulfonylation with, 磺酰化试剂 191

sulfuryl chloride (SO$_2$Cl$_2$),硫酰氯
 chlorination with,氯代试剂　140,195
 sulfonylation with,磺酰化试剂　195
Swern oxidation,斯温氧化　195

t

tele substitution,远程取代　207-209
terephthalic acid,对苯二甲酸
 hydroxymethylation,羟甲基化　7
 iodination,碘代　135
tertiary amines,叔胺
 N-arylation,N-芳基化　244,248
 Dealkylation,去烷基化　92,100,142,148,166,248
 dehydrogenation,脱氢　100
 displacement,取代　238
 nitrosation,亚硝化　142,166
tetrahydroquinolines,四氢喹啉
 halogenation,卤代　141,147
 nitration,硝化　171
tetralones,四氢萘酮
 arylation,芳基化　32
 formation,形成　15,88,101,102
1,2,4,5-tetramethylbenzene（durene）,1,2,4,5-四甲基苯(杜烯)
 acylation,酰化　93
 alkylation,烷基化　22
 nitration,硝化　164
thiazoles,噻唑
 arylation,芳基化　70,74,77
 dehalogenation,脱卤反应　145,241
 dimerization,二聚　77
 formation,形成　141
thiocyanates, formation,硫氰酸酯,形成　107,124
 from anilines,从苯胺类　124
 by S$_N$Ar,通过芳香亲核取代反应　257
thioethers,硫醚
 cleavage,断裂　150,257
 conversion to thiocyanates,转化成硫氰酸酯　107
 formation,形成　123,208,216,220,225,227,229,230,233,252
 halogenation,卤代　149-151
 isomerization,异构化　99,210
 nucleophilic displacemen,亲核取代　13,88,229-231
 oxidation,氧化　149-151
thiols,硫醇
 arylation,芳基化　216,220,225,229,252,253
 formation,形成　220,257
 halogenation,卤代　149-151
thionyl chloride（SOCl$_2$）,亚硫酰氯,氯化亚砜
 chlorinations with,氯代试剂　150,198
 dehydrogenations with,脱氢试剂　198
 reaction with anilines,与苯胺反应　140
 reaction with cinnamic acids,与肉桂酸反应　198
 reaction with pyridine,与吡啶反应　96
 reaction with sulfoxides,与亚砜反应　150,196
 sulfinylations with,次磺酰化试剂　198
thiophenes,噻吩
 acylation,酰化　102,110
 alkylation,烷基化　10,33
 arylation,芳基化　63,64,78
 cleavage,断裂　227
 dimerization,二聚　76
 formation,形成　198
 metallation,金属化　10,102
 nitration,硝化　162,163
 olefination,烯基化　51,53
thiophenols,噻吩类
 acylation,酰化　90,91
 alkylation,烷基化　13
 arylation,芳基化　230
 formation,形成　220,231,257
toluene,甲苯

bromination, 溴代 126
　　reaction with CS₂, 与二硫化碳反应 103, 104
toluenesulfonamides, halogenation, 磺酰胺类, 卤代 127
tosyl chloride, chlorination with, 对甲苯磺酰氯, 氯代试剂 123, 128
triarylmethane dyes, 三芳基甲烷染料 18
triazenes, cyclization to benzotriazoles, 三氮烯, 环化成苯并三氮唑 179
1,3,5 - triazines, arylation, 1,3,5 -三嗪, 芳基化 62
triazoles, olefination, 三唑类, 烯基化 53
(trichloromethyl) arenes, reaction with nucleophiles, 三氯甲基芳烃, 与亲核试剂反应 209
triethylamine, dehydrogenation, 三乙胺, 脱氢 100
trifluoromethylation, 三氟甲基化 8, 16
trihaloalkanes, as electrophiles, 三卤烷烃, 作为亲电试剂 88, 209
trihaloethanols, as electrophiles, 三氯乙醇, 作为亲电试剂 32
trinitrobenzene, 三硝基苯
　　arylation, 芳基化 71
　　formation, 形成 259
trioxane, 三氧杂环己烷, 三聚甲醛 7, 110
triphosgene, 三光气, 二碳酸酯 106
tritylation of arenes, 芳烃的三苯甲基化 16
trityl chloride, 三苯甲基氯 1

Troger's base, 特罗格碱 17
twofold acylation of arenes, 芳烃双重酰化 93

U

Ullmann coupling reaction, 伍尔曼偶联反应 45, 63
Urotropine, 乌洛托品 107, 108

V

valerolactone, as electrophile, 戊内酯, 作为亲电 101, 102
vanadates, catalysts for oxidations, 钒酸盐, 氧化反应的催化剂 18
vicarious nucleophilic substitution, 替代的亲核取代反应 107, 219, 221
Vilsmeier reaction, 维尔斯迈尔反应 92, 106
vinyl acetates, as electrophiles, 烯基乙酸酯, 作为亲电试剂 48
vinylic carbocations, 烯基碳正离子 47
vinylsilanes, alkylation with, 烯基硅烷, 烷基化试剂 9
von Richter rearrangement, 冯·李希特重排 220 - 223

X

xenon fluorides, 氟化氙 132
xylene, nitration, 二甲苯, 硝化 164, 171